G PROTEIN-C
RECEPTORS

Structure, Function, and Ligand Screening

 **METHODS IN SIGNAL TRANSDUCTION SERIES**

Joseph Eichberg, Jr., Series Editor

Published Titles

Lipid Second Messengers, Suzanne G. Laychock and Ronald P. Rubin

G Proteins: Techniques of Analysis, David R. Manning

Signaling Through Cell Adhesion Molecules, Jun-Lin Guan

G Protein-Coupled Receptors, Tatsuya Haga and Gabriel Berstein

Calcium Signaling, James W. Putney

G Protein-Coupled Receptors: Structure, Function, and Ligand Screening, Tatsuya Haga and Shigeki Takeda

G PROTEIN-COUPLED RECEPTORS

Structure, Function, and Ligand Screening

Edited by

Tatsuya Haga, Ph.D.

Professor and Director
Institute for Biomolecular Science
Faculty of Science
Gakushuin University
Tokyo, Japan

Shigeki Takeda, Ph.D.

Department of Nano-Material Systems
Graduate School of Engineering
Gunma University
Gunma, Japan

CRC Press
Taylor & Francis Group
Boca Raton London New York

CRC Press is an imprint of the
Taylor & Francis Group, an **informa** business

A TAYLOR & FRANCIS BOOK

Cover figure courtesy of Tetsuji Okada

CRC Press
Taylor & Francis Group
6000 Broken Sound Parkway NW, Suite 300
Boca Raton, FL 33487-2742

First issued in paperback 2019

© 2006 by Taylor & Francis Group, LLC
CRC Press is an imprint of Taylor & Francis Group, an Informa business

No claim to original U.S. Government works

ISBN-13: 978-0-8493-2771-1 (hbk)
ISBN-13: 978-0-367-39237-6 (pbk)
Library of Congress Card Number 2005040590

Library of Congress Cataloging-in-Publication Data

G protein-coupled receptors : structure, function, and ligand screening / edited by Tatsuya Haga, Shigeki Takeda.
     p. cm. – (Methods in signal transduction ; 6)
Includes bibliographical references and index.
ISBN 0-8493-2771-7
1. G proteins–Receptors. I. Haga, Tatsuya. II. Takeda, Shigeki. III. Series.

QP552.G16G175 2005
612'.01575–dc22

2005040590

Visit the Taylor & Francis Web site at
http://www.taylorandfrancis.com

and the CRC Press Web site at
http://www.crcpress.com

Series Preface

The concept of signal transduction at the cellular level is now established as a cornerstone of the biological sciences. Cells sense and react to environmental cues by means of a vast panoply of signaling pathways and cascades. While the steady accretion of knowledge regarding signal transduction mechanisms is continuing to add layers of complexity, this greater depth of understanding has also provided remarkable insights into how healthy cells respond to extracellular and intracellular stimuli and how these responses can malfunction in many disease states.

Central to advances in unraveling signal transduction is the development of new methods and refinement of existing ones. Progress in the field relies upon an integrated approach that utilizes techniques drawn from cell and molecular biology, biochemistry, genetics, immunology and computational biology. The overall aim of this series is to bring together and continually update the wealth of methodology now available for research into many aspects of signal transduction. Each volume is assembled by one or more editors who are leaders in their specialty. Their guiding principle is to recruit knowledgeable authors who will present procedures and protocols in a critical yet reader-friendly format. Our goal is to assure that each volume will be of maximum practical value to a broad audience, including students, seasoned investigators, and researchers who are new to the field.

The range of techniques used to study G protein-coupled receptor (GPCR) structure and function continues to expand at a rapid rate. The current volume edited by Haga and Takeda builds on and adds to information furnished in the previous book in the series, edited by Haga and Berstein, which dealt with this topic. While the broad areas covered remain much the same, the chapter contents reflect progress achieved in the past few years. The first portion of the book provides descriptions of recently developed screening methodologies for identification of GPCR ligands, including those for orphan GPCRs and the wealth of known odorant receptors. The next part presents a range of approaches to characterize receptors at the molecular level and to study their physiology, with particular, although not exclusive, emphasis on the muscarinic cholinergic receptor family. The last section details examples of how physical methods and computational approaches can be used to elucidate receptor and ligand structures, as well as to devise models for investigation of GPCR-ligand interactions.

Taken together, the topics covered in this volume highlight the present status of methods employed in GPCR research and point the way toward future developments in the field.

Joseph Eichberg, Ph.D.
Series Editor

Preface

G protein-coupled receptors (GPCRs) constitute one of the largest superfamilies of proteins and are major sensors of cells as well as major targets of drugs. Approximately 1000 GPCR genes appear to be present in the human genome: 350 to 400 odorant receptors, 30 to 40 taste receptors, and 350 to 450 receptors for endogenous ligands such as hormones, neurotransmitters, and chemoattractants. Rhodopsin, β adrenergic receptors, and muscarinic acetylcholine receptors are typical GPCRs for external stimulants (light), hormone/neurotransmitters (adrenaline/noradrenaline), and neurotransmitters (acetylcholine), respectively, and have been studied in detail. Many GPCRs or GPCR candidates, however, have been identified only as genes with known sequences, and their molecular and physiological functions remain to be elucidated. The molecular function of GPCRs is to recognize ligands and then activate G proteins. GPCRs for which endogenous ligands were not identified are called orphan GPCRs. It is one of the most important issues in the field of signal transduction to identify endogenous ligands for orphan GPCRs, as identification of endogenous ligands may mean discovery of novel hormones or neurotransmitters. The next step following identification of ligands is to define the function and regulation of GPCRs in terms of molecular characteristics and molecular interactions. Another important issue in GPCR studies is the determination of the tertiary structure of GPCR, which may not only lead to structural understanding of the interactions of GPCRs with ligands and G proteins but also contribute to theoretical modeling of drugs.

Tatsuya Haga and Gabriel Berstein edited a book titled *G Protein-Coupled Receptors*, which was published as part of the *CRC* Methods in Signal Transduction Series in 1999 (CRC Press). The present book may be taken as the second volume of the original *G Protein-Coupled Receptors*, not a revised edition. The present book covers current techniques in the field of GPCRs, which either were not treated in the original book or have been developed in the last few years.

The book is divided into three sections:

Section I: Screening of Ligands for GPCRs
Section II: Functions and Regulation of GPCRs
Section III: Tertiary Structure of GPCRs and Their Ligands

Section I is concerned with the methods for screening novel ligands for GPCRs, particularly endogenous ligands for orphan GPCRs. Fujino and his group have succeeded in identifying many novel endogenous ligands, and in Chapter 1 they describe the ligand screening methods that they have successfully adopted. Chapter 2 treats two topics: identification of GPCR genes from the human genome and a simple ligand screening method using the fusion protein of GPCRs with G protein alpha subunits. Chapter 3, by Kojima and Kangawa, chooses ghrelin as a representative of recently identified endogenous ligands. Ghrelin was found in the stomach and is known to regulate appetite and generation of growth hormone. Kangawa, a discoverer of ghrelin, authored papers that were most cited in 2002. Chapter 4 discusses the identification of ligands for odorant receptors. Only a few ligands for a few receptors among a thousand of odorant receptors have been identified, partly because of difficulty in expressing odorant receptors in heterologous cells. It is still a great challenge to express GPCRs, including odorant and pheromone receptors in their active states, and to identify their ligands.

In Section II, physiological and molecular characterizations of GPCRs are described. In the first three chapters, we focus on the muscarinic acetylcholine receptor as a model GPCR. Physiological functions of muscarinic receptors are being clarified by the use of knockout mice of each of five subtypes. Wess and colleagues describe a method to generate and analyze knockout mice using muscarinic receptors as a model (Chapter 5). Hulme, a pioneer of molecular characterization of muscarinic receptors, has adopted systematic mutagenesis of muscarinic receptors to reveal the structure–function relationship (Chapter 6). Desensitization, particularly internalization, of muscarinic receptors is discussed by van Koppen in Chapter 7. These methods on muscarinic receptors would also be useful for researchers working on other GPCRs. Chapter 8 by Ueda, Miyanaga and Yanagida offers a unique approach — the single molecule detection technique — that is applied to cAMP receptors involved in the chemotactic response of Dictyostelium cells. Many GPCRs have been reported to be oligomers recently, and Chapter 9 (Nakata, Yoshioka, and Kamiya) treats the methods related to this topic using adenosine and purinergic receptors as a model.

In Section III, the structures of GPCRs and their ligands are covered. Rhodopsin is the only GPCR whose atomic structure has been revealed. The critical step for determination of the atomic structure of proteins is their crystallization. It is very difficult to crystallize membrane proteins, particularly GPCRs, most parts of which are embedded in membranes. Crystallization of rhodopsin was accomplished by Okada, and his experiences and knowledge on the method are very important and useful for others working on this subject (Chapter 10). On the other hand, the steric structure of a ligand bound to a GPCR can be elucidated by using the NMR/TRNOE (transferred nuclear Overhauser effect) method without knowledge of receptor structures. As a successful application of this method, Chapter 11 describes the method using acetylcholine bound to muscarinic receptors as a model. It would be ideal for drug design if the chemical structures of antagonists or agonists could be deduced from the tertiary structure of GPCRs. In fact, we do not have the structural information, except for rhodopsin, and do not know exactly what kinds of structural changes may occur in GPCRs and ligands by their binding. Thus, it is important to make models for the interaction of GPCRs and their ligands by using a computer. This topic is treated in Chapter 12.

We hope that this book, as well as the previous work, *G Protein-Coupled Receptors*, is useful for those who wish to find endogenous ligands for orphan GPCRs, elucidate the molecular mechanisms underlying the function and regulation of GPCRs, and determine and utilize tertiary structures of GPCRs and their ligands for drug designs.

Finally, we would like to express sincere thanks to all of the contributors for taking precious time to write these excellent chapters.

Tatsuya Haga
Shigeki Takeda

The Editors

Tatsuya Haga is professor and director of the Institute for Biomolecular Science, Faculty of Science, Gakushuin University. He received his B.A. degree in biochemistry from the University of Tokyo in 1963, and his Ph.D. degree in biochemistry from the University of Tokyo, Graduate School of Science, in 1970. He was instructor and research associate in the Department of Biochemistry, Faculty of Science, and in the Department of Neurochemistry, Institute for Brain Research, Faculty of Medicine, at the University of Tokyo from 1969 to 1974, and was associate professor of biochemistry at Hamamatsu University Medical School from 1974 to 1988. Meanwhile, he served as research associate and assistant professor at the University of Virginia, Medical School, from 1975 to 1977. From 1988 to 2001, he served as professor of neurochemistry at the University of Tokyo, Faculty of Medicine, and he moved to Gakushuin University in 2001. Dr. Haga's research interests involve various aspects of neurochemistry, especially molecular characterization of muscarinic acetylcholine receptors and other G protein-coupled receptors, and of a high-affinity choline transporter. He has published more than 100 research papers, reviews, and monographs on this and other topics and delivered lectures at more than 100 national and international conferences and symposia. Dr. Haga is a member of the Japanese Biochemical Society, the American Society for Biochemistry and Molecular Biology, and the International Society for Neurochemistry. He serves, or has served, on the editorial boards of *The Journal of Biochemistry*, *The Journal of Neurochemistry*, and *Life Science*.

Shigeki Takeda is an associate professor in the Department of Nano-Material Systems, Graduate School of Engineering, Gunma University. He received his B.A. degree in pharmacology from Hokkaido University and his Ph.D. degree in biochemistry from the University of Tokyo, Graduate School of Engineering, in 1991. He was Postdoctoral Fellow of Japan Science and Technology Agency at Tsukuba and at Stanford University from 1991 to 1996, and instructor at the Tokyo Institute of Technology from 1996 to 1998. As a graduate student and postdoctoral fellow, he worked on protein engineering, structural analysis of a large protein complex using an electron microscope, NMR, and x-ray crystallography, and protein assembly of a virus. He was instructor and research associate at the University of Tokyo and Gakushuin University from 1998 to 2001, and moved to Gunma University in 2001. Since 1998, he has been working on G protein-coupled receptors, particularly on identification of endogenous ligands for orphan receptors.

Contributors

Chuxia Deng
National Institute of Diabetes and
  Digestive and Kidney Diseases
Bethesda, Maryland

Masahiko Fujino
Deceased

Chihiro Funamoto
National Institute of Advanced
  Industrial Science and Technology
Tokyo, Japan

Hiroyasu Furukawa
Columbia University
New York, New York

Dinesh Gautam
National Institute of Diabetes and
  Digestive and Kidney Diseases
Bethesda, Maryland

Tatsuya Haga
Gakushuin University
Tokyo, Japan

Toshiyuki Hamada
RIKEN Yokohama Institute
Yokohama, Japan

Sung-Jun Han
National Institute of Diabetes and
  Digestive and Kidney Diseases
Bethesda, Maryland

Shuji Hinuma
Takeda Pharmaceutical Company
  Limited
Ibaraki, Japan

Hiroshi Hirota
RIKEN Yokohama Institute
Yokohama, Japan

Edward C. Hulme
MRC National Institute for Medical
  Research
London, United Kingdom

Masaji Ishiguro
Suntory Institute for Bioorganic
  Research
Osaka, Japan

Jongrye Jeon
National Institute of Diabetes and
  Digestive and Kidney Diseases
Bethesda, Maryland

Toshio Kamiya
Tokyo Metropolitan Institute for
  Neuroscience
Tokyo, Japan

Kenji Kangawa
National Cardiovascular Center
  Research Institute
Osaka, Japan
and
Kyoto University Hospital
Kyoto, Japan

Sayako Katada
The University of Tokyo
Chiba, Japan

Masayasu Kojima
Kurume University
Fukuoka, Japan

Chris J. van Koppen
Oragnon
Oss, The Netherlands

Cuiling Li
National Institute of Diabetes and
  Digestive and Kidney Diseases
Bethesda, Maryland

Yukihiro Miyanaga
Osaka University
Osaka, Japan

Masaaki Mori
Takeda Pharmaceutical Company
  Limited
Ibaraki, Japan

Miho Muraoka
National Institute of Advanced
  Industrial Science and Technology
Tokyo, Japan

Takao Nakagawa
The University of Tokyo
Chiba, Japan

Hiroyasu Nakata
Tokyo Metropolitan Institute for
  Neuroscience
Tokyo, Japan

Tetsuya Ohtaki
Takeda Pharmaceutical Company
  Limited
Ibaraki, Japan

Yuki Oka
The University of Tokyo
Chiba, Japan

Tetsuji Okada
National Institute of Advanced
  Industrial Science and Technology
Tokyo, Japan
and
Japan Science and Technology
  Corporation
Saitama, Japan

Hinako Suga
Gunma University
Gunma, Japan

Shigeki Takeda
Gunma University
Gunma, Japan

Kazushige Touhara
The University of Tokyo
Chiba, Japan

Rumi Tsujimoto
National Institute of Advanced
  Industrial Science and Technology
Tokyo, Japan

Masahiro Ueda
Osaka University
Osaka, Japan

Jürgen Wess
National Institute of Diabetes and
  Digestive and Kidney Diseases
Bethesda, Maryland

Toshio Yanagida
Osaka University
Osaka, Japan

Kazuaki Yoshioka
Kanazawa University Graduate School
  of Medical Science
Ishikawa, Japan

Contents

PART III   *Tertiary Structure of GPCRs and Their Ligands*

Part I

Screening of Ligands for GPCRs

1 Screening of Endogenous Ligands for Orphan GPCRs

Masaaki Mori, Shuji Hinuma, Tetsuya Ohtaki, and Masahiko Fujino

CONTENTS

1.1 INTRODUCTION

G protein-coupled receptors (GPCRs) comprise one of the largest superfamilies of the human genome. The recent achievement of the human genome project has revealed that there are approximately 700 GPCR genes (excluding pseudo-genes) in the human genome. Most of these genes are identified on the basis of sequence homology to known GPCR genes. Each GPCR gene encodes a protein consisting of an extracellular N-terminal domain, seven transmembrane domains, and intracellular domains responsible for interaction with G proteins or other intracellular signaling molecules. Approximately half of GPCR genes are thought to encode sensory receptors for smell, taste, and vision. The other half encode receptors regulating cell functions. To date, natural ligands have been identified for approximately 230 of these receptors. However, the ligands of the remaining 120 receptors have not yet been identified, and they are, therefore, referred to as orphan GPCRs. The identification of ligands for orphan GPCRs is expected to lead to the discovery of new regulatory mechanisms of the human body. Furthermore, GPCRs have been historically proven to be the most successful targets in the field of drug discovery. Of the approximately 500 drugs currently on the market, more than 30% are GPCR agonists or antagonists, representing approximately 9% of global pharmaceutical sales.[1,2] Orphan GPCR research is therefore important from the perspectives of both basic and applied science. The identification of ligands for orphan GPCRs should yield important clues as to their physiological functions and will help determine whether they are suitable as drug targets.

We began our orphan GPCR research in 1994, when we isolated hGR3, an orphan GPCR, from the human pituitary.[3] This novel orphan GPCR showed low homology to known GPCRs, having at most 30% amino acid identity with the neuropeptide Y receptor. However, there were no direct clues as to its ligand. We therefore had to establish an original method of identifying ligands of orphan GPCRs.[4] Our initial approach was as follows: prepare two types of cells: the first, control cells and the second, cells expressing an orphan GPCR. Then add a ligand to these cells. This ligand would bind to the orphan GPCR and induce signal transduction only in the cells expressing the orphan GPCR, with the expectation that nothing would happen in the control cells. In this way, we could determine whether or not a sample contained a ligand by comparing signal transduction between the two types of cells. However, one problem with this approach was that each receptor has a unique signal transduction pathway; thus, it is impossible to predict exactly what kinds of signal transduction occur in each orphan GPCR. Fortunately, we found that at least one of three kinds of cellular response is invariably induced by the activation of any known regulatory GPCR: an increase of calcium ions, an increase of cyclic adenosine monophosphate (cAMP), or a decrease of cAMP. Thus, we hypothesized that we could detect the signal transduction of any orphan GPCRs using just three assays. We first applied this idea to the identification of the hGR3 ligand.

Because hGR3 showed significant homology with the neuropeptide Y receptor, we imagined that the ligand of hGR3 would be a peptide. In addition, based on the tissue distributions of hGR3, we expected that these ligands would be present in the brain. Therefore, we prepared peptide-enriched fractions from brain tissue extracts and applied these fractions to assays to detect specific signal transductions in cells expressing hGR3. Among the several different assays conducted, we successfully detected a specific response in the cells expressing hGR3 to brain tissue extracts, using an assay for arachidonic acid metabolite release. Arachidonic acid metabolite release reflects the turnover of lipid metabolism, including the activation of phospholipase A_2 induced by intracellular Ca^{2+} influx. In 1995, we purified the hGR3 ligand from bovine hypothalamic tissue extracts using a combination of various chromatographies, and we then determined its N-terminal sequence. We named it prolactin-releasing peptide (PrRP), because the hGR3 ligand could promote prolactin secretion from anterior pituitary cells. Since then, we succeeded in identifying other various orphan GPCR ligands. In this chapter, we discuss some of the ways in which we identified ligands of orphan GPCRs and what we discovered through analyses of their functions.

1.2 LIGAND SOURCES

1.2.1 Tissue Extract

The most orthodox method of ligand fishing may be to employ a tissue extract as the starting material. In this strategy, the extract of tissue is subjected to a purification procedure, while the particular responses of the cells expressing the target receptor protein are monitored. After the purification steps, which involve a combination of chromatographies, the ligand molecule is finally isolated in homogeneity, and its structure is determined. This was the main approach taken in the early attempts at ligand fishing. This method first allowed nociceptin[5] and orphanin FQ[6] to be isolated from rat and porcine brain extracts, respectively, and the discoveries of several novel peptidic ligands (orexin,[7] PrRP,[3] apelin,[8] and so on) for the orphan receptors from the tissue extracts soon followed. Many novel compounds, all of them peptidic, were discovered using this method.

Because peptide ligands in tissue generally exist at very low concentrations and are sensitive to proteolytic degradation, the final yield of target substances is usually very low. This means that ligand fishing from tissue extracts is both labor-intensive and technique-sensitive; thus, only a few research groups have been successful with this method. Nevertheless, some ligands of the orphan receptors could not have been discovered without this method. The discovery of ghrelin typified such a case. In 1999, ghrelin was isolated from rat stomach as the ligand for growth hormone secretagogue receptor.[9] It has a unique structural character, in which the hydroxyl group of the Ser residue at the third position is octanoylated. In fact, the modification of ghrelin is so extraordinary that its structure could not have been elucidated unless the ligand was purified from tissue extract.

1.2.2 KNOWN MOLECULES

Another approach to discovering ligands for the orphan receptors involves screening the library of known molecules that includes possible candidate ligands, such as the biogenic amines, peptides, chemokines, bioactive lipids, and metabolic pathway intermediates. In fact, use of this approach led to the discovery of the most ligands, including bioactive peptides, such as melanin-concentrating hormone,[10] urotensin II,[11] motilin,[12] and neuromedin U,[13] and the low molecular weight ligands, such as sphingosylphosphorylcholine,[14] lysophosphatidylcholine,[15] bile acids,[16] and free fatty acids.[17] If the target receptor is expected to possibly pair with a nonpeptidic low molecular weight ligand, this approach can identify an agonistic molecule, because only rarely will a completely novel compound be a specific ligand for such a receptor.

The success of this method depends on the size of the library, the diversity of the candidate molecules, and the throughput of the screening, all of which contribute to high financial costs. In some cases, an agonist compound fished out by this method is a surrogate ligand rather than a genuine endogenous ligand that exists and functions in the tissues. Nevertheless, the surrogate ligands are still useful for analyzing the biological functions of their receptors.

1.2.3 DATABASE SEARCH

Bioactive peptides are usually generated by cleavage at the potential processing sites (cluster of two or three successive basic amino acids) from the precursor proteins equipped with a secretory signal sequence. Accordingly, candidate genes, which possibly code the precursor proteins of novel bioactive peptides, can be discovered by searching databases using the motifs of the sequences of the secretory signal and processing site as the query. Genes encoding the preproproteins for RFamide-related peptides (RFRPs),[18] neuropeptide B,[19] and pyroglutamylated RFamide peptide[20] were discovered by this method. The predicted mature peptides were subsequently identified as the cognate ligands for orphan receptors. The application of this *in silico* approach is expected to become more effective for discovering novel bioactive peptides along with developing bioinformatics and the accumulating genomic information.

On the other hand, ghrelin possesses a unique posttranslational modification that could not be predicted from its genetic information and yet is essential for biological activity.[9] Further, angiotensin II and endothelins are known to reveal their biological activity only after processing from their pro-forms by specific convertases. These observations indicate that there are important limits to this method.

1.3 PARTICULAR EXAMPLES OF LIGAND SCREENING FOR THE ORPHAN GPCRS

1.3.1 GALANIN-LIKE PEPTIDE (GALP)

1.3.1.1 Discovery of a Novel Galanin-Family Peptide, GALP

Galanin is a regulatory peptide distributed widely in the central and peripheral organs. While it was the sole member of the "galanin-family peptide" group, the heterogeneity of galanin-like immunoreactivity in mammalian tissues was described in some of the early literature. Some immunoreactive peptides were identified with galanin precursor protein, but some remained unidentified, suggesting the possible existence of unknown galanin-family peptides. Eventually, three subtypes of galanin receptors, designated GALR1, GALR2, and GALR3, which bind galanin with this rank order of affinity, were reported. These cloned receptors allowed us to analyze galanin-family peptides on the basis of agonistic activity *in vitro*.[21]

1.3.1.2 Preparation of Tissue Extract Sample

To analyze the endogenous galanin-family peptide, a tissue extract sample was prepared from the porcine hypothalamus following a standard procedure to enrich peptide components, including heat denaturing (boiling), extraction with 1 M acetic acid, protein elimination by acetone precipitation, and lipid extraction with diethyl ether. The crude extract was further separated into 60 fractions using reversed-phase (RP) high-performance liquid chromatography (HPLC) (see Protocol 1-1).

PROTOCOL 1-1: PREPARATION OF CRUDE PEPTIDE EXTRACT

1. Cut fresh porcine hypothalamic tissues (ca. 1 kg from 30 brains) into small pieces.
2. Boil every 500 g of tissue with 2 l of water in a siliconized glass beaker for 10 min, and then cool in ice bucket.
3. Homogenate the tissues using a Polytron homogenizer (Kinematica AG, Lucerne, Switzerland) at 4°C.
4. Add 1/17 volume of acetic acid to the homogenate and stir the homogenate at 4°C overnight.

5. Spin the homogenate at $10,000 \times g$ for 30 min at 4°C. Pour the supernatant through cheesecloth into a siliconized beaker.
6. Add two volumes of chilled acetone to the extract slowly under vigorous stirring.
7. Spin down protein precipitate. Remove acetone using a rotary evaporator.
8. Mix the extract with a half volume of diethyl ether vigorously and take the water phase (repeat twice or more).
9. Load the extract onto a C18 RP column (5 × 10 cm) equilibrated with 1 M acetic acid. (Bulk C18 resins are available from Waters, YMC, etc.)
10. Wash the column with 1 M acetic acid, and elute crude peptide with 60% acetonitrile/0.1% TFA.
11. After evaporation of acetonitrile, lyophilize the crude peptide to obtain powder (0.5 to 0.8 g).
12. Fractionate the crude peptide using an RP-HPLC column (see Figure 1.1). Lyophilize an aliquot of each fraction and dissolve lyophilizate in dimethylsulfoxide (DMSO) at 1 g starting tissue equivalent/μl to make test samples

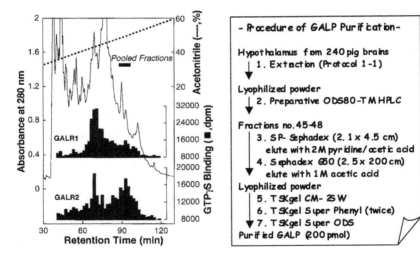

FIGURE 1.1 Screening of galanin-like agonistic activity (left) and further purification procedure (right). The lyophilized powder of porcine hypothalamic extract (every 300 to 350 mg, see Protocol 1-1) was analyzed using a TSKgel ODS80-TM (21.5 × 300 mm) with an acetonitrile concentration gradient of 20 to 60% for 120 min in 0.1% trifluoroacetic acid (TFA) at a flow rate of 4 ml/min. Fractions were collected every 2 min, and aliquots were lyophilized, dissolved in DMSO, and subjected to a [^{35}S]GTPγS binding assay.

1.3.1.3 [^{35}S]Guanosine 5'O-(γ-Thio)Triphosphate ([^{35}S]GTPγS) Binding Assay

Alpha subunits of trimeric G proteins bind guanosinediphosphate (GDP) at the resting state. The binding of the agonist to GPCR induces replacement of GDP with guanosinetriphosphate (GTP), and GTP-bound α subunits activate effecter molecules until bound GTP is hydrolyzed to GDP by the intrinsic GTPase activity of the α subunits. The [^{35}S]GTPγS binding assay measures the agonist-induced increase in the binding of [^{35}S]GTPγS (unhydrolyzable GTP analogue) to the membrane fractions carrying target GPCR, which reflects the binding of [^{35}S]GTPγS to G proteins induced by agonist-liganded GPCR.

To achieve a large increase in [^{35}S]GTPγS binding versus basal binding, the membrane fractions should contain a high concentration of target GPCR. Therefore, the membrane fractions should be of gene-transfected cell origin rather than of natural tissue origin. Stable Chinese hamster ovary (CHO) cell lines established and confirmed relevant to this assay in our laboratory have an expression level range of 5 to 15 pmol receptor/mg membrane protein. To decrease the basal binding, 1 μM GDP and 100 to 150 mM NaCl are usually included in the reaction mixtures. However, it should be noted that these additives also suppress agonist-induced [^{35}S]GTPγS binding to G proteins. Thus, this assay method is essentially suitable for detecting the activation of $G_{i/o}$ proteins, but not for detecting that of G proteins with a slow GDP/GTP exchange rate, such as in the case of G_s and $G_{q/11}$ proteins.

To detect galanin-like agonistic activity, we employed a [^{35}S]GTPγS binding assay for GALR1 and GALR2, because we knew that GALR1 was coupled to $G_{i/o}$, and GALR2 was coupled to $G_{i/o}$ and $G_{q/11}$. The GALR1 and GALR2 contents in CHO cell membranes were 13.8 and 6.6 pmol receptor/mg protein, respectively.

PROTOCOL 1-2: PREPARATION OF MEMBRANE FRACTIONS AND [^{35}S]GTPγS SOLUTION

1. Grow transformant cells expressing target GPCR in appropriate medium to subconfluency.
2. Collect the cells in phosphate-buffered saline containing 2.7 mM ethylenediaminetetraacetic acid (EDTA) (do not use trypsin).
3. Homogenize the cells in homogenizing buffer (10 mM NaHCO$_3$, 5 mM EDTA, 0.5 mM phenylmethylsulfonyl fluoride, 20 μg/ml leupeptin, 10 μg/ml pepstatin, 8 μg/ml E-64; pH 7.3) using a Polytron homogenizer at 4°C. (All of the following procedures should be done at 4°C.)
4. Spin the cell homogenate at 700 × g for 10 min and collect the supernatant.
5. Spin the supernatant at 100,000 × g for 60 min and discard the supernatant.
6. Suspend the membrane pellet at 5 to 10 mg/ml in homogenizing buffer.
7. Store aliquots of the membrane suspension frozen at −80°C until use.
8. Dilute [^{35}S]GTPγS (NEN, NEG-030H, 250 μCi) solution to 50 nM with Tris-dithiothreitol buffer. Store aliquots of the diluted solution at −80°C (−30°C is not recommended) until use (use within 1 month).

PROTOCOL 1-3:   [^{35}S]GTPγS BINDING ASSAY

1. Thaw frozen-stock membrane suspension, and dilute with GTPγS binding assay buffer [50 mM Tris, 150 mM NaCl, 5 mM MgCl$_2$, 1 μM GDP, and 1 mg/ml bovine serum albumin (BSA); pH 7.4] to 10 to 50 μg/ml. In the study of GALP, GALR1 and GALR2 membranes were diluted to 12 and 20 μg/ml, respectively.
2. Put 0.2 ml of diluted membrane suspension into 5 ml polypropylene tubes (Falcon, 2053).
3. Add 2 μl of test sample (equivalent to 2 g of starting tissue) and 2 μl of 50 nM [^{35}S]GTPγS solution to every tube.
4. Incubate reaction mixtures at 25°C for 60 min.
5. Add 1.5 ml of chilled TEM Buffer (20 mM Tris, 1 mM EDTA, 5 mM MgCl$_2$, 0.1% 3-[(3-cholamidopropyl)dimethylammonio]-1-propanesulfate (CHAPS), and 1 mg/ml BSA; pH 7.4) to reaction mixtures, and filter the mixture through a glass filter (Whatmann, GF/F).
6. Dry the filters at 50°C.
7. Subject the filters to liquid scintillation counting.

1.3.1.4 Isolation of GALP

The result of the [^{35}S]GTPγS binding assay is shown in Figure 1.1. The GALR2-agonistic activity in the porcine hypothalamus was obviously separated into two peaks by HPLC analysis. While the first peak of activity was identified with galanin on the basis of retention time, the component responsible for the second peak, which was very faint in the assay with GALR1-expressing membranes, was unknown at that time. To further characterize the active peptides in the second peak, we proceeded to conduct purification studies (see Figure 1.1 for the procedure; see Figure 1.2 for HPLC profiles).

The active peptide showing GALR2-agonistic activity was purified to a single peak (Figure 1.2) and was further subjected to mass spectrometry and peptide sequencing analysis, including chymotryptic peptide mapping (Figure 1.3). The peptide was found to be a novel peptide with 60 amino acid residues, and its amino acid sequence was finally confirmed by a cDNA cloning study. The peptide was designated "galanin-like peptide (GALP)," because it shared the same 13-amino acid sequence with galanin (Figure 1.3).[21] This sequence was conserved between the pig, rat, human, mouse, and macaque.[21–23]

1.3.1.5 Physiological Roles of GALP

GALP shows high affinity for GALR2 but lower affinity for GALR1, which is in contrast to galanin that shows high affinity for both GALR1 and GALR2. The binding affinity and agonistic activity for GALR3 remains unclear due to difficulty in the functional expression of GALR3. Histological studies have demonstrated that central GALP expression is localized to neurons in the arcuate nucleus of the hypothalamus.[24,25] This population of GALP neurons is characterized as leptin-regulated

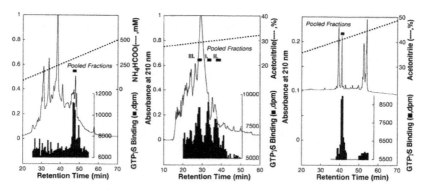

FIGURE 1.2 Purification of GALP using TSKgel CM-2SW, Super Phenyl, and Super ODS HPLC column (Steps 5, 6, and 7). Active fractions eluted from a Sephadex G50 column (Figure 1.1) were injected into a CM-2SW column (4.6 × 250 mm) equilibrated with 10 mM ammonium formate/40% acetonitrile. Elution was performed by a gradient increase of ammonium formate concentration from 10 to 500 mM for 60 min (left). Pooled fractions were next injected into a Super Phenyl column (4.6 × 100 mm) equilibrated with 0.1% TFA. Elution was performed by a gradient increase of acetonitrile concentration from 27 to 33% for 60 min (middle). Pooled fractions (I) were finally subjected to a Super ODS column (4.6 × 100 mm) equilibrated with 0.1% heptafluorobutylic acid. Elution was performed by a gradient increase of acetonitrile concentration from 33 to 48% for 60 min (right). In every step, the flow rate was set at 1 ml/min, and fractions were collected every 0.5 min.

```
APVHRGRGGW TLNSAGYLLG PVLHPPSRAE GGGKGKTALG ILDLWKAIDG LPYPQSQLAS
──────────▶──────────▶──────────────────────────────────▶──────────────▶
   Chy-f1     Chy-f2             Chy-f3                      Chy-f4
──────────────────────────────────────────────────────▶
           native peptide
```

```
                10         20         30         40         50         60
1.APAHRGRGGW TINSAGYLLG PVLHLPQMGD QDGKRETALE I IDLWKAIDG LPYSHPPQPS
2.APAHQGRGGW TINSAGYLLG PVLHLPQMGD QDRKRETALE I IDLWKAIDG LPYSHPLQPS
3.APAHRGRGGW TINSAGYLLG PVLHLSSKAN QGRKTDSALE I IDLWKAIDG LPYSRSPRMT
4.APAHRGRGGW TINSAGYLLG PVLPVSSKAD QGRKRDSALE I IDLWKIIG LPYSHSPRMT
5.APVHRGRGGW TINSAGYLLG PVLHPPSRAE GGGKGKTALG I IDLWKAIDG LPYPQSQLAS
            ** * * * * * * * * ***
6.         GWTINSAGYLLG PHAVGNHRSF SDKNGLTS
7.         GWTINSAGYLLG PHAIDNHRSP SDKHGLT-NH2
```

FIGURE 1.3 Chymotryptic peptide mapping of porcine GALP (upper) and amino acid sequence of GALP (1. human, 2. macaque, 3. rat, 4. mouse, 5. pig) and galanin (6. human, 7. rat). Arrows indicate amino acid sequences determined by a protein sequencer. The positions 44 and 45 were predicted to be $L^{44}W^{45}$ or $H^{44}Y^{45}$ from the m/z value of Chy-f3, and were later confirmed to be $L^{44}W^{45}$ from the cDNA sequence. Asterisks indicate the galanin/GALP-shared sequence.

neurons, along with the proopiomelanocortin neurons, because these GALP neurons co-express leptin receptors,[25] and leptin positively regulates the expression of GALP in the arcuate nucleus.[22,24] It is, therefore, suggested that GALP mediates the functions of leptin, such as the regulation of energy expenditure and reproduction.[26] In fact, intracerebroventricular (icv) injection of GALP reduces body weight gain, increases body temperature,[27] and increases plasma luteinizing hormone levels.[28] Because these responses are not provoked by galanin and are found in both GALR1 knockout (KO) mice and GALR2 KO mice, these responses may be mediated by some GALP-specific receptor but not by GALR1 or by GALR2.

1.3.2 METASTIN

1.3.2.1 Discovery of Metastin, the Cognate Ligand of OT7T175 (= GPR54)

OT7T175 (or GPR54, AXOR12) shows a high sequence similarity to the galanin receptor subtypes GALR1, GALR2, and GALR3. Therefore, this receptor was initially presumed to be the fourth subtype of galanin receptor or a specific receptor for GALP. However, CHO cells expressing human OT7T175 (hOT7T175) did not show any functional responses to galanin or to GALP. This finding was somewhat disappointing to us in our search for a putative GALP-specific receptor. However, this result suggested the possible existence of a novel peptide ligand for OT7T175, which we hoped would be another member of the galanin family. This prompted us to conduct a screening of a variety of tissue extracts (prepared as in Protocol 1-1) for hOT7T175-ligand activity.[29] In the beginning, because we did not know which kind of G proteins this receptor is coupled to, we employed a $[^{35}S]$GTPγS binding assay (in case of $G_{i/o}$ coupling) and fluorometric imaging plate reader (FLIPR) assay (in case of $G_{q/11}$ coupling, see below) in parallel.

1.3.2.2 FLIPR Assay

FLIPR is the fluorometric imaging plate reader system developed by Molecular Devices Co. (Sunnyvale, CA). This system is equipped with an argon laser (488 nm line) that can induce excitation of an indicator dye in each well of a 96-well (or recently 384-well) black-walled plate at once. The system also includes a charge-coupled device camera that can capture a fluorescence image of each well once per second. Each image is converted to digital data and transferred to a computer. Users can retrieve data on the time course of the fluorescence changes, which are stored as Microsoft Excel™ files. The most popular application of this system is for measuring changes in intracellular calcium ion concentration, using fluorescent Ca^{2+} dyes such as Fluo-3, Fluo-4, and Calcium Green-1. For other applications, membrane potential dyes such as DiBAC and a new dye from Molecular Devices are also available.

FLIPR is frequently used to screen for GPCR agonist and antagonist activity by detecting changes in the intracellular calcium ion concentration. This is probably because FLIPR is suitable for high-throughput screening, and is not restricted to $G_{q/11}$-coupled GPCR; it is also applicable to other GPCRs when using the co-expression system of $G\alpha_{16}$. However, we often confront the problem, especially in the screening of tissue extract samples, of high background levels of nonspecific signals that are found equally in the negative control cells. To avoid this problem, doses of tissue extract samples are limited to 1 to 2 g equivalent of tissue, or the tissue extract sample to be screened must be prepurified using ion-exchange chromatography.[30]

PROTOCOL 1-4: FLIPR ASSAY

1. Inoculate transformant cells expressing target GPCR at 30,000 cells on black-walled plates (Corning Costar, 3904) in 0.2 ml of growth medium. Grow cells at 37°C in a 5% CO_2 incubator for 1 day.
2. Dissolve 4.77 g of HEPES into 1000 ml of Hanks' balanced saline solution (HBSS), and adjust pH to 7.4 (preparation of H-HBSS).
3. Dissolve 710 mg of Probenecid (Sigma, P8761) in 5 ml of 1 N NaOH. Put the whole Probenecid solution into 1000 ml of H-HBSS (preparation of Probenecid/H-HBSS).
4. Dissolve one vial (50 µg) of Fluo 3-AM (Dojindo, 349-06961) in 20 µl of DMSO. Mix the Fluo 3-AM solution with 20 µl of 20% Pluronic F-127 (Molecular Probes, P-3000) (preparation of Fluo 3-AM/Pluronic).
5. Put 40 µl of Fluo 3-AM/Pluronic and 100 µl of fetal bovine serum (FBS) into 10 ml of Probenecid/H-HBSS (preparation of Fluo 3-AM/Pluronic/FBS/Probenecid/H-HBSS).
6. Withdraw growth medium from the cell plate, and add 0.1 ml of Fluo 3-AM/Pluronic/FBS/Probenecid/H-HBSS.
7. Incubate cells at 37°C in a 5% CO_2 incubator for 1 h.
8. Dilute 3 µl of test samples (equivalent to 3 g of starting tissues) with 150 µl of BSA/CHAPS/Probenecid/H-HBSS.
9. Put all diluted test samples into a 96-well polypropylene sample plate. Avoid use of polystyrene plates because peptides may stick to the polystyrene wall.
10. Withdraw Fluo 3-AM/Pluronic/FBS/Probenecid/H-HBSS from the cell plate. Wash the cell plate with 0.3 ml of Probenecid/H-HBSS four times (using an automatic plate washer). Put fresh 0.1 ml of Probenecid/H-HBSS into each well.
11. Set the cell plate and sample plate in the FLIPR (Molecular Devices, Inc.), and start the system. The FLIPR is programmed to transfer each 0.05 ml sample from the sample plate to the cell plate. The fluorescence intensity is measured every second for the first 60 s and every 6 s for the next 120 s.

1.3.2.3 Isolation of Metastin

During the screening of HPLC-fractionated samples prepared from various kinds of tissues, we found that human placental extract contained an active peptide that induced a marked and specific increase in the intracellular Ca^{2+} concentration of hOT7T175-expressing CHO cells (Figure 1.4). The [^{35}S]GTPγS binding assay, which

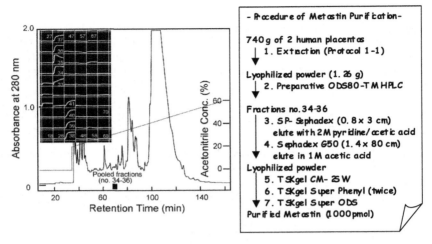

FIGURE 1.4 Screening of agonistic activity for hOT7T175 (left) and further purification procedure (right). The lyophilized powder of human placental extract was prepared following Protocol 1-1, and each 300 mg of powder was analyzed using a TSKgel ODS80-TM (21 × 300 mm) in the same conditions as in Figure 1.1. Fractions were collected every 2 min, and aliquots were lyophilized, dissolved in DMSO, and subjected to a FLIPR assay (inset). Fraction number 35 shows the peak response.

was successful for GALP, found no signals in this case. Active fractions (numbers 34 to 41) were further subjected to purification studies and finally purified to a single peak (see Figure 1.4 for procedure; see Figure 1.5 and Figure 1.6 for HPLC profiles).

1.3.2.4 Structure of Metastin

Following the N-terminal sequencing and mass spectrometric analysis, the isolated peptide was clearly identified as KiSS-1 (68–121) amide (Figure 1.7). The KiSS-1 gene was known as a metastasis-suppressor gene, but its gene product had been uncharacterized.[31] Because the KiSS-1 gene product had a potential signal sequence and two paired basic residues (cleavage sites for prohormone convertases) that flanked the sequence of the isolated peptide (Figure 1.8), we concluded that the KiSS-1 encoded a precursor protein of the regulatory peptide, which we designated "metastin." The amino acid sequences of rat and mouse metastin were later deduced from cDNA sequences.[32] Kotani et al. also reported the same peptide as kisspeptin-54.[33]

1.3.2.5 Receptor Interaction

Synthetic metastin potently induced Ca^{2+} mobilization in hOT7T175-expressing CHO cells with an EC_{50} value in the subnanomolar concentration range. In contrast, the C-terminally unamidated form of metastin was 10,000-fold less potent than metastin. The C-terminal fragment peptides, metastin (40–54) and (45–54), both amidated in the C-terminus, were threefold and tenfold more potent, respectively.

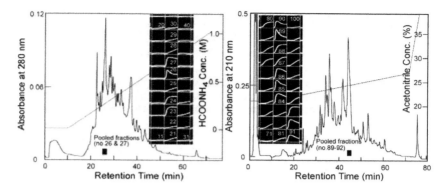

FIGURE 1.5 Purification of metastin from human placental extract (Steps 5 and 6). Active fractions eluted from a Sephadex G50 column (Figure 1.4) were injected into a CM-2SW column (4.6 × 250 mm) equilibrated with 10 mM ammonium formate/10% acetonitrile. Elution was performed by a gradient increase of ammonium formate concentration from 10 to 1000 mM for 60 min (left). Fractions were collected each 1 min and assayed by FLIPR. Peak response was found at fraction number 26 (inset). Pooled fractions were then injected into a Super Phenyl column (4.6 × 100 mm) equilibrated with 0.1% TFA. Elution was performed by a gradient increase of acetonitrile concentration from 21 to 27% for 60 min (left). Fractions were collected each 0.5 min and assayed by FLIPR. Peak response was found at fraction numbers 85 and 89 (inset). Pooled fractions were rechromatographed under the same conditions and subjected to the final purification (Figure 1.6).

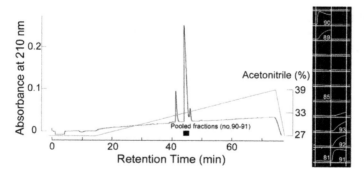

FIGURE 1.6 Purification of metastin from human placental extract (Step 7). Active fractions eluted from a Super Phenyl column (Figure 1.5) were injected into a Super ODS column (4.6 × 100 mm) equilibrated with 27% acetonitrile/0.1% heptafluorobutylic acid. Elution was performed by a gradient increase of acetonitrile concentration from 27 to 39% for 60 min (left). Fractions were collected each 0.5 min and assayed by FLIPR. Peak response was found at fraction number 90 (inset).

Further truncation beyond the Y^{45} residue decreased the agonistic activity significantly. In saturation receptor binding studies, the Kd value of $[^{125}I\text{-}Y^{45}]$ metastin (40–54) and the receptor content of the membrane fraction were determined as 95 pM and 3 pmol/mg protein, respectively. The Ki values from the competitive

1.3.3.2   Isolation and Identification of NPW as the Endogenous Ligand for the GPR7 and GPR8

We constructed an expression vector plasmid, introduced with the cloned human GPR8 gene, and transfected it into CHO cells to establish a CHO cell line expressing the human GPR8. In assays of the intracellular signal changes in the obtained GPR8-expressing CHO cells used for the test samples, including various kinds of tissue extract and known bioactive substances, fractions from an HPLC eluate of porcine hypothalamus extracts showed an inhibitory effect on cAMP accumulation induced by forskolin. It was also shown that the response of the cells was specific for GPR8.

PROTOCOL 1-5:   ASSAY FOR INHIBITION OF FORSKOLIN-INDUCED
INTRACELLULAR ACCUMULATION OF cAMP

1. Plate CHO cells expressing the receptor on 24-well plates at 5×10^4 cell/well and incubate for 48 h.
2. Wash the cells with reaction buffer (HBSS supplemented with 0.2 mM 3-isobutyl-1-methylxanthine (Wako Pure Chemical, Osaka, Japan), 0.05% BSA, and 20 mM HEPES), and expose the cells to the samples with 1 μM forskolin dissolved in the reaction buffer for 24 min.
3. Extract intracellular cAMP with 20% perchloric acid on ice for 1 h.
4. Measure the amount of extracted cAMP using an enzyme-linked immunoassay kit (Amersham Pharmacia Biotech, Piscataway, NJ).

Because the same RP-HPLC fractions also stimulated a [^{35}S]-GTPγS binding to the membrane prepared from the CHO cells expressing the GPR8 receptor, the extracts of porcine hypothalamus were subjected to purification of the ligand for GPR8 by successive chromatography monitoring of the [^{35}S]-GTPγS binding assay. The purification process consisting of four steps of HPLC purification (Scheme 1.1) yielded about 3 pmol of ligand for GPR8 with homogeneity from 500 g of porcine hypothalamus (Figure 1.9).[45]

The partial amino acid sequence of the purified peptide was determined to be WYKHTASPRYHTVGRAAXLL (X, not identified) using a protein sequencer. A cDNA encoding the precursor protein of the ligand peptide was obtained from a porcine spinal cord cDNA library utilizing this partial sequence information of the ligand peptide.[45] Subsequently, the human and rat orthologs of the genes encoding precursor proteins of GPR8 ligand peptides were also cloned utilizing information from the porcine homologue gene and *in silico* information obtained from the databases.

The deduced porcine precursor protein consisting of 152 amino acid residues was predicted to generate two mature peptides of 23 and 30 residues by alternative proteolytic processing at two pairs of basic amino acid residues (Arg-Arg) after removal of the potential signal peptide. Because the 30-residue ligand peptide was characterized as such, the first and last amino acids of the 30-residue peptide are Trp or W as single-letter abbreviations, the peptide was designated as neuropeptide W-30,

Porcine hypothalamus, 0.5 kg
↓
Extracted with 1.0 M AcOH
↓
Precipitated with 66% acetone
↓
Delipidated with diethylether
↓
ODS column (YMCgel ODS-AM 120-S50, 30 x 240 mm)
  ↓ 60% CH_3CN/0.1% TFA (batchwise)
TSKgel ODS-80T$_S$ (21.5 x 300 mm)
  ↓ 10-60% CH_3CN/0.1% TFA (80 min, 5.0 mL/min)
TSKgel SP-5PW (20 × 150 mm)
  ↓ 10-2000 mM ammonium formate/10% CH_3CN (40 min, 5.0 mL/min)
Develosil CN-UG-5 (4.6 × 250 mm)
  ↓ 21-26% CH_3CN/0.1% TFA (20 min, 1.0 mL/min)
Wakosil-II 3C18HG (2.0 × 150 mm)
  ↓ 22.5-32.5% CH_3CN/0.1% TFA (40 min, 0.2 mL/min)
Porcine NPW (3 pmole)

SCHEME 1.1 Purification procedure for NPW from porcine hypothalamus.

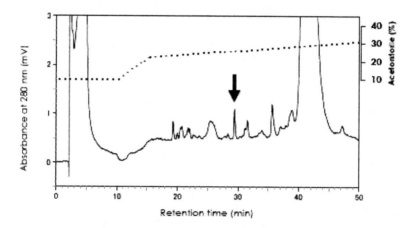

FIGURE 1.9 HPLC profile of the final purification step using a Wakosil-II 3C18HG column. Arrow marks the purified material. The eluate was manually collected, and the activity was recovered as a single peak with an elution time of 29.4 min.

Human   WYKHVASPRYHTVGRAAGLLMGLRRSPYLW
Porcine   WYKHTASPRYHTVGRAAGLLMGLRRSPYMW
Rat   WYKHVASPRYHTVGRASGLLMGLRRSPYLW

FIGURE 1.10 Amino acid sequences of human, porcine, and rat NPW.

or NPW30, and the 23-residue peptide as neuropeptide W-23, or NPW23 (Figure 1.10). Synthetic NPW23 and NPW30 were activated and bound to both GPR7 and GPR8 at similar effective doses, suggesting that both of these peptides are the endogenous ligands for GPR7 and GPR8.[45] The presence of these two peptides in the porcine hypothalamus was indicated by the coincidental retention times in HPLC of the detected agonist activity of the extract with the activity of the synthetic peptides.

1.3.3.3 Neuropeptide B (NPB) as a Paralog Peptide of NPW

After the discovery of NPW, a gene encoding a novel precursor protein, which was expected to generate a mature peptide with 61% identity to NPW, was discovered by searching the human genome database.[19] The endogenous peptide was isolated from the bovine hypothalamus and was found to be modified with bromine atom at the C-6 position of the indole ring of the amino-terminal Trp residue. Thus, this novel peptide was designated as neuropeptide B (after bromine) or NPB.[19] NPB seems to be a relatively specific ligand for GPR7 because it showed higher affinity and more intense agonistic activity to GPR7 than to GPR8. Although the structure of NPB is characterized by its unique modification by bromine, the presence of the bromine atom does not seem to affect the biological activity of the peptide either *in vitro* or *in vivo*.[19,46]

Tanaka et al.[46] also isolated and identified NPB from the extract of bovine hypothalamus as the endogenous ligand for GPR7 by monitoring a melanosome aggregation assay in *Xenopus* melanophores transfected with the human GPR7 cDNA, and they discovered that NPW was the paralog peptide through Expressed Sequence Tag database searches. The discovery of NPW and NPB from database searches based on information from a patent describing the identification of NPW as the ligand of GPR8 (M. Mori et al., WO 01/98494A1) was also reported.[47]

1.3.3.4 Biological Functions of NPW and NPB

NPW and NPB were suspected to be involved in the regulation of feeding behavior because GPR7 is expressed in brain regions responsible for feeding and energy expenditure, such as the arcuate nucleus and paraventricular nucleus of the hypothalamus in the rat brain.[44] In fact, icv-administered NPW evoked feeding behavior in the rat in the light phase.[45,48] However, it was also reported that NPW inhibited feeding in the dark phase, and, furthermore, icv-administered neutralizing antibody against NPW resulted in increased food intake.[49] An increase in energy expenditure was also observed following NPW administration.[49] These results indicated that endogenous NPW may function as an inhibitory factor for feeding behavior. In another experiment, mice centrally administered with NPB showed inhibited food intake after a transient increase of feeding.[46] Furthermore, a targeted disruption of the GPR7 gene resulted in the increase of food intake and adult-onset obesity in male mice.[50] Thus, NPW and NPB have been shown to be novel anorexigenic peptides that function through their receptors, GPR7 and GPR8.

It was also reported that icv-administered NPW increased the blood concentration of prolactin and corticosteron and decreased that of growth hormone, which suggest that these peptides may play a role in the neuroendocrine system.[45,48] Further, GPR7 and GPR8 show some similarity to opioid receptors, and it was noted that icv injection of NPB produces analgesia in rats.[46]

1.3.4 Urotensin II (UII)

1.3.4.1 GPR14

GPR14,[51] or SENR,[52] was an orphan GPCR obtained from the rat genome that showed homology to somatostatin receptor subtype 4 and to the κ, δ, or μ opioid receptors. GPR14 (SENR) was originally characterized by its unique tissue distribution; its mRNA was found to be expressed mainly in neural and sensory tissues, such as the retina, circumvallate papillae, and olfactory epithelium in the rat.[52] However, the GPR14 gene was also reported to be abundant in the heart, arteries, and pancreas in the human.[11]

1.3.4.2 Isolation and Identification of UII as the Endogenous Ligand for GPR14

We constructed an expression vector plasmid, introduced with the cloned rat GPR14 gene, and transfected it into CHO cells to establish a CHO cell line expressing the rat GPR14. Then, using the obtained cell line, we investigated the intracellular secondary signaling evoked by the extracts prepared from several tissues. Among the extracts tested, that from porcine spinal cords most intensely stimulated the release of the arachidonic acid metabolites from the GPR14-expressing cells, and the response of the cells was found to be specific for GPR14.

PROTOCOL 1-6: ASSAY FOR RELEASE OF ARACHIDONIC ACID METABOLITES

1. Plate CHO cells expressing the receptor on 24-well plates at 5×10^4 cell/well and incubate for 24 h.
2. Incorporate 9.25 kBq/well of [^{3}H]-arachidonic acid (NEN Life Science Products, Boston, MA) into the cells, and incubate the cell for 16 h.
3. Wash the cells with HBSS supplemented with 0.05% BSA, and expose the cells to the samples dissolved in 500 μl of HBSS supplemented with 0.05% BSA.
4. After incubation for 30 min, mix 350 μl of culture supernatant with scintillation cocktail, and measure the released radioactivity using a scintillation counter.

Porcine spinal cord, 1.0 kg

↓

Extracted with 70% acetone/20 mM HCl/1.0 M AcOH

↓

Delipidated with diethylether

↓

ODS column (YMCgel ODS-AM 120-S50, 30 × 240 mm)

↓ 60% CH_3CN/0.1% TFA (batchwise)

TSKgel ODS-80T$_S$ (21.5 × 300 mm)

↓ 10-60% CH_3CN/0.1% TFA (80 min, 5.0 mL/min)

TSKgel SP-5PW (20 × 150 mm)

↓ 10-300 mM ammonium formate/10% CH_3CN (30 min, 5.0 mL/min)

Vydac 219-TP54 (diphenyl) (4.6 x 250 mm)

↓ 26-31% CH_3CN/0.1% TFA (20 min, 1.0 mL/min)

Develosil CN-UG-5 (4.6 × 250 mm)

↓ 28.5-33.5% CH_3CN/0.1% TFA (20 min, 1.0 mL/min)

Wakosil-II 3C18HG (2.0 × 150 mm)

↓ 30-37.5% CH_3CN/0.1% heptafluorobutyric acid (30 min, 0.2 mL/min)

Porcine UII-1 and UII-2 (10 pmole)

SCHEME 1.2 Purification procedure for UII from porcine spinal cord.

Starting with 50 porcine spinal cords (about 1.0 kg), the two active substances were purified using a combination of HPLC processes. In the first step, using a semipreparative C18 column, the activity was recovered in two fractions, and they were separately subjected to the same subsequent chromatographic procedure that included five steps of HPLC purification (Scheme 1.2). Finally, we obtained two active substances that showed different behaviors in each purification step (Figure 1.11). The isolated materials, estimated at about 10 pmol, were subjected to N-terminal amino acid sequence analysis using a protein sequencer. The sequences of active substances were determined as GPTSECFWKYCV and GPPSECF-WKYCV, which showed homology to fish or human urotensin II (UII)[53,54] (Figure 1.12); thus, the ligands of the GPR14 obtained from the porcine spinal cord were identified as two molecular species of porcine homologue peptides of UII (porcine UII-1 and UII-2).[55]

Porcine UII-1 and UII-2 were chemically synthesized using a solid-phase peptide synthesizer. The chromatographic behaviors of the active substances isolated from the spinal cords were indistinguishable from those of the synthetic peptides. The synthetic porcine UII-1 and UII-2 evoked the release of arachidonic acid metabolites from the GPR14-expressing CHO cells in a dose-dependent manner, with an estimated EC_{50} value of 1.0 nM. Both somatostatin and cortistatin, which share a partial amino acid sequence with UII, Phe-Trp-Lys in a Cys-Cys ring structure, failed to show any activity.

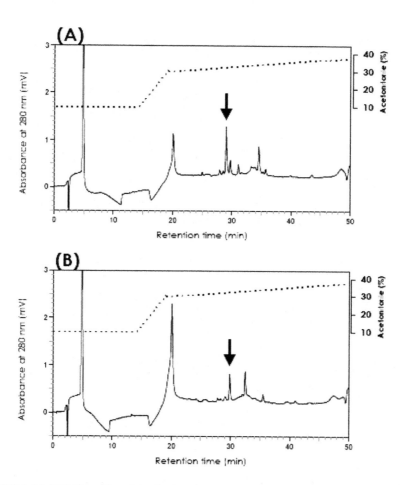

FIGURE 1.11 HPLC profiles of the final purification step on a Wakosil-II 3C18HG column. Arrow marks the purified material. The eluate was manually fractionated, and activity was recovered as peaks at 32.2% (A) or 32.5% (B) of acetonitrile.

Although three other groups independently and almost simultaneously reported the identification of UII as the cognate ligand for GPR14,[11,56,57] and several reports revealed the presence of the genes encoding the mammalian homologs of UII precursor protein,[54,58,59] ours was the only group to report the presence of UII in mammalian tissue in the functional mature form generated from its precursor protein.

We isolated UII-like peptide, designated UII-related peptide or URP, as the sole UII-immunoreactive material in the rat brain and demonstrated that URP also possessed a highly specific agonist activity against the GPR14 receptor.[60] It is likely that URP is the molecular species responsible for the biological functions through its receptor, GPR14, in both the rat and mouse.

Human	ET –PDCFWKYCV
Porcine –1	GPT– SECFWKYCV
Porcine –2	GPP –SECFWKYCV
Frog	AGNLSECFWKYCV
Goby	AGT –ADCFWKYCV
Sucker A	GSG–AECFWKYCV
Carp α	GGG– ADCFWKYCV
Lamptera	NNF – SDCFWKYCV

URP (human, rat, mouse)

ACFWKYCV

FIGURE 1.12 Amino acid sequences of human, porcine, and fish UII, and human, rat, and mouse URP.

1.3.4.3 Pathophysiological Significance of UII

UII is a piscine neuropeptide originally isolated from the teleost urophysis. It is involved in the cardiovascular regulation, osmoregulation, and regulation of lipid metabolism in fish.[53] It has been reported that intravenously administrated UII caused myocardial contractile dysfunction in nonhuman primates,[11] suggesting that UII might function as a novel cardiovascular peptide in the pathophysiology of circular diseases, such as ischemic heart failure in mammals. However, UII did not necessarily show significant vasoconstrictive activity to isolated human vessels,[61–64] and, furthermore, the *in vivo* effects of UII by systemic infusion in the human were not remarkable.[65,66] These results demonstrate that the action of UII in the human cardiovascular system is not obvious. However, there are reports of the strong expression of UII mRNA in cardiomyocytes of patients with congestive heart failure[67] and of the upregulation of UII-immunoreactivity and UII binding sites in the right ventricle in the experimental pulmonary hypertensive rat[68] and in the left ventricle in a rat model of heart failure after myocardial infarction.[69] Furthermore, UII was observed to activate extracellular signal-regulated kinases in cultured rat cardiomyocytes.[70] These results indicate that UII and its receptor participate in cardiomyocyte hypertrophy and are involved in the development of heart failure by promoting cardiac remodeling.

1.3.5 FREE FATTY ACIDS (FFAS)

1.3.5.1 FFAs Are Signaling Molecules

Although free fatty acids (FFAs) are known as an important energy source as nutrients, recent studies have suggested that they also work as signaling molecules, for example, as ligands for peroxisome proliferators-activated receptor α, which regulates target genes involved in lipid metabolism, and that they are also closely linked to insulin resistance and diabetes. In fact, plasma FFA levels are frequently elevated in diabetes. In pancreatic β cells, which are the only cells capable of secreting insulin, FFAs acutely enhance insulin secretion and support the glucose response.[71] On the other hand, prolonged exposure to FFAs reduces insulin secretion and impairs β cell functions (lipotoxicity).[72] In any case, the molecular mechanism behind these effects of FFAs on pancreatic β cells remains unclear.

1.3.5.2 GPR40

GPR40 was an orphan GPCR identified originally from a human genomic DNA fragment.[73] The GPR40 gene is located downstream of CD22 on chromosome 19q13.1. Three other GPCR genes (GPR41, GPR42, and GPR43) cluster in close to the GPR40 gene in the human genome. GPR40 has 28 to 30% amino acid identity with GPR41, GPR42, and GPR43. In rats, the highest level of expression of GPR40 mRNA was detected in the pancreas, suggesting that GPR40 functions in the pancreas.[18,74,75] To more precisely determine the distribution of GPR40 mRNA in the pancreas, we prepared islets from rats and examined its expression in the islets. The expression level of GPR40 mRNA and insulin mRNA were approximately 80 times greater in the islets than in the pancreas as a whole, suggesting that GPR40 mRNA is predominantly expressed in the β cells of the islets. Additionally, GPR40 mRNA was expressed markedly in pancreatic β cell lines, with its highest level detected in MIN6 cells (mouse β cells). However, our group found no evidence of GPR40 mRNA expression in any other cell lines, including a pancreatic α cell line. Furthermore, to confirm the cell types in which GPR40 mRNA is expressed, we performed in situ hybridization in rat islets. Hybridization signals were detected in the insulin-immunoreactive regions of the islets, indicating that GPR40 mRNA is expressed in the β cells of the islets.[74]

1.3.5.3 Identification of FFAs as Ligands for GPR40

CHO cells transiently expressing human GPR40 cDNA were exposed to over 1000 kinds of chemical compounds, and their intracellular Ca^{2+} influx was assessed using a FLIPR assay system. Long-chain FFAs were found to evoke specific Ca^{2+} mobilization in these cells (Figure 1.13a).[74] No significant responses were detected to these FFAs in CHO cells expressing other GPCRs, including GPR41, GPR42, and GPR43 (data not shown). CHO cells stably expressing human GPR40 (CHO-hGPR40), mouse GPR40 (CHO-mGPR40), and rat GPR40 (CHO-rGPR40) were used for precise determination of the Ca^{2+} influx-inducing activities of various FFAs (Table 1.1).[74] Apparent stimulatory activities were detected in C12- to 16-length saturated FFAs and in both C18- and C22-length unsaturated FFAs. Among the FFAs tested, docosahexaenoic acid (DHA), linoleic acid, α-linolenic acid, oleic acid, γ-linolenic acid, and elaidic acid were highly potent to human GPR40. In contrast, methyl linoleate did not show stimulatory activity, suggesting that the carboxyl group is indispensable for stimulating GPR40. The stimulatory activities of FFAs to GPR40 were generally consistent among the human, rat, and mouse, although DHA showed

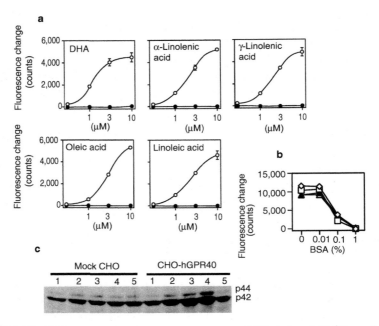

FIGURE 1.13 (a) Representative dose-response curves of FFA-induced $[Ca^{2+}]_i$ rise in CHO-hGPR40. ○,CHO-hGPR40; ●, control cells (CHO cells expressing human histamin H1 receptor). Data represent the mean values ± S.E.M. of the maximal fluorescent changes induced by FFAs in three separate experiments with a FLIPR. (b) Effects of BSA on the FFA-induced $[Ca^{2+}]_i$ rise in CHO-hGPR40. □, γ-linolenic acid; ◇, linoleic acid; △, oleic acid; ▲, arachidonic acid. Each trace represents the mean value in duplicate assays at 10 μM of FFAs. (c) MAP kinase activation in CHO-hGPR40 induced by FFAs. Lanes 1, no FFA; 2, linoleic acid; 3, oleic acid; 4, DHA; 5, ML. Treatments were for 10 min.

a lower potency to mouse and rat GPR40 than to human GPR40. We found that BSA at more than 0.1% inhibited the Ca^{2+} influx-inducing activities of FFAs on CHO-hGPR40, suggesting that BSA may mask the agonistic potency of chemicals, such as FFAs, that bind to BSA (Figure 1.13b).[74] To demonstrate that CHO-hGPR40 specifically responded to FFAs, we prepared CHO cells expressing a fusion protein of human GPR40 and green fluorescent protein (GFP) (CHO-hGPR40-GFP) and then examined the Ca^{2+} influx in these cells. We confirmed that the fusion protein was localized at the plasma membrane, and that the CHO-hGPR40-GFP cells responded to DHA, while the control CHO cells not expressing the fusion protein did not (data not shown). In addition, we found that long-chain FFAs slightly decreased cAMP production in forskolin-stimulated CHO-hGPR40 cells (data not shown). These results suggest that GPR40 couples to $G\alpha_{q/i}$. We subsequently examined the activation of mitogen-activated protein (MAP) kinase in CHO-hGPR40 cells after stimulation with DHA. Treatment with DHA resulted in a specific and rapid increase of phosphorylated MAP kinase (i.e., p44/42) after 5 to 20 min, which returned to its base level after 120 min (Figure 1.13c).[74]

TABLE 1.1
The Potency of Fatty Acids to Induce Ca²⁺ Influx in CHO Cells Expressing GPR40

FFA	(EC_{50}, µM)		
	Human	Mouse	Rat
Acetic acid (C2)	Inactive	Inactive	Inactive
Butyric acid (C4)	Inactive	Inactive	Inactive
Caproic acid (C6)	Inactive	Inactive	>300
Caprylic acid (C8)	>300	Inactive	>300
Capric acid (C10)	43 ± 2.2	>100	>100
Lauric acid (C12)	5.7 ± 1.4	5.6 ± 1.6	13 ± 3.3
Myristic acid (C14)	7.7 ± 1.4	6.0 ± 0.8	7.3 ± 0.5
Palmitic acid (C16)	6.8 ± 0.5	4.6 ± 1.2	6.6 ± 0.4
Stearic acid (C18)	>300	>300	>300
Oleic acid (C18:1)	2.0 ± 0.3	2.7 ± 0.5	3.4 ± 0.4
Elaidic acid (C18:1)	4.7 ± 0.4	6.5 ± 1.5	11 ± 1.5
Linoleic acid (C18:2)	1.8 ± 0.1	2.9 ± 0.3	4.1 ± 0.5
Methyl linoleate	Inactive	Inactive	Inactive
α-Linolenic acid (C18:3)	2.0 ± 0.3	3.6 ± 0.3	4.0 ± 0.7
γ-Linolenic acid (C18:3)	4.6 ± 1.6	5.2 ± 0.6	5.4 ± 1.1
Arachidonic acid (C20:4)	2.4 ± 0.6	5.4 ± 0.8	8.0 ± 0.6
Eicosapentaenoic acid (C20:5)	2.3 ± 0.4	4.9 ± 0.8	9.8 ± 0.6
Docosahexaenoic acid (C22:6)	1.1 ± 0.3	16 ± 4.7	13 ± 1.7

Note: Ca²⁺ influx induced in CHO-hGPR40, CHO-mGPR40 and CHO-rGPR40 cells by the indicated samples was measured with a fluorometric imaging plate reader (FLIPR). EC_{50} (concentration of a sample that produces 50% of the maximal response) was calculated from dose–response curves. "Inactive" indicates no response at 300 µM. Data represent mean values ± S.E.M. in three to six assays.

1.3.5.4 Role of GPR40 Expressed in β Cells

To examine the role of GPR40 expressed in β cells, we treated MIN6 cells with FFAs and examined the effect of each FFA on insulin secretion. Oleic acid, linolenic acid, α-linolenic acid, and γ-linolenic acid stimulated insulin secretion from the MIN6 cells, but methyl linoleate and butyric acid did not (data not shown). The stimulatory activities of oleic acid and linoleic acid on insulin secretion from MIN6 cells were also detected under high-glucose conditions (11 and 22 mM), indicating that FFAs amplified glucose-stimulated insulin secretion from pancreatic β cells (Figure 1.14).[74] We next performed experiments to inhibit the expression of GPR40 in MIN6 cells by small interfering RNA (siRNA). The previously observed increase of insulin secretion from MIN6 cells after stimulation with linoleic acid and γ-linolenic acid was apparently eliminated by treatment with a siRNA specific for mouse GPR40.[74] These results suggest that at least part of the stimulation of insulin secretion by these FFAs is via GPR40.

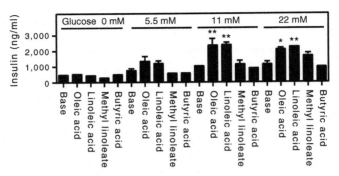

FIGURE 1.14 FFA-induced insulin secretion from MIN6. Data represent the mean values ± S.E.M. in quadruplicate assays at 10 µM of FFAs.

Several studies have shown that FFAs modulate insulin secretion. However, our results suggest that GPR40 may play a role in at least one of the mechanisms responsible for the enhancement of glucose-dependent insulin secretion by FFAs. The discovery of a cell-surface FFA receptor on pancreatic β cells will help to clarify the relationship between FFAs and insulin secretion, and thus may lead to the development of new antidiabetic drugs.

1.3.6 Pyroglutamylated RFamide Peptide (QRFP)

1.3.6.1 RFamide Peptides in Mammals

The first report on a peptide with RFamide involved the isolation of FMRFamide from bivalve mollusks. Since then, a number of bioactive peptides with the same structure have been found throughout the animal kingdom — these have been designated RFamide peptides.[76] In mammals, four RFamide peptide genes have been identified, namely, neuropeptide FF (NPFF),[77,78] PrRP,[3] RFRP,[19] and metastin.[29] In addition, all of their receptors have been identified through orphan GPCR research. Based on the identification of PrRP and RFRP, we proposed that a variety of RFamide peptides exist and have physiological functions, even in mammals. Here, we show the identification of pyroglutamylated RFamide peptide (QRFP) utilizing a human genome database.[20]

1.3.6.2 Identification of a Novel RFamide Peptide Gene

We searched for unknown members of the RFamide peptide family in the human genome database using queries to detect repetitive patterns generating RFamide peptide (i.e., RFGR or RFGK, where RF is followed by G as an amide donor and by R or K as a proteolytic cleavage site) and a secretory signal peptide sequence upstream of the patterns, as reported previously.[19] This search revealed a human genomic sequence that possibly encoded an RFamide peptide (i.e., QRFP). On the basis of the sequences detected, we isolated human, bovine, rat, and mouse cDNAs with full coding regions (Figure 1.15). Two RFGR motifs were found in the human preproprotein. The motif at the C-terminal side was conserved among the different species, but that at the N-terminal side was not. Based on these sequence analyses,

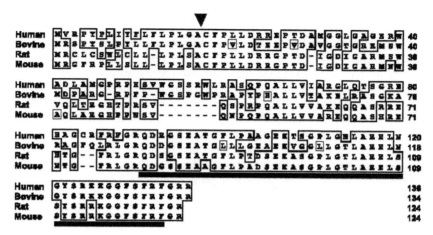

FIGURE 1.15 Amino acid sequences of human, bovine, rat, and mouse QRFP prepropro-teins. The closed arrowhead shows the predicted cleavage site of the N-terminal secretory signal peptide sequence. The fully active structure of QRFP is underlined. Residues that are identical in at least two of the species are boxed. Amino-acid numbers are shown on the right.

we predicted that an RFamide peptide (QRFP) would be produced from the C-terminal motif in the human preproprotein.

1.3.6.3 Identification of a Receptor for QRFP

AQ27 is a novel GPCR that we isolated from human brain poly(A)+RNA based on public genome information (accession No. AQ270411), and that is identical to GPR103.[79] We isolated its rat and mouse counterparts from their respective brain poly(A)+RNA by reverse transcription polymerase chain reaction (RT-PCR). Among the ligand-known GPCRs, AQ27 showed 30% and 32% amino acid identity with OT7T022 (the receptor for RFRP) and HLWAR77 (the receptor for NPFF), respec-tively. We therefore inferred that QRFP might act as a ligand for AQ27. To investigate this, we synthesized a short peptide with an amino acid length of seven (GGFSFR-Famide). We then subjected this short peptide to an assay with HEK293 cells transiently expressing AQ27 and a reporter gene (CRE-luciferase). Because AQ27 coupled to $G\alpha_q$, we monitored the activation of AQ27 treated with the peptide by increased luciferase activity. However, the agonistic activity of this peptide was very weak. We then considered that a longer form of the peptide would show full activity. To investigate this, we expressed the human QRFP cDNA in CHO cells and examined whether more effective peptidic ligands were secreted in the culture supernatant. As we were able to detect specific stimulatory activity on HEK293 cells expressing AQ27 in the culture supernatant, we purified the ligand for AQ27 from the culture supernatant. As a result, we determined the structure of the purified peptide to be <EDEGSEATGFLPAAGEKTSGPLGNLAEELNGYSRKKGGFSFRFamide (<E indicates pyroglutamic acid). We therefore designated it QRFP, for pyroglutamylated RFamide peptide. In the reporter gene assays, the purified peptide was estimated to show agonistic activity even at 10^{-10} to 10^{-9} M.

1.3.6.4   Interaction between QRFP and AQ27

We next determined the intracellular Ca^{2+} and cAMP changes in a stable CHO cell line expressing AQ27 (CHO-hAQ27) after stimulation with QRFP. We detected a rapid rise of intracellular Ca^{2+} concentrations in CHO-hAQ27 after the stimulation. We also detected the suppression of cAMP production in these cells (Table 1.2). These results suggest that AQ27 couples to the G proteins, $G\alpha_{i/o}$ and $G\alpha_q$, in CHO cells. Because the purified QRFP had a long N-terminal region, we chemically synthesized various forms of QRFP and determined their cAMP-production-inhibitory activities. QRFP inhibited the cAMP production of CHO-hAQ27 in a dose-dependent manner. QRFP(43), which is a 43-amino-acid-length QRFP, was found to be the most potent. Serial deletion of the N-terminal region gradually attenuated the inhibitory activity of QRFP. QRFP(26OH), which is the nonamidated form of the peptide, showed no evident inhibitory activities. To examine the binding of QRFPs with AQ27, we labeled human QRFP(43) with ^{125}I. Scatchard plot analysis indicated that the membrane fractions of CHO-hAQ27 had a single class of high-affinity binding sites for [^{125}I-Tyr^{32}] QRFP at the dissociation constant (*Kd*) of 5.3×10^{-11} M and maximal binding sites (*Bmax*) of 0.51 pmol mg^{-1} protein. In competitive binding experiments, various lengths of QRFP inhibited the binding of [^{125}I-Tyr^{32}] QRFP with AQ27 (Table 1.2). QRFP(43) proved to be the most potent, even in the competition assays.

1.3.6.5   Functions of QRFP and Its Receptor

QRFP mRNA was highly expressed in the hypothalamus, optic nerve, trachea, and mammary gland in rats. On the other hand, high levels of AQ27 mRNA were detected in the central nervous system, adrenal gland, and testis. The very high expression of AQ27 mRNA in the adrenal gland, especially in the zona glomerulosa of the cortex, suggested that QRFP affected the function of the adrenal glands in rats. We found that intraveneous injection of QRFP resulted in an increase of plasma aldosterone in rats. These results suggest that QRFP and its receptor play a regulatory role in aldosterone secretion from the adrenal cortex in rats. Future studies on QRFP and its receptor will contribute to our understanding of the physiological significance of RFamide peptides in mammals.

ACKNOWLEDGMENTS

The authors wish to thank Drs. Yasuaki Itoh and Shoji Fukusumi for their contributions to the preparation of this chapter.

TABLE 1.2
Interaction of Various Lengths of QRFP with AQ27

Name	Sequence	Binding Assay IC_{50} (nM)	cAMP Assay EC_{50} (nM)
QRFP43	<EDEGSEATGFLPAAGEKTSGPLGNLAEELNGYSRKKGGFSFRF-NH$_2$	0.52	2.7
QRFP26	TSGPLGNLAEELNGYSRKKGGFSFRF-NH$_2$	3.2	3.9
QRFP26OH	TSGPLGNLAEELNGYSRKKGGFSFRF-OH	1200	>1000
QRFP23	PLGNLAEELNGYSRKKGGFSFRF-NH$_2$	15	8.9
QRFP(19)	LAEELNGYSRKKGGFSFRF-NH$_2$	49	440
QRFP(15)	LNGYSRKKGGFSFRF-NH$_2$	37	460
QRFP(10)	RKKGGFSFRF-NH$_2$	220	>1000
QRFP(8)	KGGFSFRF-NH$_2$	2900	>1000
QRFP(7)	GGFSFRF-NH$_2$	>10,000	>1000

Note: Pyroglutamic acid is shown as <E.

ABBREVIATIONS

BSA	bovine serum albumin
CHAPS	3-[(3-cholamidopropyl)dimethylammonio]-1-propanesulfate
CHO	Chinese hamster ovary
DHA	docosahexaenoic acid
DMSO	dimethylsulfoxide
EDTA	ethylenediaminetetraacetic acid
FBS	fetal bovine serum
FFA	free fatty acid
FLIPR	fluorometric imaging plate reader
GALP	galanin-like peptide
GDP	guanosinediphosphate
GFP	green fluorescent protein
GPCR	G protein-coupled receptor
GTP	guanosinetriphosphate
GTPγS	guanosine $5'O$-(γ-thio)triphosphate
HBSS	Hanks' balanced saline solution
HPLC	high-performance liquid chromatography
icv	intracerebroventricular, intracerebroventricularly
IHH	idiopathic hypogonadotropic hypogonadism
MAP	mitogen-activated protein
NPB	neuropeptide B
NPFF	neuropeptide FF
NPW	neuropeptide W
PrRP	prolactin-releasing peptide
QRFP	pyroglutamylated RFamide peptide
RFRP	RFamide-related peptide
RP	reversed phase
siRNA	small interfering RNA
TFA	trifluoroacetic acid
UII	urotensin II
URP	urotensin II-related peptide

REFERENCES

1. Drews, J., Drug discovery: a historical perspective, *Science*, 287, 1960, 2000.
2. Wise, A., Jupe, S.C., and Rees, S., The identification of ligands at orphan G-protein coupled receptors, *Ann. Rev. Pharmacol.* Toxicol., 44, 43, 2004.
3. Hinuma, S. et al., A prolactin-releasing peptide in the brain, *Nature*, 393, 272, 1998.
4. Hinuma, S., Onda, H., and Fujino, M., The quest for novel bioactive peptides utilizing orphan seven-transmembrane-domain receptors, *J. Mol. Med.*, 77, 495, 1999.
5. Meunier, J.C. et al., Isolation and structure of the endogenous agonist of opioid receptor-like ORL1 receptor, *Nature*, 377, 532, 1995.
6. Reinscheid, R.K. et al., Orphanin FQ: a neuropeptide that activates an opioid like G protein-coupled receptor, *Science*, 270, 792, 1995.

7. Sakurai, T. et al., Orexins and orexin receptors: a family of hypothalamic neuropeptides and G protein-coupled receptors that regulate feeding behavior, *Cell*, 92, 573, 1998.
8. Tatemoto, K. et al., Isolation and characterization of a novel endogenous peptide ligand for the human APJ receptor, *Biochem. Biophys. Res. Commun.*, 251, 471, 1998.
9. Kojima, M. et al., Ghrelin is a growth-hormone-releasing acylated peptide from stomach, *Nature*, 402, 656, 1999.
10. Chambers, J. et al., Melanin-concentrating hormone is the cognate ligand for the orphan G-protein-coupled receptor SLC-1, *Nature*, 400, 261, 1999.
11. Ames, R.S. et al., Human urotensin-II is a potent vasoconstrictor and agonist for the orphan receptor GPR14, *Nature*, 401, 282, 1999.
12. Feighner, S.D. et al., Receptor for motilin identified in the human gastrointestinal system, *Science*, 284, 2184, 1999.
13. Szekeres, P.G. et al., Neuromedin U is a potent agonist at the orphan G protein-coupled receptor FM3, *J. Biol. Chem.*, 275, 20,247, 2000.
14. Xu, Y. et al., Sphingosylphosphorylcholine is a ligand for ovarian cancer G-protein-coupled receptor 1, *Nat. Cell Biol.*, 2, 261, 2000.
15. Kabarowski, J.H. et al., Lysophosphatidylcholine as a ligand for the immunoregulatory receptor G2A, *Science*, 293, 702, 2001.
16. Maruyama, T. et al., Identification of membrane-type receptor for bile acids (M-BAR), *Biochem. Biophys. Res. Commun.*, 298, 714, 2002.
17. Briscoe, C.P. et al., The orphan G protein-coupled receptor GPR40 is activated by medium and long chain fatty acids, *J. Biol. Chem.*, 278, 11,303, 2003.
18. Hinuma, S. et al., New neuropeptides containing carboxy-terminal RFamide and their receptor in mammals, *Nat. Cell Biol.*, 2, 703, 2000.
19. Fujii, R. et al., Identification of a neuropeptide modified with bromine as an endogenous ligand for GPR7, *J. Biol. Chem.*, 277, 34,010, 2002.
20. Fukusumi, S. et al., A new peptidic ligand and its receptor regulating adrenal function in rats, *J. Biol. Chem.*, 278, 46,387, 2003.
21. Ohtaki, T. et al., Isolation and cDNA cloning of a novel galanin-like peptide (GALP) from porcine hypothalamus, *J. Biol. Chem.*, 274, 37,041, 1999.
22. Kumano, S. et al., Changes in hypothalamic expression levels of galanin-like peptide in rat and mouse models support that it is a leptin-target peptide, *Endocrinology*, 144, 2634, 2003.
23. Cunningham, M.J., Scarlett, J.M., and Steiner, R.A., Cloning and distribution of galanin-like peptide mRNA in the hypothalamus and pituitary of the macaque, *Endocrinology*, 143, 755, 2002.
24. Jureus, A. et al., Galanin-like peptide (GALP) is a target for regulation by leptin in the hypothalamus of the rat, *Endocrinology*, 141, 2703, 2000.
25. Takatsu, Y. et al., Distribution of galanin-like peptide in the rat brain, *Endocrinology*, 142, 1626, 2001.
26. Gottsch, M.L., Clifton, D.K., and Steiner, R.A., Galanin-like peptide as a link in the integration of metabolism and reproduction, *Trends Endocrinol. Metab.*, 15, 215, 2004.
27. Hansen, K.R. et al., Activation of the sympathetic nervous system by galanin-like peptide: a possible link between leptin and metabolism, *Endocrinology*, 144, 4709, 2003.
28. Matsumoto, H. et al., Stimulation effect of galanin-like peptide (GALP) on luteinizing hormone-releasing hormone-mediated luteinizing hormone (LH) secretion in male rats, *Endocrinology*, 142, 3693, 2001.

29. Ohtaki, T. et al., Metastasis suppressor gene KiSS-1 encodes peptide ligand of a G-protein-coupled receptor, *Nature*, 411, 613, 2001.
30. Masuda, Y. et al., Isolation and identification of EG-VEGF/prokineticins as cognate ligands for two orphan G-protein-coupled receptors, *Biochem. Biophys. Res. Commun.*, 293, 396, 2002.
31. Lee, J.H. et al., KiSS-1, a novel human malignant melanoma metastasis-suppressor gene, *J. Natl. Cancer Inst.*, 88, 1731, 1996.
32. Terao, Y. et al., Expression of KiSS-1, a metastasis suppressor gene, in trophoblast giant cells of the rat placenta, *Biochim. Biophys. Acta*, 1678, 102, 2004.
33. Kotani, M. et al., The metastasis suppressor gene KiSS-1 encodes kisspeptins, the natural ligands of the orphan G protein-coupled receptor GPR54, *J. Biol. Chem.*, 276, 34,631, 2001.
34. Hori, A. et al., Metastin suppresses the motility and growth of CHO cells transfected with its receptor, *Biochem. Biophys. Res. Commun.*, 286, 958, 2001.
35. Bilban, M. et al., Kisspeptin-10, a KiSS-1/metastin-derived decapeptide, is a physiological invasion inhibitor of primary human trophoblasts, *J. Cell. Sci.*, 117, 1319, 2004.
36. Horikoshi, Y. et al., Dramatic elevation of plasma metastin concentrations in human pregnancy: metastin as a novel placenta-derived hormone in humans, *J. Clin. Endocrinol. Metab.*, 88, 914, 2003.
37. de Roux, N. et al., Hypogonadotropic hypogonadism due to loss of function of the KiSS1-derived peptide receptor GPR54, *Proc. Natl. Acad. Sci. USA*, 100, 10,972, 2003.
38. Seminara, S.B. et al., The GPR54 gene as a regulator of puberty, *New Engl. J. Med.*, 349, 1614, 2003.
39. Funes, S. et al., The KiSS-1 receptor GPR54 is essential for the development of the murine reproductive system, *Biochem. Biophys. Res. Commun.*, 312, 1357, 2003.
40. Gottsch, M.L. et al., A role for kisspeptins in the regulation of gonadotropin secretion in the mouse, *Endocrinology*, 145, 4073, 2004.
41. Navarro, V.M. et al., Developmental and hormonally regulated messenger ribonucleic acid expression of KiSS-1 and its putative receptor GPR54 in rat hypothalamus and potent LH releasing activity of KiSS-1 peptide, *Endocrinology*, 145, 4565, 2004.
42. Matsui, H. et al., Peripheral administration of metastin induces marked gonadotropin release and ovulation in the rat, *Biochem. Biophys. Res. Commun.*, 320, 383, 2004.
43. O'Dowd, B.F. et al., The cloning and chromosomal mapping of two novel human opioid-somatostatin-like receptor genes, GPR7 and GPR8, expressed in discrete areas of the brain, *Genomics*, 28, 84, 1995.
44. Lee, D.K. et al., Two related G protein-coupled receptors: the distribution of GPR7 in rat brain and the absence of GPR8 in rodents, *Mol. Brain Res.*, 71, 96, 1999.
45. Shimomura, Y. et al., Identification of neuropeptide W as the endogenous ligand for orphan G-protein-coupled receptors GPR7 and GPR8, *J. Biol. Chem.*, 277, 35,826, 2002.
46. Tanaka, H. et al., Characterization of a family of endogenous neuropeptide ligands for the G protein-coupled receptors GPR7 and GPR8, *Proc. Natl. Acad. Sci. USA*, 100, 6251, 2003.
47. Brezillon, S. et al., Identification of natural ligands for the orphan G protein-coupled receptors GPR7 and GPR8, *J. Biol. Chem.*, 278, 776, 2003.
48. Baker, J.R. et al., Neuropeptide W acts in brain to control prolactin, corticosterone, and growth hormone release, *Endocrinology*, 144, 2816, 2003.

49. Mondal, M.S. et al., A role for neuropeptide W in the regulation of feeding behavior, *Endocrinology*, 144, 4729, 2003.
50. Ishii, M. et al., Targeted disruption of GPR7, the endogenous receptor for neuro-peptides B and W, leads to metabolic defects and adult-onset obesity, *Proc. Natl. Acad. Sci. USA*, 100, 10,540, 2003.
51. Marchese, A. et al., Cloning and chromosomal mapping of three novel genes, GPR9, GPR10, and GPR14, encoding receptors related to interleukin 8, neuropeptide Y, and somatostatin receptors, *Genomics*, 29, 335, 1995.
52. Tal, M. et al., A novel putative neuropeptide receptor expressed in neural tissue, including sensory epithelia, *Biochem. Biophys. Res. Commun.*, 209, 752, 1995.
53. Conlon, J.M. et al., Distribution and molecular forms of urotensin II and its role in cardiovascular regulation in vertebrates, *J. Exp. Zool.*, 275, 226, 1996.
54. Coulouarn, Y. et al., Cloning of the cDNA encoding the urotensin II precursor in frog and human reveals intense expression of the urotensin II gene in motoneurons of the spinal cord, *Proc. Natl. Acad. Sci. USA*, 95, 15,803, 1998.
55. Mori, M. et al., Urotensin II is the endogenous ligand of a G-protein-coupled orphan receptor, SENR (GPR14), *Biochem. Biophys. Res. Commun.*, 265, 123, 1999.
56. Liu, Q. et al., Identification of urotensin II as the endogenous ligand for the orphan G-protein-coupled receptor GPR14, *Biochem. Biophys. Res. Commun.*, 266, 174, 1999.
57. Nothacker, H.P. et al., Identification of the natural ligand of an orphan G-protein-coupled receptor involved in the regulation of vasoconstriction, *Nat. Cell Biol.*, 1, 383, 1999.
58. Coulouarn, Y. et al., Cloning, sequence analysis and tissue distribution of the mouse and rat urotensin II precursors, *FEBS Lett.*, 457, 28, 1999.
59. Elshourbagy, N.A. et al., Molecular and pharmacological characterization of genes encoding urotensin-II peptides and their cognate G-protein-coupled receptors from the mouse and monkey, *Br. J. Pharmacol.*, 136, 9, 2002.
60. Sugo, T. et al., Identification of urotensin II-related peptide as the urotensin II-immunoreactive molecule in the rat brain, *Biochem. Biophys. Res. Commun.*, 310, 860, 2003.
61. Hillier, C. et al., Effects of urotensin II in human arteries and veins of varying caliber, *Circulation*, 103, 1378, 2001.
62. MacLean, M.R. et al., Contractile responses to human urotensin-II in rat and human pulmonary arteries: effect of endothelial factors and chronic hypoxia in the rat, *Br. J. Pharmacol.*, 130, 201, 2000.
63. Paysant, J. et al., Comparison of the contractile responses of human coronary bypass grafts and monkey arteries to human urotensin-II, *Fundam. Clin. Pharmacol.*, 15, 227, 2001.
64. Stirrat, A. et al., Potent vasodilator responses to human urotensin-II in human pulmonary and abdominal resistance arteries, *Am. J. Physiol. Heart Circ. Physiol.*, 280, H925, 2001.
65. Affolter, J.T. et al., No effect on central or peripheral blood pressure of systemic urotensin II infusion in humans, *Br. J. Clin. Pharmacol.*, 54, 617, 2002.
66. Wilkinson, I.B. et al., High plasma concentrations of human urotensin II do not alter local or systemic hemodynamics in man, *Cardiovasc. Res.*, 53, 341, 2002.
67. Douglas, S.A. et al., Congestive heart failure and expression of myocardial urotensin II, *Lancet*, 359, 1990, 2002.
68. Zhang, Y. et al., Effect of chronic hypoxia on contents of urotensin II and its functional receptors in rat myocardium, *Heart Vessels*, 16, 64, 2002.

69. Tzanidis, A. et al., Effect of chronic hypoxia on contents of urotensin II and its functional receptors in rat myocardium, *Cir. Res.*, 93, 246, 2003.

70. Zou, Y., Nagai, R., and Yamazaki, T., Urotensin II induces hypertrophic responses in cultured cardiomyocytes from neonatal rats, *FEBS Lett.*, 508, 57, 2001.

71. Haber, E.P. et al., Pleiotropic effects of fatty acids on pancreatic beta-cells, *J. Cell Physiol.*, 194, 1, 2003.

72. Bergman, R.N. and Ader, M., Free fatty acids and pathogenesis of type 2 diabetes mellitus, *Trends Endocrinol. Metab.*, 11, 351, 2000.

73. Sawzdargo, M. et al., A cluster of four novel human G protein-coupled receptor genes occurring in close proximity to CD22 gene on chromosome 19q13.1, *Biochem. Biophys. Res. Commun.*, 239, 543, 1997.

74. Itoh, Y. et al., Free fatty acids regulate insulin secretion from pancreatic β cells through GPR40, *Nature*, 422, 173, 2003.

75. Kotarsky, K. et al., A human cell surface receptor activated by free fatty acids and thiazolidinedione drugs, *Biochem. Biophys. Res. Commun.*, 301, 406, 2003.

76. Price, D.A. and Greenberg, M. J., Structure of a molluscan cardioexcitatory neuropeptide, *Science*, 197, 670, 1977.

77. Perry, S.J. et al., A human gene encoding morphine modulating peptides related to NPFF and FMRFamide, *FEBS Lett.*, 409, 420, 1997.

78. Elshourbagy, N.A. et al., Receptor for the pain modulatory neuropeptides FF and AF is an orphan G protein-coupled receptor, *J. Biol. Chem.*, 275, 25,965, 2000.

79. Lee, D.K. et al., Discovery and mapping of ten novel G protein-coupled receptor genes, *Gene*, 275, 83, 2001.

2 Screening of Ligands for Human GPCRs by the Use of Receptor-Gα Fusion Proteins

Shigeki Takeda, Hinako Suga, and Tatsuya Haga

CONTENTS

2.1 INTRODUCTION

G protein-coupled receptors (GPCRs) are integral proteins in the cell membranes; receive extracellular signals, such as light, odorants, and taste substances, as well as most hormones and neurotransmitters; and interact with and activate G proteins in the intracellular surface. GPCRs constitute one of the largest protein superfamilies and are targets for 30 to 60% of present drugs.[1] Thus, identification of GPCRs in the human genome and their functional characterization should give a great impetus not only to the understanding of human physiology, but also to the development of novel drugs.

There are a variety of GPCRs ranging from well-characterized, classical GPCRs, such as muscarinic acetylcholine receptors and adrenergic receptors, to those that have been identified only from their sequences from the genome data. Endogenous ligands are known for more than 100 GPCRs. It is common for there to be multiple receptor subtypes for an endogenous ligand: for example, there are five subtypes M1 through M5 for muscarinic acetylcholine receptors and nine subtypes, α_{1A}, α_{1B}, α_{1D}, α_{2A}, α_{2B}, α_{2C}, β_1, β_2, and β_3, for adrenergic receptors. Although muscarinic acetylcholine receptors and adrenergic receptors are the most characterized GPCRs, subtype-specific ligands have not yet been fully developed. The development of strictly subtype-specific ligands is a key step in elucidating physiological functions of each receptor subtype and also in developing drugs with fewer side effects. On the other hand, over a few hundred putative GPCR genes are predicted from human genome analysis.[2-5] Natural ligands for these GPCRs and their functions remain unknown. These "orphan GPCRs" and their ligands should be involved in unique biological events that have not yet been investigated, and could be potential targets for novel drug discovery. It is now routine practice in pharmaceutical industries to screen many thousands of synthetic and natural compounds for each target GPCR in order to identify its endogenous and synthetic ligands. This situation has induced a number of excellent assay systems that are suitable for high-throughput screening.

In this chapter, we cover two topics. The first is concerned with the screening of sequence information for GPCR genes from the human genome and other databases. There are a few reports that attempt to prepare a list of human GPCR genes. Although the process of genome annotation seems to be close to the final stage, the list of GPCRs is still temporary, mostly because of the difficulty in predicting pseudogenes and alternative splicings. The second topic is concerned with a ligand screening method using GPCR-Gα fusion proteins with a cell-based assay *in vivo* or with a filter-binding assay *in vitro*. This method could be useful for middle-scale screening in a laboratory with the 96-well plate format and also for large-scale high-throughput screening with 386-well formant using an automatic system.

2.2 LIST OF HUMAN GPCR GENES

GPCRs comprise one of the largest superfamilies throughout various species. The number of GPCR genes was estimated to be 1050 for the *Caenorhabditis elegans* genome[6] and 160 for *Drosophila melanogaster*,[7] which corresponds to 5.5% and 1% of the total genes, respectively. In mice, the presence of 500 to 1000 odorant receptors[8] and 40 taste receptors[9] was suggested. Based on draft sequences of the human genome, Venter et al. reported the presence of 616 GPCR genes that belonged to the rhodopsin-class, secretin receptor-class, or metabotropic glutamate receptor-class,[10] and the public sequencing consortium reported the presence of 569 class A (rhodopsin-class) GPCR genes.[11] Both analyses suggested that the number of GPCR genes in the human genome would be approximately 1000.

2.2.1 SCREENING OF GPCRS FROM THE HUMAN GENOME

As the draft sequence of the human genome was announced in 2000, we attempted to identify human GPCRs from the sequence. Taking advantage of the fact that many mammalian GPCR genes do not have introns in their coding region,[12] we focused on GPCR genes with no introns in the coding regions. Thus, we could avoid the difficult and risky task of predicting the exon–intron boundaries in the genome sequence. We translated the human genome data without predicting the exon–intron boundary and extracted 335,570 open reading frames (ORFs), which had 200 to 1500 amino acid residues. The length of 200 to 1500 amino acid residues corresponds to those for most of the GPCRs. Then we took advantage of a common characteristic of GPCRs that they have seven transmembrane segments, and adapted the program SOSUI, which was developed to identify transmembrane segments in protein sequences and to discriminate membrane proteins (http://sosui.proteome.bio.tuat.ac.jp/sosuiframe0.html). Using SOSUI, we selected 7780 ORFs, which were judged to have six to eight transmembrane segments, as GPCR candidates. We removed some candidates as redundant ORFs when they were more than 95% homologous to each other as determined by a BLAST search. We then analyzed the homology of obtained ORFs to protein sequences in databases and excluded those with homology to known proteins different from GPCRs. We found 322 putative odorant receptors, 41 of which had been registered as such, and 281 of which were 40 to 85% homologous to registered odorant receptors. Glusman et al. independently reported the presence of 322 odorant receptor genes in the human genome.[13] More recently, human odorant receptors were reported to be composed of 339 genes.[14] Putative taste receptors were detected, including 11 registered ones and 11 novel ones with homology of 30 to 70% to registered bitter receptors of mice or rat. We also found nonodorant and nontaste GPCRs: 128 of the GPCRs for endogenous ligands had already been registered on the "nr" database as of August 20, 2000, and 50 of the GPCR candidates were not registered on the database. The homology search indicated that out of the 50 GPCR candidates, eight were homologous to amine receptors, four to peptide receptors, six to nucleotide receptors, four to lipid receptors, six to putative pheromone receptors, nine to various kinds of GPCRs, and thirteen to orphan receptors.

We then estimated the total number of GPCRs by correcting for the proportion of intron-less GPCRs, the success rate for SOSUI, and the completeness of the draft sequence. We first checked the proportions of intron-less GPCRs for 152 of the registered GPCRs, for which both cDNA and genomic clones had been sequenced. We found no genes with introns in the coding regions among 43 human odorant receptor genes in the public database, 3 sweet receptor genes with introns,[15–17] and 11 bitter receptor genes without introns[9,18] among 14 taste receptor genes, and 63 intron-less ORFs among 95 known GPCR genes that have endogenous ligands. Correct prediction by SOSUI analysis was obtained for 77% of the 43 odorant receptors, 100% of the 11 taste receptors, and 95% of the 207 receptors for endogenous ligands. In the draft sequence that we used, we could find 79% of the 43 odorant receptors, 100% of the 11 taste receptors, and 90% of the 207 receptors for endogenous ligands. Subjecting registered GPCR genes to the present system independently checked the efficiency of the present screening method. We obtained recoveries of 67% of 43 odorant receptors, 79% of 14 taste receptors, and 54% of 207 receptors for endogenous ligands. The recovery of 54% for receptors for endogenous ligands was essentially the same as the product (56%) of three factors — the proportion of intron-less (66%), the performance of SOSUI (95%), and the

TABLE 2.1
Number of Identified GPCRs and Estimation of Total Number of GPCRs in the Human Genome in 2000

	Number of Identified Receptors (Already-Registered Receptors)	Estimated Number of Total Receptors
Odorant receptors	322 (41)	481
Taste receptors	22 (11)	28
Receptors for endogenous ligands	178 (128)	330
Open reading frames without homology to registered GPCRs	59 (0)	109
Total	581 (128)	948

Note: See the text for the strategy to identify GPCRs.

Source: Takeda, S. et al., *FEBS Lett.*, 520, 97, 2002.

completeness of the genome (90%). A similar correspondence was also observed for odorant receptors (67% vs. 61%) and taste receptors (79% vs. 79%). These results indicate that the efficiency of this screening strategy depends solely on the above three factors. On the assumption that the efficiency when screening novel GPCRs is the same for registered and nonregistered GPCRs, we estimated the total number of GPCRs in the human genome to be 948 (Table 2.1). This consists of 481 odorant receptors, 28 taste receptors, and 330 receptors for endogenous ligands. In addition, approximately 100 ORFs might represent a new class of GPCRs, which are not homologous to registered GPCRs (Table 2.1).

During the last few years, several groups published an advanced, but still temporary, list of human GPCRs for endogenous ligands (Table 2.2).[3–5] These estimates are primarily based on the sequence homology and might not cover GPCRs that do not have homology to established GPCRs. In fact, a progestin receptor, which is not included in the above lists, was functionally identified as a GPCR that does not have significant homology with other GPCRs. This suggests that there might be a new class of GPCRs with no sequence homology to registered GPCRs in the human genome. We reasoned that 59 genes might be GPCR genes because they contained seven putative transmembrane segments and no introns in the coding region, although there was no observed sequence homology with registered GPCRs. The progestin receptor gene is included in these 59 genes, and 26 other genes remain as GPCR gene candidates, although several others were found not to be included in the most recent human genome data and might be artifacts due to errors in the draft sequence data.

2.2.2 LIST OF USEFUL DATABASES FOR GPCRs

Internet Web sites/databases for GPCRs are frequently updated. Because several GPCRs are registered under different names or accession numbers in different

TABLE 2.2
Summary of Human GPCRs Predicted by Genome Sequencing

	Vassilatis et al.	Fredriksson et al.	Inoue et al.
Class A (Rhodopsin family)	284	241	226
Class B (Secretin receptor family)	50	15	18
Class C (Glutamate receptor family	17	15	9
Class F/S (Frizzled receptor family	11	24	8
Adhesion receptor family	—	24	—
Other	5	23	8
Total	367	342	269

Source: Vassilatis, D.K. et al., *Proc. Natl. Acad. Sci. USA*, 100, 4903-8, 2003; Fredriksson, R. et al., *Mol. Pharmacol.*, 63, 1256, 2003; Inoue, Y., Ikeda, M., and Shimizu, T., *Comput. Biol. Chem.*, 28, 39, 2004.

databases, investigators should use more than one database and compare them. This section describes Web sites useful for obtaining and examining GPCR sequences. Sequence homology search is a powerful tool for classifying GPCRs. Phylogenetic analysis is helpful in evaluating structural homology and sometimes in predicting ligand structure for the given orphan GPCR. These data are also useful for computer modeling and three-dimensional structural prediction and for ligand–receptor binding simulation for computer-aided drug design (see Chapter 12).

1. For a general guide and portal site of the human genome:
   Human Genome Browser at the Sanger Center
       (http://www.ensembl.org/Homo_sapiens/)
   Human Genome Browser in National Center for Biotechnology Information
       (http://www.ncbi.nlm.nih.gov/genome//guide/human/)
   DDBJ (DNA Data Bank of Japan)/CIB Human Genomics Studio
       (http://studio.nig.ac.jp/index.htm)
2. Frequently updated database for GPCRs:
   GPCRDB (http://www.gpcr.org/)
3. Download sites for files and data of recent publications:
   Temporary list of human GPCRs reported in Reference 4
       (http://www.neuro.uu.se/medfarm/schiothArt.html)
   Temporary list of mammalian GPCRs reported in Reference 5
       (ftp://bioinfo.si.hirosaki-u.ac.jp/GPCRbyBTP)
4. Computational prediction for GPCR genes:
   SEVENS (http://sevens.cbrc.jp/)
5. Mutation database for GPCRs:
   GRAP Mutant Databases (http://tinygrap.uit.no/)
6. Database for olfactory receptors:
   ORDB (http://senselab.med.yale.edu/senselab/ordb/)

7. Database for chemokine receptors:
   dbCRF (http://crf.medic.kumamoto-u.ac.jp/CSP/Receptor.html)
8. General receptor database including other types of receptors:
   RDB (http://impact.nihs.go.jp/RDB.html)
9. SNP database on general proteins detected in Japanese species:
   SNPs (http://snp.ims.u-tokyo.ac.jp/index.html)
10. Database for genetic disorder for general use:
    Genes, locations, and genetic disorders
       (http://www.gdb.org/gdbreports/GeneticDiseases.html)
11. Prediction of the coupling specificity of GPCRs to G proteins from their
    primary structures:
    GPCR coupling specificity (http://ep.ebi.ac.uk/GPCR/)
12. Prediction of GPCR subfamilies from their primary structures:
    GPCR Subfamily Classifier (http://www.soe.ucsc.edu/research/comp-
    bio/gpcr-subclass/)

2.2.3 LIST OF USEFUL PROGRAMS FOR GPCR AND OTHER PROTEIN RESEARCH

1. Sequence homology search:
   Blast (http://www.ncbi.nlm.nih.gov/blast/)
   FTP site for Blast program and Blast database for downloading
      (ftp://ftp.ncbi.nlm.nih.gov/blast/executables/) and
      (ftp://ftp.ncbi.nlm.nih.gov/blast/db/)
2. Prediction of protein function from primary structure and sequence homology search:
   GTOP (http://spock.genes.nig.ac.jp/~genome/gtop.htm)
3. A large collection of multiple sequence alignments and protein families:
   Pfam (http://pfam.wustl.edu/ and several mirror sites)
4. Prediction of transmembrane segments from primary structure (also see
   text):
   SOSUI (http://sosui.proteome.bio.tuat.ac.jp/sosuiframe0.html)
5. For drawing and generation of snake-like diagram for GPCRs from primary structure:
   RbDg (http://icb.med.cornell.edu/crt/RbDg/index.xml)
6. For drawing and generation of multiple sequence alignment:
   ClastalW (http://www.genome.ad.jp/SIT/CLUSTALW.html)

2.3 LIGAND SCREENING SYSTEM FOR GPCRS USING RECEPTOR-Gα FUSION PROTEINS

In this section, we describe the ligand screening system for GPCRs using nociceptin
and nociceptin receptor (NR) as a model system. NR is structurally similar to opioid
receptors, and nociceptin is a typical example identified as a ligand for the orphan

GPCR that is NR. Nociceptin is a peptide with a structure similar to that of opioid peptides.[19,20]

Several protocols were already developed for screening ligands for GPCRs, including NR. If a specific radioligand is available for a given GPCR, the binding experiment with receptor-expressing membranes is a useful method for finding compounds that interact with the receptor and then inhibit the radioligand binding. This is one of the simplest methods for screening ligands, but the drawback is that agonists and antagonists are not discriminated by this method in general. The measurement of second messengers in receptor-expressing cells is another typical protocol for the ligand screening. For example, the activity of nociceptin in a brain extract was monitored by measuring the inhibition of forskolin-stimulated cAMP accumulation in cells expressing nociceptin receptors.[19,20] This protocol enables nociceptin to be detectable with an EC_{50} of 1 nM. One of the most popular methods for ligand screening is measurement of intracellular Ca^{2+} concentrations in receptor-expressing cells. This system was originally used for G_q-coupled GPCRs, but it can be used for G_i/G_o-coupled GPCRs by coexpressing $G_{q/i}$ chimera or promiscuous G protein G_{16}, which may enable G_i/G_o-coupled receptors to initiate G_q-mediated signal cascades.[21,22] The EC_{50} for nociceptin is reported to be 8 nM by measuring a nociceptin-stimulated Ca^{2+} increase in cells expressing G_i/G_o-coupled NR and $G_{q/i}$ or G_{16}.[23] The lower sensitivity of this method compared with the cAMP monitoring method may be due to weak coupling between NR and G_{16} or $G_{q/i}$. A reporter gene assay is a useful method as well as is measurement of intracellular Ca^{2+} concentrations for the ligand screening, and this was applied to NR with an EC_{50} of 0.8 nM.[24] The activity was measured by inhibition of forskolin-induced luciferase expression using a stable Chinese hamster ovary (CHO) clone that was transfected with the reporter plasmid carrying six cAMP-responsive elements upstream of an SV40 promoter and luciferase gene. The agonist-dependent change in pH for receptor-expressing cells appears to provide the most sensitive method, and the EC_{50} for nociceptin is reported to be 0.06 nM.[24]

Receptor-Gα fusion has been used to examine interactions between GPCR and Gα. In this chapter, we will show that the receptor-Gα fusion proteins are also useful for ligand screening, and we describe how to prepare and employ them for the ligand screening in membrane or cell preparations. The membrane preparations expressing the receptor-Gα fusion proteins can be used to detect both agonists and antagonists by measuring agonist-dependent [^{35}S]GTPγS binding activity. The cells expressing receptor-Gα fusion proteins can also be used for the ligand screening by measuring intracellular Ca^{2+} ions or other markers. The interaction between GPCR and Gα in fusion proteins is more efficient than that between independent GPCR and Gα, and then the ligand is expected to be detected with the higher sensitivity for membranes or cells expressing the fusion proteins as compared to those expressing receptors and Gα separately. An important advantage of [^{35}S]GTPγS binding assay using membrane preparations is that false-positive reactions are negligible, whereas the cell-based assay may be affected by false-positive responses mediated by endogenous receptors. Another benefit for the former system is that a large amount of membranes expressing the fusion protein can be easily prepared by using the baculovirus-Sf9 system. The disadvantage for [^{35}S]GTPγS binding assay using membrane preparations is that the sensitivity is not as high as that of the methods using cell preparations, and it is not easy to apply for G_q-coupled receptors.[25]

2.3.1 Gene Construction for Receptor-Gα Fusion Proteins

In this section, we describe the construction of receptor-Gα fusion proteins using the rat CX_3C chemokine receptor 1 (CX_3CR1)-human $G\alpha_{16}$ fusion protein as a model system.[26] CX_3CR1 couples to G_i family G proteins. G_{16} is promiscuously activated by different kinds of GPCRs that are intrinsically coupled to G_s, G_i, or G_q family G proteins.[27,28] Activated G_{16} then activates phospholipase Cβ, as G_q does. Therefore, fusion proteins of G_i-coupled receptors with $G\alpha_{16}$ can convert the coupling of the G_i-coupled receptors to the PLC   pathway. Ligands for the G_i-coupled receptor-$G\alpha_{16}$ fusion protein can be screened relatively easily by expressing the fusion protein in cultured cells and measuring the downstream signals mediated by PLC. The method using receptor-$G\alpha_{16}$ fusion proteins is particularly useful for orphan receptors because it is generally difficult to determine the G protein coupled to a given orphan receptor. It should be noticed, however, that G_{16} is not omnipotent — it is likely to be activated by G_s-coupled receptors as well as G_q-coupled receptors more easily than by G_i-coupled receptors, and it may be inhibitory for some receptors.[29]

The fusion of a receptor with a Gα is performed by the polymerase chain reaction (PCR).[26,30,31] The 3' end of cDNA encoding the receptor, from which the stop codon is removed, is fused directly to the 5' end of the Gα cDNA. Alternatively, the stop codon is modified to add restriction enzyme sites, which introduce a few amino acids between the receptor and Gα. In some cases, epitope tags are added to the 5' end of a fusion protein cDNA or between a receptor cDNA and a Gα cDNA, or the green fluorescent protein cDNA is added between a receptor cDNA and a Gα cDNA.[32]

Co-expression of G_i-coupled receptors and $G\alpha_{16}$ followed by measurements of intracellular Ca^{2+} ions is widely used for the ligand screening of G_i/G_o-coupled receptor. We found, however, that cells expressing receptor-$G\alpha_{16}$ fusion proteins showed greater responses and sensitivity than those expressing receptor and $G\alpha_{16}$ together but as independent entities.[26]

Protocol 2-1:   Gene Construction for Receptor-Gα Fusion Proteins

2-1-1   PCR

Figure 2.1 shows a schematic diagram of how to generate a gene for a receptor-Gα fusion protein. Primer 1 carries the recognition sequence for a restriction enzyme with some nucleotides at its 5' termini, and approximately 18 nucleotides corresponding to the 5' end of a receptor at its 3' termini. The restriction enzyme should be chosen from those with restriction sites, which are present in multicloning sites of the expression vector and are not present in genes for the receptor and Gα. Many restriction enzymes need some nucleotides in addition to the recognition sequence for effective cutting, and these nucleotides may be chosen from the reference appendix "cleavage close to the end of DNA fragments" in the New England Biolabs

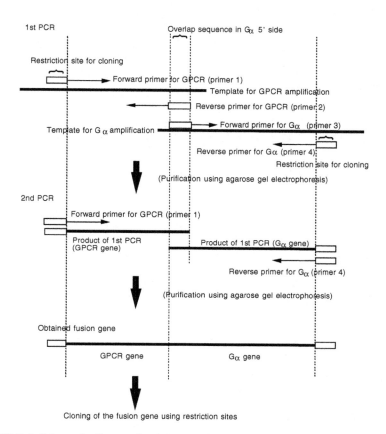

FIGURE 2.1 Schematic diagram for the construction of receptor-Gα fusion gene.

(Beverly, MA) catalog. Nucleotides corresponding to the 5' end of the receptor should have Tm = 55 to 60°C. Primer 2 carries about 18 nucleotides corresponding to the opposite strand of the receptor 3' end from which the stop codon was removed. Its Tm should be 55 to 60°C. Primer 3 carries about 24 nucleotides corresponding to the 3' end of the receptor from which the stop codon was removed at its 5' terminus, and about 18 nucleotides corresponding to the 5' end of a Gα at its 3' terminus. Nucleotides corresponding to the 3' end of the receptor should have Tm = 65 to 70°C. Nucleotides corresponding to the 5' end of the Gα should have Tm = 55 to 60°C. Primer 4 is for the Gα 3' end with same criteria for Primer 1. All four primers should not include GC, CG, AT, or TA at their 3' termini, because these sequences generate unexpected dimerization of the primers.

1. Primers are described as follows:
    Primer 1 (sense primer for CX₃CR1): 5-ACGC<u>GTCGAC</u>**ATG**CCTAC CTCCTTCCCGG-3 (restriction site for *Sal*I underlined; initiation Met bold)
    Primer 2 (reverse primer for CX₃CR1): 5-GAGCAGGAGAGATCCCTC-3

Primer 3 (sense primer for $G\alpha_{16}$): 5-GAGGGAGAGGGATCTCTCCT-GCTC**ATG**GCCCGCTCGCTGACC-3 (initiation Met bold)

Primer 4 (reverse primer for $G\alpha_{16}$): 5-GG<u>ACTAGT</u>**TCA**CAGCAGGT TGATCTCGT-3 (restriction site for *Spe*I underlined; termination codon bold)

2. PCR for receptor, for the $G\alpha$, and conditions:

First PCR for the receptor: Mix the following on ice:

100 ng/µl Template (pBluescript-human CX₃CR1):	1 µl
20 µM Primer 1	1 µl
20 µM Primer 2	1 µl
2 mM dNTPs	5 µl
25 mM MgCl₂	2 µl
10× Buffer (Toyobo)	5 µl
KOD DNA polymerase (Toyobo)	1 µl
Sterile water	34 µl
Total	50 µl

First PCR for the $G\alpha$: Mix the following on ice:

100 ng/µl Template (pCIS-human $G\alpha_{16}$):	1 µl
20 µM Primer 3	1 µl
20 µM Primer 4	1 µl
2 mM dNTPs	5 µl
25 mM MgCl₂	2 µl
10 × Buffer (Toyobo)	5 µl
KOD DNA polymerase (Toyobo)	1 µl
Sterile water	34 µl
Total	50 µl

PCR conditions: Carry out "cycle (1)" of KOD DNA polymerase according to the manufacturer's protocols.

98°C 2 min

25 cycles of 98°C 15 s, 55°C 2 s, 74°C 30 s

74°C 5 min

4°C

3. The PCR products (10 µl) are mixed with 2 µl of 6 × gel-loading buffer. Load into the slots of a 1% agarose gel in the electrophoresis tank (Mupid-2). Load size markers too. Run the gel according to the manufacturer's protocols. Apply a constant voltage of 100 V for 20 to 25 min. Examine the gel by ultraviolet light. Confirm the bands are the expected sizes.

4. The remainder of the PCR products are electrophoresed as described above. The PCR products (40 µl) are mixed with 8 µl of 6× gel-loading buffer. Load into the slots of a 1% agarose gel in the electrophoresis tank. Load size markers too. Apply a constant voltage of 100 V for about 25 min. Examine the gel by ultraviolet light. Cut out the bands using a scalpel. Put them in the microfuge tubes.

5. Purify the PCR products using the Ultra clean 15 DNA purification kit (Mo Bio Laboratories, Carlsbad, CA) according to the manufacturer's protocols. Use 6 μl of Ultra Bind. Elute the purified DNA with 12 μl of 10 mM Tris/0.1 mM EDTA.

6. Tethering reaction of the receptor gene with the Gα gene:
Mix the following on ice:

PCR product of the receptor	1 to 6 μl
PCR product of the Gα	1 to 6 μl
2 mM dNTPs	5 μl
25 mM MgCl$_2$	2 μl
10 × Buffer (Toyobo)	5 μl
KOD DNA polymerase (Toyobo)	1 μl
Sterile water	to 48 μl
Total	48 μl

Reaction conditions:
98°C 2 min
5 cycles of 98°C 15 s, 65°C 2 s, 74°C 30 s
4°C

7. Second PCR for the amplification of the receptor-Gα fusion gene:
Add the primers to the tethering reaction mixture on ice:

Tethering reaction mixture: 48 μl	
20 μM Primer 1	1 μl
20 μM Primer 4	1 μl
Total	50 μl

PCR conditions:
98°C 2 min
25 cycles of 98°C 15 s, 55°C 2 s, 74°C 30 s
74°C 5 min
4°C

8. The PCR product (10 μl and 40 μl) is subjected to electrophoresis as described above (2-1-1, 3, and 4).

9. Purify the PCR product using the Ultra clean 15 DNA purification kit.

Note: An appropriate restriction site may be put between the receptor and Gα genes, and the receptor gene may be inserted upstream of the G α subunit gene instead of performing the second PCR.

2-1-2   Subcloning

1. Digest the PCR product with 5 U of the restriction enzymes, SalI and SpeI, overnight.

2. Digest 1 μg of pEF-BOS-KS[33], a derivative of the mammalian expression vector pEF-BOS[34] that contains the multiple cloning site of pBluescript KS (Stratagene), with 5 U of the restriction enzymes, SalI and SpeI.

3. Subject the digests to electrophoresis as described above (2-1-1, 3, and 4).

4. Purify the DNAs using the Ultra clean 15 DNA purification kit. Use 6 µl of Ultra Bind. Elute the purified DNA with 12 µl of 10 mM Tris/0.1 mM EDTA.

5. Mix 6 µl of the digested PCR product with 1 µl of the digested vector. Ligate them using the ligation high kit (Toyobo) according to the manufacturer's protocols.

6. Add half of the mixture to competent DH5α cells (Toyobo) on ice. Transform the cells using the Competent high DH5α kit (Toyobo) according to the manufacturer's protocols. Spread the cells on LB agar Amp (+) plates. Incubate overnight at 37°C.

7. PCR for the insert check:
   Mix the following:

	One Tube	Nine Tubes
20 µM Primer 1	1 µl	9 µl
20 µM Primer 4	1 µl	9 µl
2 mM dNTPs	5 µl	45 µl
10 × PCR Buffer (PE Applied Biosystems)	5 µl	45 µl
AmpliTaq Gold DNA polymerase (PE Applied Biosystems)	0.3 µl	2.7 µl
Sterile water	37.7 µl	339.3 µl
Total	50 µl	450 µl

   Distribute 50 µl aliquots of the mixture to eight numbered microfuge tubes for PCR. Number eight separate colonies on the back of the LB agar Amp (+) plate. Pick a colony with a sterile toothpick. Shake the toothpick several times in the appropriately numbered PCR mixture. Pick up another seven colonies in the same manner.

   PCR conditions:
   95°C 10 min
   30 cycles of 95°C 30 s, 50°C 30 s, 72°C 2 min
   72°C 5 min
   4°C

8. Subject the PCR product (15 µl) to electrophoresis as described above (2-1-1, 3, and 4).

9. Choose two colonies that harbor the receptor-Gα fusion gene inserted in the plasmid. Put 3 ml of LB Amp (+) medium into two centrifuge tubes. Pick each colony with a sterile toothpick. Rub the wall of one LB Amp (+) medium centrifuge tube with the toothpick several times. Repeat this process with the other tube and colony. Shake overnight at 37°C.

10. Check the turbidity of the LB Amp (+) medium to determine whether the transformant cells have grown enough. Add 1.5 ml of the medium (3 ml) to a microfuge tube. Centrifuge at 15,000 rpm for 30 s. Discard the supernatant. Add the other 1.5 ml to the microfuge tube, and repeat the centrifugation. Discard the supernatant. Purify the plasmid using the Qiagen plasmid mini kit (Qiagen) according to the manufacturer's protocols. Dissolve the purified plasmid in 50 µl of 10 mM Tris, 0.1 mM EDTA.

11. Add 99 μl of distilled water to a microfuge tube. Add 1 μl of the purified plasmid. Measure the absorbance at 260 nm (1 OD_{260} = 50 μg/ml). Calculate the concentration of the plasmid using the following equation:

$$\text{Measured value} \times 100 \times 50/1000 = \text{Real concentration (μg/μl)}$$

12. Verify the construct by restriction endonuclease mapping.
13. Confirm the sequence by DNA sequencing.

2.3.2 IN VIVO CELL-BASED ASSAY

The constructed plasmids encoding fusion proteins are transfected in CHO cells. To estimate the expression level of fusion proteins, ligand-binding activity is measured when applicable. The expression of fusion proteins is also measured by immunostaining, Northern blot, or reverse transcription polymerase chain reaction (RT-PCR). In these cases, however, it is not known whether receptors are functionally expressed on cell membranes or not. Agonist-stimulated increases in intracellular Ca^{2+} in cultured cells are most commonly utilized to screen for ligands of receptors. The prostaglandin E_2 (PGE_2) generation is also useful for detecting the activation of receptor-$G\alpha_{16}$ fusion proteins (Figure 2.2). These assay systems are amenable to high-throughput screening of a large number of drug candidates, channel a broad spectrum of downstream pathways to a single measurable endpoint, do not require knowledge of an orphan receptor's G protein-coupling mechanism, and are adaptable to a wide range of previously characterized and novel receptors.

PROTOCOL 2-2: TRANSFECTION AND EXPRESSION OF RECEPTOR-Gα FUSION PROTEINS IN CHO CELLS

2-2-1   Stable Transfection

1. Maintain CHO-K1 cells (Health Science Research Resources Bank, Tokyo, Japan) in Ham's F-12 medium (Sigma-Aldrich, St. Louis, MO) supplemented with 10% heat-inactivated fetal bovine serum (JRH Biosciences, Lenexa, KS) and penicillin/streptomycin (50 U/ml and 50 μg/ml, respectively; Life Technologies, Inc., Gaithersburg, MD) at 37°C under humidified atmosphere of 5% CO_2. Warm the medium at 37°C before it is added to the cells. Use 0.05% trypsin containing 0.53 mM EDTA (trypsin-EDTA) for passages.
2. Co-transfect the constructed plasmids with pEF-neo (a gift from Dr. S. Nagata), conferring neomycin resistance to the cells, using the TransFast kit (Promega, Madison, WI) according to the manufacturer's protocols.
3. Transfect the cells grown on 100 mm dishes (Falcon) with 5 μg of fusion cDNA and 2 μg of pEF-neo. The ratio of fusion cDNA to pEF-neo is 2.5:5.
4. After 1 h, add the normal medium to the cells. Incubate for 48 h to recover the cells before replating and selecting.

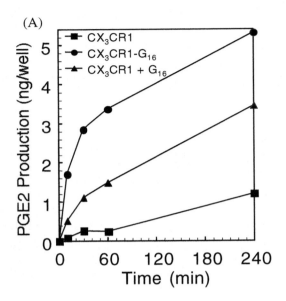

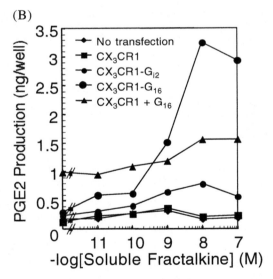

FIGURE 2.2 PGE_2 generation in stably transfected CHO cells. (A) Time course of PGE_2 generation. Cells expressing CX_3CR1 (squares), CX_3CR1-$G\alpha_{16}$ (circles), and CX_3CR1 plus $G\alpha_{16}$ (triangles) were stimulated with agonist, 10^{-8} M sFKN, for the times indicated, and PGE_2 released into the culture supernatants was measured using the PGE_2-enzyme immunoassay kit. A representative result of three experiments is shown. (B) Dose–response curves of PGE_2 generation. Cells expressing CX_3CR1 (squares) , CX_3CR1-$G\alpha_{i2}$ (open circles), CX_3CR1-$G\alpha_{16}$ (closed circles), CX_3CR1 plus $G\alpha_{16}$ (triangles), and parental CHO cells (diamonds) were stimulated with various concentrations of sFKN for 4 h, and PGE_2 released into the culture supernatants was measured.

5. Discard the medium. Add 5 ml of Dulbecco's phosphate-buffered saline (PBS; Sigma-Aldrich) to the cells. Discard the PBS.

6. Add 1 ml of trypsin-EDTA. Incubate for 1 to 2 min at 37°C. Shake and tap the dish gently by hand to detach the cells. Add 10 ml of culture medium. Suspend by pipetting.

7. Transfer the cell suspension to a 15 ml centrifuge tube. Centrifuge at 800 to 1000 rpm for 5 min. Discard the supernatant. Resuspend the pellet by pipetting with 10 ml of medium.

8. Add 10 ml of medium containing 400 μg/ml G418 (Nacalai Tesque, Inc., Kyoto, Japan) to three 100-mm dishes. Distribute 0.1, 0.05, 0.02 ml of the cells to each 100-mm dish (100, 200, 500 × dilution, respectively). Spread the cells by shaking the dish by hand.

9. Note that the cells should begin to die after 3 to 4 days. Culture for about 8 to 9 days, until the cells make 2 to 3 mm resistant colonies. Change medium containing 400 μg/ml G418 when the medium turns yellow, or every 2 to 3 days.

10. Choose one or two dishes that have separate colonies. Mark 10 separate and large colonies per dish from the bottom of the dish. Distribute 0.5 ml of medium containing 400 μg/ml G418 into 10 wells of a 24-well culture plate (Costar) per dish.

11. Discard the medium in the dish. Add 5 ml of PBS to the cells. Discard the PBS.

12. Push a cloning cylinder onto silicon grease a few times using forceps, and attach silicon grease to the bottom of the cloning cylinder. Put the cloning cylinder on a colony. Push the cloning cylinder to stick to the dish. Repeat for the other nine colonies. Add three drops of trypsin-EDTA to the inside of the cylinder. Incubate for 2 to 3 min at 37°C. Suspend by pipetting four times using a pipetman P200 set to 50 μl. Transfer the cell suspension to the medium in a well of the 24-well culture plate. Repeat for the other nine colonies. Spread the cells by shaking the dish by hand.

13. Culture for several days until the cells become confluent. Change medium containing 400 μg/ml G418 when it turns yellow.

14. Add 0.5 ml of Cell Banker (Nihon Zenyaku Kogyo, Fukishima, Japan) to the same number of cryogenic vials as confluent wells. Add 5 ml of medium containing 400 μg/ml G418 to the same number of 60-mm dishes (Falcon) as confluent wells.

15. Wash confluent cells grown in the 24-well culture plate with 0.5 ml of PBS.

16. Add 0.2 ml of trypsin-EDTA. Incubate for 1 to 2 min at 37°C. Suspend by pipetting.

17. Transfer 0.1 ml of the cell suspension to Cell Banker in the cryogenic vials. Suspend by pipetting. Store at 80°C. Transfer 0.1 ml of the cell suspension to the medium containing 400 μg/ml G418 in the 60-mm dish. Spread the cells by shaking the dish by hand.

18. Culture and use the cells for various assays. Stably transfected cells were maintained in a medium containing 200 μg/ml G418 until they were plated into multiwell plates or dishes for experiments.

2-2-2   Transient Transfection

1. Transfect the constructed plasmids using the LipofectAMINE 2000 (Invitrogen, Carlsbad, CA) according to the manufacturer's protocols.
2. Transfect the cells grown on 60 mm dishes (Falcon) with 8 μg of fusion cDNA.
3. After 18 h, wash the cells with 2 to 3 ml of PBS.
4. Add 0.5 ml of trypsin-EDTA. Incubate for 1 to 2 min at 37°C. Shake and tap the dish gently by hand to detach the cells. Add 5 ml of culture medium. Suspend by pipetting.
5. Transfer the cell suspension to a 15-ml centrifuge tube. Centrifuge at 800 to 1000 rpm for 5 min. Discard the supernatant. Resuspend the pellet by pipetting with 5 ml of medium.
6. For intracellular Ca^{2+} measurements, add 17 ml of medium to a 50-ml centrifuge tube. Add 4 ml of the cell suspension (total 21 ml). Suspend by pipetting. Distribute every 0.2 ml of the cells to a well of a 96-well culture plate (Costar). For PGE_2 measurements, add 19.5 ml of medium to the 50-ml centrifuge tube. Add all (5 ml) of the cell suspension (total 24.5 ml). Suspend by pipetting. Distribute every 1 ml of the cells to a well of a 24-well culture plate. The concentration of the cells should be 50% confluent.
7. At 42 to 48 h after transfection, initiate the measurements. Protocol 2-3: Detection of Receptor-$G\alpha_{16}$ Fusion Protein

2-3-1   Ligand-Binding Assay

Buffer

Hepes binding buffer: 50 mM Hepes-NaOH buffer, pH 7.4, supplemented with 150 mM NaCl, 5 mM $MgCl_2$, 0.01% (w/v) bovine serum albumin (BSA).

1. Wash confluent CHO cells stably expressing fusion proteins grown in a 60 mm dish with 2 to 3 ml of PBS.
2. Add 0.5 ml of trypsin-EDTA. Incubate for 1 to 2 min at 37°C. Shake and tap the dish gently by hand to detach the cells. Add 5 ml of culture medium. Suspend by pipetting.
3. Transfer the cell suspension to a 15 ml centrifuge tube. Centrifuge at 800 to 1000 rpm for 5 min. Discard the supernatant. Resuspend the pellet by pipetting with 5 ml of medium.
4. Add 23.5 ml of medium to a 50-ml centrifuge tube. Add 1.1 ml of the cell suspension (total 24.6 ml). Suspend by pipetting. Distribute every 1 ml of the cells to a well of a 24-well culture plate. The concentration of the cells should be 10% confluent.
5. Culture for 3 days until the cells become confluent. Change medium when it turns yellow.
6. For total binding, prepare 0.2 ml/well of Hepes binding buffer containing increasing concentrations of radiolabeled agonist, e.g., 1 to 100 nM [^{125}I]

human fractalkine chemokine domain (soluble fractalkine, sFKN; specific activity 2000 Ci/mmol; Amersham Pharmacia Biotech, Piscataway, NJ) for CX3CR1-Gα_{16} fusion protein.

7. Wash the confluent CHO cells grown in the 24-well plate twice with 0.5 ml of PBS.

8. Initiate the experiments by adding 0.19 ml of Hepes binding buffer containing radiolabeled agonist with or without excess amount of unlabeled agonist. Incubate the cells for 1 h at 37°C.

9. Wash twice with 0.5 ml of PBS. Add 0.25 ml of 1% Triton X-100, shake using a plate mixer to lyse the cells. Add 0.25 ml of 1 N NaOH, shake using a plate mixer to lyse the cells. Transfer the lysate (total 0.5 ml) to glass tubes.

10. Count the radioactivity with a γ-counter (Cobra, Packard, Chicago, IL). Specific binding is determined by subtraction of the nonspecific binding from the total binding.

11. Count the cell numbers or measure the protein concentrations to analyze the number of agonist binding sites. To measure the protein concentrations, wash the cells in the 24-well plate with PBS. Add 0.2 ml of PBS. Scrape the cells using a rubber policeman. Transfer to a microfuge tube. Sonicate in ice water with a probe- or bath-type sonicator for 10 s (total protein). Centrifuge at $150,000 \times g$ for 30 min at 4°C. Discard the supernatant. Add 0.2 ml of PBS. Sonicate in ice water by probe- or bath-type sonicator for 10 s (membrane protein). Measure the protein concentrations using micro assay method of BCA protein assay (Pierce, Rockford, IL) according to the manufacturer's protocols.

2-3-2 Western Blot (See also Protocol 9-4)

Buffers

1 × SDS sample buffer: 50 mM Tris-HCl (pH 6.8), 6% 2-mercaptoethanol, 2% SDS, 0.1% bromophenol blue, 10% glycerol.

Blotting buffer: 100 mM Tris, 192 mM glycine, 10% methanol, 0.02% SDS.

Blocking solution: PBS containing 5% powdered skim milk and 0.1% Tween 20 (polyoxyethylene sorbitan monolaurate).

T-PBS: PBS containing 0.1% Tween 20.

1. Wash confluent cells grown in 60 mm dishes with 5 ml of PBS.

2. Add 0.25 ml of 1 × SDS sample buffer. Scrape the cells using a rubber policeman. Transfer the cell suspension to a microfuge tube.

3. Sonicate it in ice water with a bath-type sonicator (Tomy, Tokyo, Japan) for 10 s. Centrifuge at 15,000 rpm for 5 min.

4. Load 0.05 ml of the supernatant into a well of a 12% SDS-polyacrylamide gel (~14 cm × ~11 cm × 1 mm). Load 0.005 ml of the prestain marker low range (Bio-Rad, Hercules, CA) to another well. Apply a constant

current of 25 mA. Run until the bromophenol blue has migrated to about 1 cm from the bottom of the gel.

5. Prepare six sheets of Whatman 3MM paper and a sheet of polyvinylidene difluoride membrane (0.2 μm, Fluorotrans; Pall Corp., East Hills, NY) cut to the same size as the running gel (~14 cm × ~8 cm). Immerse the membrane in methanol. Immerse the membrane in the blotting buffer. Shake gently for 5 min. Handle the membrane using gloves and forceps.

6. Remove the one side of the glass plate from the gel. Cut off the stacking gel from the running gel. Immerse the running gel in the blotting buffer. Shake gently.

7. Wipe the positive electrode of the semidry blotting machine using a sheet of Kimwipe soaked in the blotting buffer. Immerse the three sheets of the Whatman 3MM paper in the blotting buffer. Place the papers on the positive electrode. Place the membrane on the papers. Place the gel on the membrane. Immerse the other three sheets of the Whatman 3MM paper in the blotting buffer. Place the papers on the gel. Wipe the negative electrode using a sheet of Kimwipe soaked in the blotting buffer. Place the negative electrode on the paper. Make sure that there are no air bubbles between the papers, the gel, and the membrane. Apply a constant current of 0.5 A for 30 min.

8. Immerse the membrane in blocking solution. Shake gently overnight at room temperature or at 4°C.

9. Add T-PBS, and shake gently. Repeat three times to wash the membrane.

10. Add T-PBS containing 0.5% powdered skimmed milk and an appropriate concentration of antiserum or antibody against receptor-$G\alpha_{16}$ fusion. Shake gently at room temperature for 1 h.

11. Discard the solution. Add T-PBS, and shake gently for 5 min. Repeat three times to wash the membrane.

12. Discard the solution. Add T-PBS containing 0.5% powdered skim milk and 1:5000 dilution of anti-rabbit IgG antibodies cross-linked with horseradish peroxidase (BioRad). Shake gently at room temperature for 1 h.

13. Discard the solution. Add T-PBS, and shake gently for 5 min. Repeat three times to wash the membrane.

14. Discard the solution. Add PBS.

15. Visualize the bands with the ECL chemiluminescence Western blot detection kit (Amersham Pharmacia Biotech) according to the manufacturer's protocols.

PROTOCOL 2-4: INTRACELLULAR Ca²⁺ MEASUREMENTS (SEE ALSO PROTOCOL 4-4)

1. Wash confluent CHO cells stably expressing fusion proteins grown in a 60 mm dish with 2 to 3 ml of PBS.

2. Add 0.5 ml of trypsin-EDTA. Incubate for 1 to 2 min at 37°C. Shake and tap the dish gently by hand to detach the cells. Add 5 ml of culture medium. Suspend by pipetting.

3. Transfer the cell suspension to a 15-ml centrifuge tube. Centrifuge at 800 to 1000 rpm for 5 min. Discard the supernatant. Resuspend the pellet by pipetting with 5 ml of medium.

4. Add 19 ml of medium to a 50-ml centrifuge tube. Add 2 ml of the cell suspension (total 21 ml). Suspend by pipetting. Transfer the cells to a 100-mm dish. Distribute every 0.2 ml of the cells to a well of a 96-well culture plate (Packard) using a multichannel pipette. The concentration of the cells should be 25% confluent.

5. Culture for 2 days until the cells become confluent. Change medium when it turns yellow. The multiwell plate washer manifold (Sigma-Aldrich) is convenient.

6. Discard the medium. Add 0.03 ml/well of Hank's balanced salt solution (Life Technologies) containing 5 μM fura-2/AM (Molecular Probes, Eugene, OR) for 2 h at room temperature in the dark. If fura-2/AM precipitates, sonicate briefly in a bath-type sonicator, centrifuge, and use the supernatant.

7. Prepare Hank's balanced salt solution containing 2 × concentrations of various stimulants and 0.1 mg/ml BSA. Add 0.1 ml of each solution in a well of a 96-well plate (Costar). Use 0.01 mM ionomycin (final 0.005 mM) as the positive control.

8. Wash the cells with 0.1 ml/well of Hank's balanced salt solution.

9. Add 0.03 ml/well of Hank's balanced salt solution.

10. Using a fluorescence imaging plate reader FDSS (Hamamatsu Photonics, Shizuoka Pref., Japan), record the baseline of intracellular Ca^{2+}-dependent fluorescence, with excitation at 340 and 380 nm and emission at 510 nm. Calculate the fluorescence intensity ratio I_{340}/I_{380} to visualize intracellular Ca^{2+} mobilization. After 1.5 min, add 0.03 ml/well of the stimulants prepared in Step 7, and record the changes in intracellular Ca^{2+}-dependent fluorescence.

PROTOCOL 2-5: PGE$_2$ MEASUREMENTS

1. Wash confluent CHO cells stably expressing fusion proteins grown in a 60-mm dish with 2 to 3 ml of PBS.

2. Add 0.5 ml of trypsin-EDTA. Incubate for 1 to 2 min at 37°C. Shake and tap the dish gently by hand to detach the cells. Add 5 ml of culture medium. Suspend by pipetting.

3. Transfer the cell suspension to a 15-ml centrifuge tube. Centrifuge at 800 to 1000 rpm for 5 min. Discard the supernatant. Resuspend the pellet by pipetting with 5 ml of medium.

4. Add 23.5 ml of medium to a 50-ml centrifuge tube. Add 1.1 ml of the cell suspension (total 24.6 ml). Suspend by pipetting. Distribute every 1 ml of the cells to a well of a 24-well culture plate. The concentration of the cells should be 10% confluent.

5. Culture for 3 days until the cells become confluent. Change medium when it turns yellow.
6. Prepare 0.2 ml/well of medium containing various stimulants.
7. Wash confluent CHO cells grown in the 24-well plate with 0.5 ml of culture medium.
8. Initiate the experiments by adding 0.19 ml of culture medium containing various stimulants. The cells were incubated for 10 min to 4 h at 37°C, being flushed with 5% CO_2 in humidified air.
9. Collect the culture medium. Store at 4°C. Centrifuge briefly to precipitate debris before use.
10. Measure the PGE_2 generation using an enzyme immunoassay kit (Cayman Chemical, Ann Arbor, MI) according to the manufacturer's instructions. At first, dilute the medium sample to 1/10 or 1/20 with EIA buffer, and measure the PGE_2 generation. Then change the dilution appropriately (Figure 2.2).

2.3.3 IN VITRO [35S]GTPγS BINDING ASSAY

It is generally difficult to measure the agonist-induced activation of G protein using membrane preparations because the expression levels of receptors and G proteins are usually not high enough to enable detection of their interaction. On the other hand, the interaction of receptor and G proteins can be measured rather easily by using membrane preparations containing expressed receptor-Gα fusion proteins. The receptor-Gα fusion protein was first reported for β_2 adrenergic receptor and $G\alpha_s$ by Bertin et al.[35] The agonist-bound receptor interacts with GDP-bound Gα and induces release of GDP followed by binding of GTP and its breakdown to GDP and phosphate. The interaction of receptor and Gα may be monitored as agonist-induced increase in the [35S]GTPγS binding or in the GTPase activity.

We examined the interaction between receptor and Gα in detail by using Sf9 cell membranes expressing muscarinic M_2 receptor-$G\alpha_{i1}$ or NR-$G\alpha_{i2}$ fusion proteins as model systems. In both systems, the effect of agonists on [35S]GTPγS binding was much more clearly observed in the presence of GDP rather than in its absence. The displacement curves of [35S]GTPγS binding by GDP shifted to the right in the presence of agonists as compared with its absence or in the presence of antagonists, and the smaller extent of shift was observed in the presence of partial agonists (Figure 2.3). Agonists and partial agonists gave similar but smaller effects on the displacement curves by GTP but did not affect the displacement curves by cold GTPγS. The extent of the shift for the displacement curves by GDP in the presence of agonists was increased by increasing $MgCl_2$ concentrations from 0 to 10 mM, although the displacement curve in the presence of antagonists was not affected by concentrations of $MgCl_2$: the displacement curves in the absence of $MgCl_2$ were not affected by the presence of agonists or antagonists. These results indicate that the apparent affinity for GDP of the fusion protein is determined by the species of bound ligands, and those that have the lowest, medium, and highest affinity in the presence of full agonists, partial agonists, and antagonists, respectively. The results also indicate that Mg^{2+} is required for the agonist-bound receptor to interact with Gα, reducing its affinity for GDP.[31,36] Thus, the stimulation by agonists of [35S]GTPγS binding is not due to the increase in affinity for GTPγS but is due to the decrease in affinity for GDP. Therefore, the presence of an appropriate concentration of GDP is essential to achieve the high signal-to-noise (S/N) ratio (Figure 2.3).

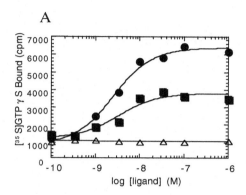

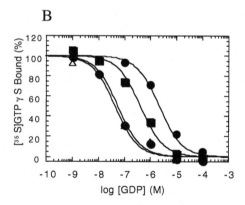

FIGURE 2.3 (A) Stimulation by nociceptin and other ligands of [35S]GTPγS binding to a fusion protein of rat nociceptin receptor and bovine G_{i2} protein α subunit (NR-Gα$_{i2}$). The membrane preparations expressing NR-Gα$_{i2}$ (20 μg of protein) were incubated with indicated concentrations of ligands {circles, nociceptin (full agonist); squares, [Phe^{1}Ψ(CH$_2$-NH)Gly^2]nociceptin(1-13)NH$_2$ (partial agonist); triangles, naloxone benzoylhydrazone (antagonist)} at 30°C for 30 min in 100 μl of 20 mM HEPES-KOH (pH 8.0), 1 mM EDTA, 160 mM NaCl, 1 mM DTT, 100 pM [35S]GTPγS, 1 μM GDP, and 10 mM MgCl$_2$ in 96-well microplates. The membranes were trapped on a GF/B glass filter that was washed three times with 300 μl each of cold 20 mM potassium phosphate buffer (pH 7.0). Radioactivity was then counted with a liquid scintillation counter. (B) Displacement by GDP of [35S]GTPγS binding to NR-Gα$_{i2}$. The same assay was performed in the presence of 1 μM nociceptin (full agonist, closed circles), [Phe^{1}Ψ(CH$_2$-NH)Gly^2]nociceptin(1-13)NH$_2$ (partial agonist, squares), naloxone benzoylhydrazone (antagonist, triangles), and in the absence of ligand (open circles) under the same conditions as in (a) except for the concentrations of GDP and ligands.

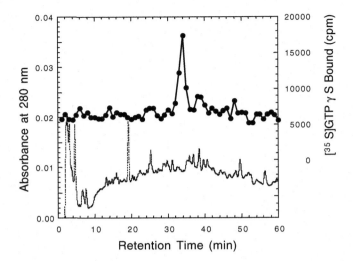

FIGURE 2.4 Identification of the active substance that stimulates [^{35}S]GTPγS binding to the NR-Gα$_{i2}$ in fractions in ion-exchange chromatography (TSKgel SP-2SW, 4.6 mm × 250 mm, Tosoh). For [^{35}S]GTPγS binding assay, a part of each fraction was lyophilized with 50 μg of bovine serum albumin, then dissolved in 100 μl of 20 mM HEPES-KOH (pH 8.0), 1 mM EDTA, and 160 mM NaCl. An aliquot (50 μl) of each fraction was assayed for [^{35}S]GTPγS binding under the conditions described in the legend to Figure 2.3, using eluates from the column instead of synthetic ligands. The solvents used were 10 mM ammonium formate (pH3.8)/10% acetonitrile to 1 M ammonium formate (pH 3.8)/10%. Solid and dotted lines represent [^{35}S]GTPγS binding activity and absorbance at 280 nm, respectively.

As a model system to screen endogenous ligands for G$_i$-coupled receptors, we examined nociceptin binding to the NR-Gα$_{i2}$ fusion proteins and could detect the stimulation of [^{35}S]GTPγS binding by nociceptin with an EC$_{50}$ of 2.0 nM and a gain of approximately five times. We examined the activity to stimulate [^{35}S]GTPγS binding to the fusion protein in various fractions of gel filtration or cation exchange column chromatographies, to which the brain extract was applied. The activity was eluted in the same fraction as one where authentic nociceptin is eluted, and it could be detected by using the fraction derived from 2 to 3 g wet weight tissue with high S/N ratio (Figure 2.4). Thus, the sensitivity of the fusion protein is comparable to the cell-based assay systems (see above) and is high enough to be applied for detection of endogenous ligands.

We also constructed a series of orphan receptor-Gα fusion proteins — most of the genes for these were identified from the human genome sequence by a computational search.[37] We employed a chemical compound library with 640 compounds and a bioactive lipid library with 240 lipids for ligand screening for them. Among them, the fusion protein of hGPCR48 with Gα$_{i1}$ was found to be activated by 5-oxo-ETE (5-oxo-eicosatetraenoic acid), 5(S)-hydroperoxy-eicosatetraenoic acid (5(S)-HPETE), and arachidonic acid with EC$_{50}$ of 5.5 nM, 19 nM, and 1.4 μM, respectively (Figure 2.5).[37] 5-Oxo-ETE, a metabolite of arachidonic acid, had been known to act as a chemoattractant for neutrophils and eosinophils and was thought

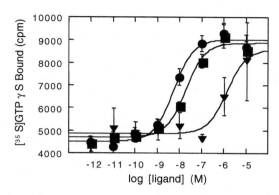

FIGURE 2.5 Stimulation by 5-oxo-ETE and other lipids of [^{35}S]GTPγS binding to the hGPCR48-Gα$_{i1}$ fusion protein. Sf9 cell membranes expressing the hGPCR48-Gα$_{i1}$ fusion protein were incubated with 0.1 nM [^{35}S]GTPγS, 1 μM GDP, and various concentrations of 5-oxo-ETE (circles), 5(S)-HPETE (squares), and arachidonic acid (triangles) for 30 min at 30ºC. The [^{35}S]GTPγS bound to the membranes was collected and counted with a liquid scintillation counter. Data shown are the averages and standard deviations for three experiments.

to be involved in inflammation and allergy. The resultant EC$_{50}$ value of hGPCR48 for 5-oxo-ETE was essentially the same as the reported values for its binding to plasma membranes of neutrophils and for its effect to induce the increase of the intracellular Ca^{2+} ions and the chemotaxis of neutrophils and eosinophils. The expression of hGPCR48 in neutrophils was evidenced by RT-PCR. These results indicate that hGPCR48 is a receptor for 5-oxo-ETE that is responsible for 5-oxo-ETE-induced chemotaxis of neutrophils. Because RT-PCR indicated its expression in liver and kidney at high levels and in the central nervous system at low levels, hGPCR48 might exert other physiological functions. For the other receptors, eleven compounds were found to act as surrogate agonists for a receptor-Gα$_s$ and four receptor-Gα$_i$ fusion proteins, one compound as an inverse agonist for two receptor-Gα$_s$ fusion proteins.[38] These achievements show the usefulness of the fusion protein for the ligand screening.

The most important advantage of the [^{35}S]GTPγS binding assay using the receptor-Gα fusion protein over cell-based assays is that false-positive reactions are negligible (see Figure 2.4). The cell-based assays may be affected by false-positive responses mediated through endogenous receptors. Another benefit of the fusion protein is the high S/N ratio in the presence and absence of agonist, which is roughly five (Figure 2.3). In contrast, the measurement of cAMP decrease, which is caused by activation of G$_i$-coupled receptors, is characterized by a low S/N ratio between the presence and absence of agonist. In addition, a large amount of membranes expressing a fusion protein can be easily prepared using the baculovirus-Sf9 system. One liter of cultured Sf9 cells is estimated to be enough for more than 100,000 assays. It is, therefore, not difficult to use the same batch of membrane preparations with the same specific activity from start to end of a large-scale ligand screening.

One of the possible disadvantages in this system is that the effective dose of agonists might be higher for fusion proteins than for cultured cells. The cells that express excess amounts of receptors can respond to lower concentrations of ligands than those expected from mass action, because effectors may be fully activated when only a small fraction of receptors is bound with agonist, whereas the sensitivity for the fusion proteins is determined by mass action between the ligand and receptor. Another possible disadvantage for the fusion protein is that the agonist-dependent increase in [^{35}S]GTPγS binding is clearly observed for G_i- and G_s-coupled receptors but not G_q-coupled receptors[25] (but see also Reference 39). Co-expression of βγ subunits does not resolve this problem.

PROTOCOL 2-6:   GENERAL METHOD FOR EXPRESSION OF THE RECEPTOR-Gα FUSION PROTEIN

2-6-1   Gene Construction of Fusion Proteins

The gene construction takes advantage of the use of the PCR (Figure 2.1).

2-6-2   Expression of the Fusion Protein Using Sf9 Cells

1. The recombinant baculoviruses are usually produced using Bac-to-Bac expression system (Life Technologies, Inc.) according to the manufacturer's instructions. Original viruses are amplified at a multiplicity of infection (m.o.i) of 0.01 to 0.1 for 72 or 96 h. To express the fusion protein, Sf9 cells are infected with the recombinant viruses at an m.o.i of 5 to 10 and are then cultured for 48 or 60 h. See also Protocol 2-1 for details. Optimal infection conditions may vary depending on the virus species. In general, the receptors or fusion proteins appear to start to degrade 48 h after infection. A longer culture after infection may cause higher expression yield with a higher risk of protein degradation. The Protocol 2.7 without virus titration may be adopted if cultured cells are used only to check the responses of a fusion protein to ligand candidates on a scale that is not too large.
2. Membrane fractions are prepared by homogenization of cultured cells on ice in 20 mM Hepes-KOH (pH 8.0), 1 mM EDTA, 2 mM MgCl$_2$, 0.5 mM phenylmethylsulfonyl fluoride, 5 μg/ml leupeptin, 5 μg/ml pepstatin, 5 mM benzamidine using a Dounce homogenizer followed by centrifugation at 150,000 × g for 30 min at 4°C.
3. The pellet was then homogenized on ice in the same buffer and stored in aliquots at 80°C in 0.5 to 1.5 ml tubes. Avoid repetitions of freeze and thaw. Protein concentrations are determined by the BCA protein assay (Pierce) using bovine serum albumin as the standard.

PROTOCOL 2-7:   QUICK (BUT NOT OPTIMIZED) EXPRESSIONS OF THE FUSION PROTEINS WITHOUT VIRUS TITRATION

1. Transfect Sf9 cells with recombinant Bacmid vector that is obtained using Bac-to-Bac system (Life Technologies, Inc.), and incubate for 96 h. See details of the manufacturer's instructions. Recover and keep medium as an original virus stock.
2. Seed Sf9 cells into a 75-cm² flask with 10 ml IPL-41 medium, and cultivate them to 60 to 70% confluence at 28°C.
3. Add 1 ml of the original virus stock to the above flask. Keep the flask at 28°C for 96 h, and recover the medium as an amplified virus solution.
4. Add 5 to 10 ml of the amplified virus solution to 200 ml of Sf9 cells in 3 l spinner culture flask equipped with a rotating paddle with a density of over 3×10^6 cells/ml. Incubate the vessel for 96 h at 28°C, and recover its medium by centrifugation (5000 rpm, 10 min) as the second amplified virus solution.
5. Add 40 to 50 ml of the second amplified virus solution to 200 ml of Sf9 cells in a vessel at a density of over 3×10^6 cells/ml. Incubate 48 h at 28°C, and recover the cells by centrifugation (5000 rpm, 10 min).
6. Prepare cell membrane fractions as described in Protocol 2-6-2 and 11-2, and keep them in 0.5 to 1.5-ml tubes at –80°C. Avoid repetitions of freeze and thaw. From 200 ml cultivation, approximately 10 to 15 ml of membrane suspensions with 5 mg protein/ml and 0.1 to 1 pmole receptor/mg protein is obtained.

Notes

1. Check the cells with an optical microscope, and confirm virus infection in every step and every day. Pay attention to cell size, shape of membrane, and nuclear size before and after infection.
2. Keep the depth of the culture medium less than 3 cm after virus infection unless a specific oxygen supplier is used. Because Sf9 cells consume a lot of oxygen after infection, cultivation under deep medium easily causes exhaustion of oxygen and death of cells.

PROTOCOL 2-8:   CONFIRMING EXPRESSION OF THE FUSION PROTEIN

1. Ligand-binding assay with a radioisotope labeled ligand is the direct and accurate method to measure the expression level. See also Protocol 2-3 for details.
2. Western blotting is a useful method to confirm the expression of the fusion proteins for which radiolabeled ligands are not available, although the estimation is not as quantitative as the ligand-binding assay. Mix 25 µl of membrane fraction (~5 mg protein/ml) and 25 µl of digitonin solution

(4%). Keep the mixture on ice for 1 h and centrifuge it for 10 min at 12,000 rpm. Add the SDS-PAGE buffer to the supernatant, and apply it to the SDS-PAGE gel: be sure that the sample is not boiled; the applied sample should contain approximately 100 μg of membrane protein. We use rabbit polyclonal anti-Gα$_i$ antibody against synthetic C-terminal peptide of Gα$_i$. It should be noted that even nonfunctional receptors can be detected with the Western blotting assay in contrast with the ligand-binding assay.

PROTOCOL 2-9:   [^{35}S]GTPγS BINDING ASSAY

This protocol is for high-throughput screening and thus saves reagents.

1. Prepare 10× assay buffer (200 mM HEPES pH 8.0, 100 mM MgCl$_2$, 1.6 M NaCl, 10 mM EDTA, 10 mM dithiothreitol). Mix 10 μl of 10× assay buffer, 20 μl of membrane fraction (1 mg/ml, 20 μg of protein), 10 μl of 10 μM GDP, 10 μl of 1 nM [^{35}S]GTPγS (DuPont–New England Nuclear, Boston, MA), ligand candidates, or standard agonist. Add water, and adjust the total volume to 100 μl.
2. Mix gently, and incubate at 30°C for 30 min.
3. Terminate the reactions with filtration through GF/B Unifilter plates (Packard) followed by rinsing three times with 200 μl of cold 20 mM potassium phosphate buffer (pH 7.0) in a cell harvester: membranes, together with the fuison protein and the bound [^{35}S]GTPγS, are trapped on the filter, separated from the free [^{35}S]GTPγS in the reaction mixture.
4. Measure the radioactivity on the Unifilter plates using Topcount (Packard) or appropriate liquid scintillation counter.
5. Note that several computer software packages, for example KaleidaGraph, are available to analyze the obtaned data. The equation below may be used to fit the curve and estimate EC50 or IC50.

$$y = a + b/(1 + (c/x)^d) \qquad (23.1)$$

where y is the amount of bound [^{35}S] GTPγS; x is the concentration of agonist or antagonist; a and b are parameters representing the amount of bound [^{35}S] GTPγS in the absence of agonist and the agonist-induced increase in bound [^{35}S] GTPγS in the presence of maximal agonist; c is EC$_{50}$ or IC$_{50}$; d is 1 with experiments with different concentrations of agonists, and −1 with experiments with a certain concentration of agonist and different concentrations of antagonists.

Notes

1. Ten percent of dimethylsulfoxide is the first-choice solvent for water-insoluble reagents. The presence of 10% diemthysulfoxide in a reaction mixture causes a reduction of efficiency of radioactivity counting but does

not appear to affect the activation of the fusion protein. The addition of 0.1% bovine serum albumin is recommended to reduce nonspecific adsorption of sample reagents to microplates.

2. To optimize assay conditions, various concentrations for proteins and GDP should be examined using fusion proteins of receptors with known radioligand before applying to orphan GPCRs. The radiocount ratio of resting and activated state, that is, the ratio of $(a + b)/a$ in Equation 2.1, should be around two to five.

3. The assay using SPA technology (wheat germ agglutinin scintillation proximity assay, Amersham) is convenient for high-throughput screening and is applicable to this assay instead of filtration through filter plates.

ACKNOWLEDGMENTS

We sincerely thank Dr. Melvin Simon for human Gα16 cDNA, Dr. Shigekazu Nagata for pEF-BOS and pEF-neo, Dr. Hiroshi Takeshima for rat nociceptin receptor gene, and Ms. Michiko Okamura for her help in the initial attempt to prepare receptor-Gα fusion proteins. This study was supported in part by Japan Science and Technology Corporation (CREST) and research grants from the Ministry of Education, Culture, Sports, Science and Technology (Intellectual Infrastructure Research and Development for Its Comprehensive Promotion System, Special Coordination Funds for the Promotion of Science and Technology).

ABBREVIATIONS

5-oxo-ETE	5-oxo-6E,8Z,11Z,14Z-eicosatetraenoic acid
5(S)-HETE	5(S)-hydoxy-6E,8Z,11Z,14Z-eicosatetraenoic acid
5(S)-HPETE	5(S)-hydroperoxy-6E,8Z,11Z,14Z-eicosatetraenoic acid
[^{35}S]GTPγS	guanosine 5-O-(3–[^{35}S]thiotriphosphate)
CHO	Chinese hamster ovary
CX$_3$CR1	CX$_3$C chemokine receptor 1
GPCRs	G protein-coupled receptors
NR-Gα_{i2}	a fusion protein of rat nociceptin receptor and bovine G$_{i2}$ protein α subunit
PGE$_2$	prostaglandin E$_2$
Sfkn	human fractalkine chemokine domain (soluble fractalkine)

REFERENCES

1. Howard, A.D. et al. Orphan G-protein-coupled receptors and natural ligand discovery, *Trends Pharmacol Sci.*, 22, 132, 2001.
2. Takeda, S. et al., Identification of G protein-coupled receptor genes from the human genome sequence, *FEBS Lett.*, 520, 97, 2002.
3. Vassilatis, D.K. et al., The G protein-coupled receptor repertoires of human and mouse, *Proc. Natl. Acad. Sci. USA*, 100, 4903-8, 2003.

4. Fredriksson, R. et al., The G-protein-coupled receptors in the human genome form five main families, Phylogenetic analysis, paralogon groups, and fingerprints, *Mol. Pharmacol.*, 63, 1256, 2003.

5. Inoue, Y., Ikeda, M., and Shimizu, T., Proteome-wide classification and identification of mammalian-type GPCRs by binary topology pattern, *Comput. Biol. Chem.*, 28, 39, 2004.

6. Bargmann, C.I., Neurobiology of the Caenorhabditis elegans genome, *Science*, 282, 2028, 1998.

7. Adams, M.D. et al., The genome sequence of Drosophila melanogaster, *Science*, 287, 2185, 2000.

8. Buck, L. and Axel, R. A novel multigene family may encode odorant receptors: a molecular basis for odor recognition, *Cell*, 65, 175, 1991.

9. Matsunami, H., Montmayeur, J.P., and Buck, L.B., A family of candidate taste receptors in human and mouse, *Nature*, 404, 601, 2000.

10. Venter, J.C. et al., The sequence of the human genome, *Science*, 291, 1304, 2001.

11. Lander, E.S. et al., Initial sequencing and analysis of the human genome, *Nature*, 409, 860, 2001.

12. Gentles, A.J. and Karlin, S., Why are human G-protein-coupled receptors predominantly intronless? *Trends in Genet.*, 15, 47, 1999.

13. Glusman, G. et al., The complete human olfactory subgenome, *Genome Res.*, 11, 685, 2001.

14. Malnic, B., Godfrey, P.A., and Buck, L.B., The human olfactory receptor gene family, *Proc. Natl. Acad. Sci. USA*, 101, 2584, 2004.

15. Montmayeur, J.P. et al., A candidate taste receptor gene near a sweet taste locus, *Nat. Neurosci.*, 4, 492, 2001.

16. Kitagawa, M. et al., Molecular genetic identification of a candidate receptor gene for sweet taste, *Biochem. Biophys. Res. Commun.*, 283, 236, 2001.

17. Max, M. et al., Tas1r3, encoding a new candidate taste receptor, is allelic to the sweet responsiveness locus Sac, *Nat. Genet.*, 28, 58, 2001.

18. Adler, E. et al., A novel family of mammalian taste receptors, *Cell*, 100, 693, 2000.

19. Meunier, J.C. et al., Isolation and structure of the endogenous agonist of opioid receptor-like ORL1 receptor, *Nature*, 377, 532, 1995.

20. Reinscheid, R.K. et al., Orphanin FQ: a neuropeptide that activates an opioidlike G protein-coupled receptor, *Science*, 270, 792, 1995.

21. Brauner-Osborne, H. and Brann, M.R., Pharmacology of muscarinic acetylcholine receptor subtypes (m1-m5): high throughput assays in mammalian cells, *Eur. J. Pharmacol.*, 295, 93, 1996.

22. Milligan, G. and Rees, S., Chimaeric G alpha proteins: their potential use in drug discovery, *Trends Pharmacol. Sci.*, 20, 118, 1999.

23. Coward, P. et al., Chimeric G proteins allow a high-throughput signaling assay of Gi-coupled receptors, *Anal. Biochem.*, 270, 242, 1999.

24. Wnendt, S. et al., Agonistic effect of buprenorphine in a nociceptin/OFQ receptor-triggered reporter gene assay, *Mol. Pharmacol.*, 56, 334, 1999.

25. Guo, Z.D. et al., Receptor-Gα fusion proteins as a tool for ligand screening, *Life Sci.*, 68, 2319, 2001.

26. Suga, H. et al., Stimulation of increases in intracellular calcium and prostaglandin E2 generation in Chinese hamster ovary cells expressing receptor-Gα_{16} fusion proteins, *J. Biochem. (Tokyo)*, 135, 605, 2004.

27. Offermanns, S. and Simon, M.I., Gα_{15} and Gα_{16} couple a wide variety of receptors to phospholipase C, *J. Biol. Chem.*, 270, 15175, 1995.

28. Milligan, G., Marshall, F., and Rees, S., G_{16} as a universal G protein adapter: implications for agonist screening strategies, *Trends Pharmacol. Sci.*,17, 235, 1996.
29. Kostenis, E., Is $G\alpha_{16}$ the optimal tool for fishing ligands of orphan G-protein-coupled receptors? *Trends Pharmacol. Sci.*, 22, 560, 2001.
30. Milligan, G., The use of receptor G-protein fusion proteins for the study of ligand activity, *Receptors Channels*, 8, 309, 2002.
31. Zhang, Q. et al., Effects of partial agonists and Mg^{2+} ions on the interaction of M2 muscarinic acetylcholine receptor and G protein $G\alpha_{i1}$ subunit in the M2-$G\alpha_{i1}$ fusion protein, *J. Biochem. (Tokyo)*, 135, 589, 2004.
32. Bevan, N. et al., Functional analysis of a human A(1) adenosine receptor/green fluorescent protein/$G_{i1}\alpha$ fusion protein following stable expression in CHO cells, *FEBS Lett.*, 462, 61, 1999.
33. Zhang, L. and Saffen, D., Muscarinic acetylcholine receptor regulation of TRP6 Ca^{2+} channel isoforms. Molecular structures and functional characterization, *J. Biol. Chem.*, 276, 13,331, 2001.
34. Mizushima, S. and Nagata, S., pEF-BOS, a powerful mammalian expression vector, *Nucleic Acids Res.*, 18, 5322, 1990.
35. Bertin, B. et al., Cellular signaling by an agonist-activated receptor/$G_s\alpha$ fusion protein, *Proc. Natl. Acad. Sci. USA*, 91, 8827, 1994.
36. Takeda, S. et al., The receptor-$G\alpha$ fusion protein as a tool for ligand screening: a model study using a nociceptin receptor-$G\alpha_{i2}$ fusion protein, *J. Biochem. (Tokyo)*, 135, 597, 2004.
37. Takeda, S., Yamamoto, A., and Haga, T., Identification of a G protein-coupled receptor coupled for 5-oxo-eicosatetraenoic acid, *Biomed. J.*, 23, 101, 2002.
38. Takeda, S. et al., Identification of surrogate ligands for orphan G protein-coupled receptors, *Life Science*, 74, 367, 2003.
39. Liu, S. et al., Effective information transfer from the $\alpha_1\beta$-adrenoceptor to $G\alpha_{11}$ requires both β/γ interactions and an aromatic group four amino acids from the C terminus of the G protein, *J. Biol. Chem.*, 277, 25,707, 2002.

3 Screening and Identification of Ghrelin, an Endogenous Ligand for GHS-R

Masayasu Kojima and Kenji Kangawa

CONTENTS

3.1 INTRODUCTION

Growth hormone (GH), a multifunctional hormone secreted from somatotrophs of the anterior pituitary, regulates overall body and cell growth, carbohydrate–protein–lipid metabolism, and water–electrolyte balance.[1] Production and release of GH are properly controlled because GH excess results in acromegaly and gigantism, while GH deficiency in children results in impaired growth and short stature. GH is controlled by many factors, in particular, by two hypothalamic neuropeptides: GH release is stimulated by hypothalamic growth-hormone releasing hormone (GHRH) and inhibited by somatostatin.[2] However, a third independent pathway regulating GH release has sprung from studies of growth-hormone secretagogues (GHSs).[3,4] GHSs are synthetic compounds that are potent stimulators of GH release, working through a novel G protein-coupled receptor, the GHS-receptor (GHS-R). Because GHSs are a group of artificial compounds and do not exist in nature, it has been speculated that there must exist an endogenous ligand that binds to GHS-R and mimics the action of GHSs, that is, until the discovery of ghrelin.[5] We survey here the purification and structure of ghrelin.

3.2 REVERSE PHARMACOLOGY

Recent progress in genomic research revealed the nucleotide sequences of many orphan G protein-coupled receptors (GPCRs).[6] Figure 3.1 presents a so-called "orphan receptor strategy" to search for endogenous ligands.[7,8] In this method, first, a cell line was established to express an orphan GPCR stably. Then, peptide extract was applied to the cell, and a second messenger was measured. If a target orphan GPCR is functionally expressed on the cell surface and the extract contains the endogenous ligand, the second messenger level will increase or decrease in response to a coupled G protein. Usually, cyclic adenosine monophosphate (cAMP) concentration or intracellular calcium change was measured as an indicator for the second messenger. Using this assay system and several steps of chromatographies, searches for novel ligands using orphan GPCR-expressing cells have resulted in the discovery of several novel bioactive peptides, such as nociceptin/orphanin FQ,[9,10] orexin/hypocretin,[11] prolactin-releasing peptide,[12] apelin,[13] metastin,[14] neuropeptide B,[15,16] and neuropeptide W.[16,17] Therefore, orphan receptors represent important new tools for the discovery of novel bioactive molecules and for in drug development.

   In these many orphan GPCR receptors, GHS-R attracted many researchers and pharmaceutical companies, because a putative ligand for the GHS-R should be used directly for treatment of GH deficiency.[18] Unlike in the case of other orphan GPCRs, GHS-R was known to bind artificial ligands, such as GHRP-6 or Hexarelin, providing a convenient positive control for screening assay. The search for a GHS-R endogenous ligand was attempted in brain and hypothalamic extracts, because GHS-R is known to be mainly expressed in the pituitary and hypothalamus,[19–21] thus suggesting that its ligand should be produced in the brain. However, numerous attempts failed to find the ligand in the brain, and unexpectedly, we succeeded in the purification and identification of the endogenous ligand for the GHS-R, ghrelin, from the stomach.[5] Ghrelin is derived from "ghre," a word root in Proto-Indo-European languages

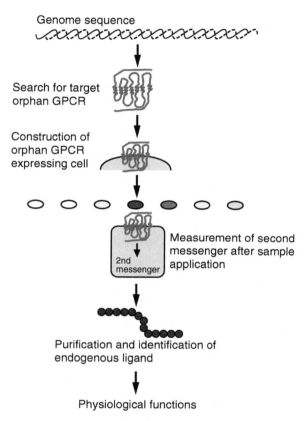

Genome sequence

Search for target orphan GPCR

Construction of orphan GPCR expressing cell

2nd messenger

Measurement of second messenger after sample application

Purification and identification of endogenous ligand

Physiological functions

FIGURE 3.1 Orphan receptor strategy. A number of GPCRs are found in genome sequences of mammals, fishes, and even in *Caenorhabditis elegans*. These GPCRs are expressed in culture cells and are used for an assay to measure second messenger changes. By using these assay systems, the endogenous ligands are purified, and their structures are determined. After these steps, physiological functions of these ligands are examined. Thus, the orphan receptor strategy is completely reverse to the classical strategy for hormone research. The classical strategy uses physiological functions as an assay system to purify endogenous ligands. After the identification of the ligand, its receptor is searched.

meaning "grow," reflecting its role in stimulating GH release. Ghrelin is a novel growth-hormone-releasing and appetite-stimulating peptide.

3.3 GHS (GROWTH-HORMONE SECRETAGOGUE) AND ITS RECEPTOR

3.3.1 HISTORY

Figure 3.2 shows the structures of typical GHSs. In 1976, Bowers et al. found that some opioid peptide derivatives had weak GH-releasing activity, although they did not exhibit opioid activity.[22] The structure of the first GHS was Tyr-D-Trp-Gly-Phe-

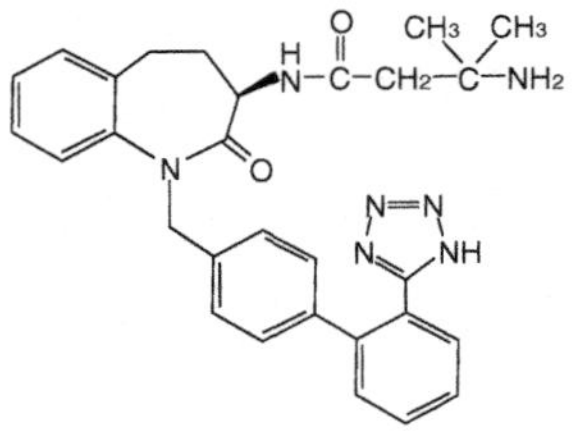

L-692,429 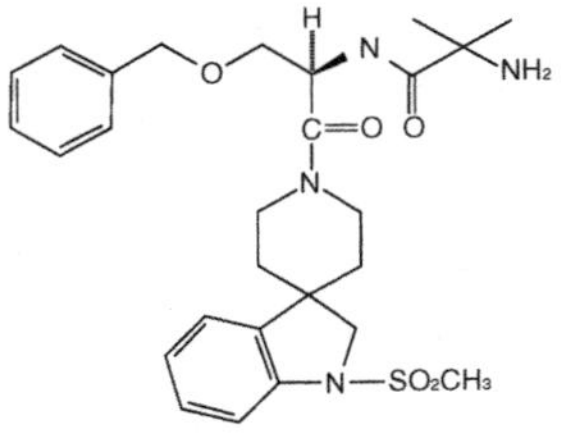   L-163,191 (MK-0677)

His-D-Trp-Ala-Trp-D-Phe-Lys-NH2
GHRP-6

Tyr-D-Trp-Gly-Phe-Met-NH2   His-D-2-Methyl-Trp-Ala-Trp-D-Phe-Lys-NH2
The first GHS   Hexarelin

FIGURE 3.2 Typical structures of GHSs. GHSs are grouped into nonpeptidyl GHSs (like L-692,429 and MK-0677) and peptidyl GHSs (like GHRP-6 and Hexarelin). The first GHS was a methionine enkephalin derivative.

Met-NH$_2$, which released GH by a direct action on the pituitary. This synthetic peptide was a methionine enkephalin derivative, which replaced the second Gly with D-Trp and had C-terminal amide structure. The GH-releasing activities of early GHSs were so weak that it was found only *in vitro*. After this observation, many peptidyl derivatives were synthesized in the search for more potent growth-hormone secretagogues.

It was in 1984 that a potent GHS, GHRP-6, was synthesized based on conformational energy calculations in conjunction with peptide chemistry modifications and biological activity assay.[23] Hexapeptide GHRP-6 was shown to be active both *in vitro* and *in vivo*, which suggested a possible application for clinical use.

In 1993, the first nonpeptide GHS, L-692,429, was synthesized by Smith et al.[24] This nonpeptidyl GHS was a milestone for clinical use of GHSs, because L-692,429 was active enough even when orally administered.

During this period, researchers investigated the mechanisms of GHS action. While GH release from the pituitary was known to be stimulated by the hypothalamic GHRH, exogenous GHSs were thought to induce GH release through a different pathway from that of GHRH.[25] GHRH acts on the GHRH receptor to increase intracellular cAMP, which serves as a second messenger. On the other hand, GHSs act on a different receptor, increasing intracellular calcium concentration via inositol 1,4,5-trisphosphate (IP$_3$) signal transduction (Figure 3.3).

In 1996, the growth-hormone secretagogue receptor (GHS-R) was identified by expression cloning.[21] This strategy was based on the findings that GHSs stimulated phospholipase C, resulting in an increase in inositol triphosphate and intracellular calcium. Xenopus oocytes were injected with *in vitro* transcribed complementary ribonucleic acids (cRNAs) derived from swine pituitary, supplemented simultaneously with various G subunit mRNAs. A MK0677-stimulated calcium increase could be detected by bioluminescence of the jellyfish photoprotein aequorin, which was incorporated in the Xenopus oocytes. The identified GHS-R is a typical G protein-coupled seven-transmembrane receptor. *In situ* hybridization analyses showed that GHS-R is expressed in the pituitary, hypothalamus, and hippocampus. This receptor

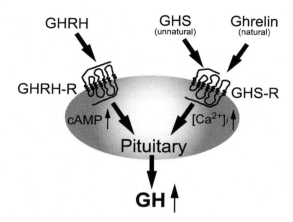

FIGURE 3.3 The regulation of growth hormone release from the pituitary. In the pituitary somatotroph cells, GHRH stimulates GH release through binding to the GHRH receptor and increasing cAMP levels. By contrast, GHSs stimulate GH release through the GHS receptor (GHS-R, ghrelin receptor) to increase intracellular Ca^{2+} ($[Ca^{2+}]i$) levels. Because GHSs are artificial molecules and do not exist naturally, the endogenous ligand for the GHS-R has been postulated. The endogenous ligand for the GHS-R was not known until the discovery of ghrelin.

was, for some time, an example of an orphan GPCR; that is, a GPCR with no known natural ligand. After the identification of the GHS-R, a search for its endogenous ligand had been actively undertaken, using the orphan receptor strategy (Figure 3.1).

3.3.2 GHS RECEPTOR (GHS-R) SUPERFAMILY

Figure 3.4 shows a dendrogram of GHS-R and its related receptor superfamily. In 1996, the GHS-R was identified by expression cloning and was shown to be a typical G protein-coupled seven-transmembrane receptor. The receptor most closely related to GHS-R is the motilin receptor (GPR38); the human forms of these receptors possess 52% amino acid homology.[26,27] Two receptors for NMU (NMU-R1 and R2), a neuropeptide that promotes smooth muscle contraction and suppresses food intake, are also homologous to GHS-R.[28–31] Because motilin and NMU are found mainly in gastrointestinal organs, it was speculated that the endogenous ligand for GHS-R may be another gastrointestinal peptide. This speculation was confirmed by the isolation of ghrelin from stomach tissue.

3.4 PURIFICATION OF GHRELIN

3.4.1 CONSTRUCTION OF GHS-R EXPRESSING CELLS

Because most orphan GPCRs have no ligand for positive control in the assay, we have to use orphan GPCR expressing cells without checking the proper functional expression of the receptor. We can check only the messenger RNA (mRNA) level of an expressed orphan receptor by Northern blot analysis. However, a high mRNA expression level does not mean that the expressed orphan receptor is functional.

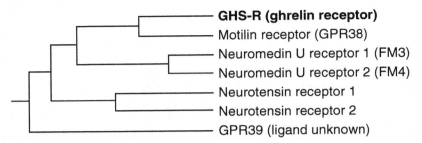

FIGURE 3.4 Dendrogram alignment of ghrelin receptor (GHS-R) and other G protein-coupled receptors. The ghrelin receptor forms a group of the GPCR superfamily, which contains motilin, neuromedin U, and neurotensin receptors. Ghrelin receptor (GHS-R) is most homologous to the motilin receptor. Because their endogenous ligands, ghrelin and motilin, have in part, a homologous amino-acid sequence, ghrelin and motilin systems may be evolved from an ancestral common peptide system. This superfamily also contains an orphan receptor, GPR39, with an endogenous ligand that should be a peptide(s).

In contrast to many orphan GPCRs, GHS-R was known to bind artificial ligands, such as GHRP-6 or hexarelin, providing a convenient positive control for constructing the assay system used to identify the endogenous ligand. A cultured cell line expressing the GHS-R was established and used to identify tissue extracts that could stimulate the GHS-R as monitored by increases in intracellular Ca²⁺ levels.

PROTOCOL 3-1: CONSTRUCTION OF GHS-R EXPRESSING CELL LINE

1. A full-length complementary deoxyribonucleic acid (cDNA) of rat GHS-R was obtained by reverse transcription polymerase chain reaction (RT-PCR) from a rat brain cDNA template. Sense and antisense primers were synthesized based on the reported sequence of rat GHS-R. The sense primer is 5′-ATGTGGAACGCGACCCCCAGCGA-3′, and the antisense primer is 5′-ACCCCCAATTGTTTCCAGACCCAT-3′.
2. PCR conditions were 35 cycles at 94°C for 1 min, 63 °C for 2 min, and 72 °C for 3 min. After amplification, the PCR products were resolved by electrophresis in 1% agarose gel, and the band that matched the predicted length of GHS-R cDNA was cut and purified.
3. The purified cDNA of rat GHS-R was ligated into the pcDNA3.1 vector (Invitrogen, Carlsbad, CA).
4. The expression vector, GHSR-pcDNA3.1, was transfected into Chinese hamster ovary (CHO) cells using FuGENE6 transfection reagent (Roche, Indianapolis, IN).
5. Stable transfectants were selected in 1 mg/ml G418 and isolated as a single clone cell.
6. The isolated cell lines were screened for intracellular calcium concentration ([Ca²⁺]i) changes induced by GHRP-6 (Peninsula Laboratories, Belmont, CA).

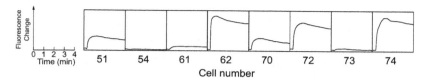

FIGURE 3.5 Ca^{2+} mobilization responses of the GHS-R expressing CHO cells. Each square represents responses of isolated stable cells as determined by FLIPR. Intracellular calcium concentrations were indicated by fluorescence changes using Fluo-4 as an indicator dye and measuring for 4 min. GHRP-6 (10^{-9} M) was used for activating the cells.

3.4.2 INTRACELLULAR CALCIUM MEASUREMENT ASSAY BY FLIPR

Figure 3.5 shows intracellular calcium concentration changes by fluorometric imaging plate reader (FLIPR) assays in several GHS-R expressing cell lines. Isolated GHS-R expressing cell lines are activated by GHRP-6, an artificial ligand to GHS-R. The calcium changes are varied in each cell line in relation to the expression levels of GHS-R mRNA.

PROTOCOL 3-2: FLIPR ASSAY

1. The GHS-R expressing cells were plated in flat-bottom, black-wall 96-well plates (Corning Corstar Corporation, Cambridge, MA) at 4×10^4 cells/well for 12 to 15 h before the assay.
2. Cells were loaded with 4 mM Fluo-4-AM fluorescent indicator dye (Molecular Probes, Inc., Eugene, OR) for 1 h in assay buffer [Hanks Balanced Salts Solution (HBSS), 10 mM HEPES (pH 7.4), 2.5 mM probenecid, 1% fetal calf serum (FCS)] and were washed four times in assay buffer without FCS.
3. $[Ca^{2+}]i$ changes were measured by a FLIPR (Molecular Devices, Sunnyvale, CA). We used the maximum change in fluorescence over baseline to determine agonist responses.
4. The clone (CHO-GHSR62) that showed the highest response was used in the following assays.

3.4.3 PURIFICATION OF RAT GHRELIN

The purification steps of ghrelin are summarized in Figure 3.6. Cells contain many proteases for digestion, processing, and protection from endogenous or exogenous molecules or organisms. In intact cells, these many proteases are compartmented into limited parts of cellular components, such as lysosome, microsome, Golgi apparatus, and mitochondria; bioactive peptides in secretary granules are not digested. However, when cells are homogenized for extracting peptide fractions, intracellular proteases are released into solutions, and sometimes inactive forms of proteases are activated. These proteases digest and inactivate endogenous peptides very easily. It is, thus, important to inactivate intracellular proteases before taking purification steps. Our simple method by boiling the tissues in water for 5 to 10 min before homogenization is very good in keeping peptide intact during purification.[32,33]

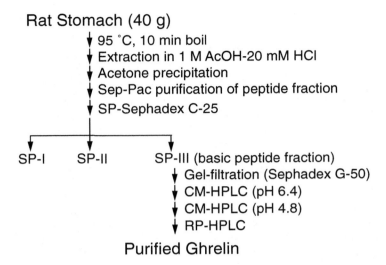

FIGURE 3.6 Purification steps of ghrelin from rat stomach tissues. AcOH, acetic acid; CM, carboxymethyl; RP-HPLC, reverse-phase high-performance liquid chromatography.

PROTOCOL 3-3: PURIFICATION OF GHRELIN FROM RAT STOMACH TISSUE (FIGURE 3.7)

1. Fresh rat stomach (40 g) was minced and boiled for 10 min in 5 × volumes of water to inactivate intrinsic proteases. Boiling in acid solution like acetic acid should be avoided because boiling of tissues in acid solution sometimes results in cleavage of peptides. In the case of ghrelin, boiling in acetic acid cleaves the ester bond between modified fatty acid and Ser3 and results in inactivation of ghrelin.
2. The solution was adjusted to 1 M acetic acid (AcOH)-20 mM HCl.
3. Boiled stomach tissue was homogenized with a Polytron mixer.
4. The supernatant of the extracts was concentrated to approximately 40 ml by evaporation.
5. A 2 × volume of acetone was added to the concentrate for acetone precipitation in 66% acetone.
6. Acetone was removed from the separated supernatants by evaporation.
7. The supernatant was loaded onto a 10 g Sep-Pak C18 column (Waters, Milford, MA), pre-equilibrated in 0.1% trifluoroacetic acid (TFA). The Sep-Pak cartridge was washed with 10% acetonitrile (CH_3CN)/0.1% TFA, and the peptide fraction was eluted in 60% CH_3CN/0.1% TFA.
8. The eluate was evaporated and lyophilized.
9. The residual materials were redissolved in 1 M AcOH and adsorbed on a SP-Sephadex C-25 column (H^+-form, Pharmacia, Uppsala, Sweden) pre-equilibrated with 1 M AcOH.

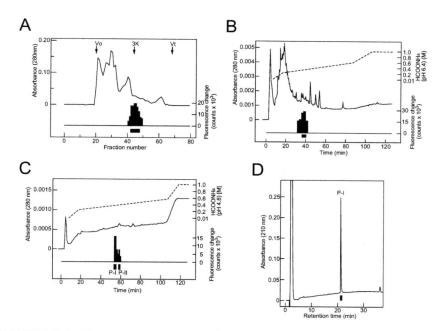

FIGURE 3.7 Purification of ghrelin from rat stomach. Black bars indicate the fluorescence changes due to $[Ca^{2+}]i$ increase on CHO-GHSR62 cells. The gradient profiles are indicated by the dotted lines. (A) Gel filtration chromatography of the basic peptide fraction extracted from rat stomach (40 g). Active fractions (43 to 48) were eluted at approximately Mr 3000 (3 K). Vo, void volume; Vt, total volume. (B) CM-ion-exchange HPLC (pH 6.4) of the active fractions derived from gel filtration. Active fractions, indicated by solid bars, were processed by a second CM-ion-exchange HPLC. (C) The second CM-ion-exchange HPLC (pH 4.8) of the active fractions from the first CM-HPLC. Two active fractions, P-I and P-II, are indicated by solid bars. (D) Final purification of P-I by RP-HPLC.

10. Successive elutions with 1 M AcOH, 2 M pyridine, and 2 M pyridine-AcOH (pH 5.0) yielded three fractions: SP-I, SP-II, and SP-III (basic peptide fraction).

11. The SP-III fraction was lyophilized (dry weight: 39 mg).

12. The lyophilized sample was dissolved in 1 ml of 1M AcOH and applied to a Sephadex G-50 gel-filtration column (1.8 × 130 cm) (Pharmacia, Uppsala, Sweden).

13. Five-milliliter fractions were collected, and a portion of each fraction was subjected to the $[Ca^{2+}]i$ change assay using CHO-GHSR62 cells (Figure 3.7A).

14. Active fractions were separated by carboxymethyl (CM) ion-exchange high-performance liquid chromatography (HPLC) on a TSK CM-2SW column (4.6 × 250 mm, Tosoh, Tokyo, Japan) using an ammonium formate gradient $(HCOONH_4)$ (pH 6.4) of 10 mM to 1 M in the presence of 10% CH_3CN at a flow rate of 1 ml/min for 100 min. Two-milliliter fractions were collected and assayed for $[Ca^{2+}]i$ change (Figure 3.7B).

15. Active fractions were further fractionated by a second round of CM-HPLC on the same column at pH 4.8, yielding two active peaks (P-I and P-II) (Figure 3.7C). Ghrelin was purified from P-I, and des-Gln14-ghrelin was isolated from P-II.

16. The P-I was further purified manually using a C18 reverse-phase HPLC (RP-HPLC) column (Symmetry 300, 3.9 × 150 mm, Waters, Milford, MA), and finally purified as a single peak (Figure 3.7D).

3.4.4 STRUCTURAL DETERMINATION OF RAT GHRELIN

Ghrelin is a 28-amino-acid peptide. The sequence, except for the third amino acid, was determined by Edman degradation. An expressed-sequence-tag clone containing the coding region of the peptide then revealed that the third residue is a Ser, which was then confirmed in cDNA clones encoding the peptide precursor isolated from a rat stomach cDNA library.

A 28-amino-acid peptide based on the Ser-containing cDNA sequence was synthesized, and its characteristics were compared with those of purified ghrelin. This comparison revealed the following: (1) the synthetic peptide, unlike the purified peptide, did not increase $[Ca^{2+}]i$ levels in GHS-R-expressing cells; (2) the retention time of purified ghrelin on RP-HPLC was longer than that of the synthetic peptide; and (3) by mass spectrometric analysis, the molecular weight (Mr) of purified ghrelin (3315) was 126 Da greater than that of the synthetic peptide (3189).

The most probable modification was that of a fatty acid, n-octanoic acid (Figure 3.8). When the hydroxyl group of Ser3 of the synthetic peptide (Mr = 3189) was esterified by n-octanoic acid (Mr = 144), the resulting modified peptide was of the same molecular weight as purified ghrelin (3315). Moreover, the modification by fatty acid appeared to increase the hydrophobicity of the peptide, explaining the increase in retention time when it was subjected to HPLC.

The peptide was then synthesized with a Ser3 n-octanoyl modification, and its characteristics were compared with those of purified ghrelin. This synthetic modified peptide eluted at the same retention time as did purified ghrelin on RP-HPLC, and mass spectrometric fragmentation patterns of the synthetic modified peptide and purified ghrelin were the same. Moreover, the synthetic modified peptide had the same effect as purified ghrelin on the GHS-R-expressing cells. These results confirmed the primary structure of ghrelin, an octanoyl-modified peptide (Figure 3.9).[5]

3.4.5 MINOR MOLECULAR FORMS OF HUMAN GHRELIN

The major form of active human ghrelin in stomach is, like rat ghrelin, a 28-amino-acid peptide with an octanoyl modification at its third amino acid, serine. During the course of purification, we isolated several minor forms of ghrelin peptides.[34] These peptides were classified into four groups by the type of acylation observed at Ser3: nonacylated, octanoylated (C8:0), decanoylated (C10:0), and possibly decenoylated (C10:1). All peptides found are either 27 or 28 amino acids in length, the former lacking the C-terminal Arg28, and are derived from the same ghrelin precursor through two alternative pathways. Synthetic octanoylated and decanoy-

n-octanoic acid
O=C-(CH2)6-CH3
O
H
Mr 144
H
O
1
10
20
28
GSSFLSPEHQKAQQRKESKKPPAKLQPR
Mr 3,189
H2O
O=C-(CH2)6-CH3
O
GSSFLSPEHQKAQQRKESKKPPAKLQPR
Mr 3,315
3,189 + 144 - 18 = 3,315

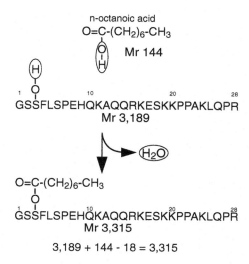

FIGURE 3.8 Ghrelin modification by n-octanoic acid.

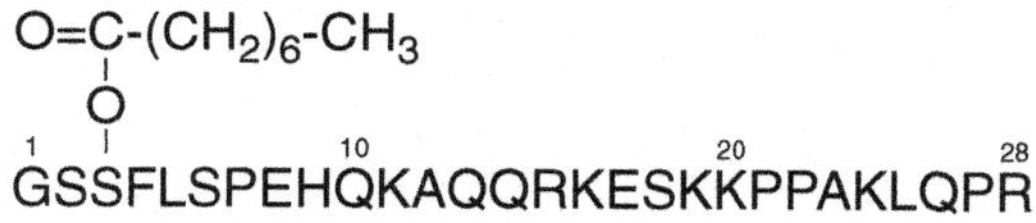

O=C-(CH2)6-CH3
O
1
10
20
28
GSSFLSPEHQKAQQRKESKKPPAKLQPR

FIGURE 3.9 Structure of rat ghrelin. The third amino acid, serine, is modified by a fatty acid, n-octanoic acid, and this modification is essential for ghrelin's activity.

lated ghrelins induce intracellular calcium increases in GHS-R-expressing cells and stimulate GH release in rats to similar degrees.

3.4.6 Ghrelin Precursor and Des-Gln14-Ghrelin

The rat and human ghrelin precursors are both composed of 117 amino acids.[5] In these precursors, the ghrelin sequence immediately follows that of the signal peptide. At the C-terminus of the ghrelin sequence, processing occurs at an uncommon Pro-Arg recognition site. In the rat stomach, two isoforms of mRNA-encoding pro-ghrelin are produced from the gene by an alternative splicing mechanism.[35] One mRNA encodes the ghrelin precursor, and another encodes a precursor for des-Gln14-ghrelin, a peptide identical to ghrelin, but with a deletion of Gln14. This deletion results from the use of the CAG codon, which encodes Gln14 as well as acts as a splicing signal. Thus, two types of active ghrelin peptide are produced in rat stomach — ghrelin and des-Gln14-ghrelin (Figure 3.10). However, des-Gln14-ghrelin is only present in low amounts in the stomach, and full-length ghrelin is the major active form.

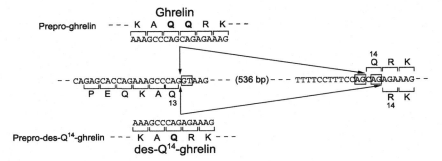

FIGURE 3.10 Schematic diagram of the splice junction of rat ghrelin gene. The genomic sequences of the exon–intron boundaries of the intron 2 of the ghrelin gene can be seen here. The splicing signals, GT for the 5'-side and AGs for the 3'-side of the intron 2, are boxed. When the first AG is used for the splicing signal, prepro-ghrelin mRNA is produced, and the second CAG is translated into Gln14. On the other hand, when the second AG is used, des-Gln14-ghrelin mRNA is created to produce des-Gln14-ghrelin, accordingly.

PROTOCOL 3-4: PURIFICATION OF DES-GLN14-GHRELIN FROM RAT STOMACH TISSUE (FIGURE 3.11)

Des-Gln14-ghrelin was purified by the same method as previously described for ghrelin.

Steps 1 to 13 are the same as in Protocol 3-3.

1. First, CM-HPLC active fractions were further fractionated by a second CM-HPLC on the same column at pH 4.8 to give two active peaks (P-I and P-II) (Figure 3.11A).
2. P-II was purified manually using a C18 RP-HPLC column (Symmetry 300, 3.9 × 150 mm, Waters, Milford, MA) and identified to be des-Gln14-ghrelin (Figure 3.11B).

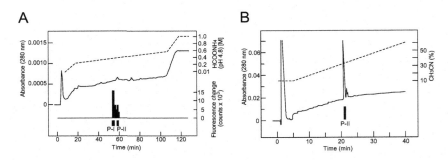

FIGURE 3.11 Purification of des-Gln14-ghrelin from rat stomach. Black bars indicate the fluorescence changes due to $[Ca^{2+}]i$ increase on CHO-GHSR62 cells. The gradient profiles are indicated by the dotted lines. (A) The second CM-ion-exchange HPLC (pH 4.8) of the active fractions from the first CM-HPLC (Figure 3.7b). Two active fractions, P-I and P-II, are indicated by solid bars. (B) Final purification of P-II by RP-HPLC.

3.5 NONMAMMALIAN GHRELIN

Figure 3.12 shows a sequence comparison of nonmammalian ghrelins.

```
                     *
Human:     GSSFLSP-EHQRVQQRKESKKPPAKLQPR
Rat:       GSSFLSP-EHQKAQQRKESKKPPAKLQPR
Bullfrog:  GLTFLSPADMQKIAERQSQNKLRHGNMN
Chicken:   GSSFLSP-TYKNIQQQKDTRKPTARLH
Goldfish:  GTSFLSPA--QKPQGRRPPRM-NH2
Eel:       GSSFLSP-S-QRPQG-KD-KKPPRV-NH2
```

FIGURE 3.12 Alignment of amino-acid sequences of human, rat, bullfrog, chicken, goldfish, and eel ghrelins. Sets of residues identical in at least three species are shaded. The asterisk indicates the acylated third amino acid, serine or threonine. Eel, and possibly goldfish, ghrelin has an amide structure at its C terminus.

3.5.1 BULLFROG GHRELIN

Amphibian ghrelin has been identified from the stomach of bullfrog.[36] The three forms of ghrelin identified in bullfrog stomach, each comprising 27 or 28 amino acids, possess 29% sequence identity to the mammalian ghrelins. A unique threonine at the third position of the sequence (Thr3) in bullfrog ghrelin differs from the serine present in the mammalian ghrelins; this Thr3 is acylated by either n-octanoic or n-decanoic acid.

PROTOCOL 3-5: PURIFICATION OF BULLFROG GHRELIN BY USING ANTI-GHRELIN IgG-AFFINITY CHROMATOGRAPHY

1. A frozen bullfrog stomach (10.75 g) was pulverized and boiled for 10 min in 5 volumes of water to inactivate intrinsic proteases.
2. The steps of extraction, sample preparation, and chromatographies are the same as those used in rat ghrelin purification, except that the gel-filtration chromatography step was skipped.
3. The active peak after the second CM-ion-exchange HPLC was diluted in an equal volume of 0.1% TFA, was desalted by using a Sep-Pak C18 cartridge, and was then lyophilized.
4. The lyophilized sample was subsequently dissolved in 500 μl of 100 mM phosphate buffer (pH 7.4) and applied on anti-rat ghrelin [1-11] immunoglobulin G (IgG) immunoaffinity column.
5. Absorbed substances were eluted in 1 ml of 60% CH_3CN/0.1% TFA.
6. The eluate was concentrated by evaporation and then subjected to RP-HPLC using a μBondasphare C18 column (2.1 × 150 mm, Waters) at a flow rate of 0.2 ml/min under a linear gradient from 10 to

60% CH₃CN/0.1% TFA for 40 min. The eluate corresponding to each absorption peak was collected.

7. A part of each fraction (100 mg tissue equivalent) was assayed for activity on GHS-R expressing cells by FLIPR.

8. The active fractions (numbers 9 to 11) were combined and further purified by RP-HPLC using a diphenyl column (2.1 × 150 mm, 219TP5215, Vydac, Hesperia, CA) at a flow rate of 0.2 ml/min with a linear gradient from 10 to 60% CH₃CN/0.1% TFA for 80 min. Each absorption peak was collected; a part of each fraction (100 mg tissue equivalent) was assayed for activity by FLIPR.

9. The active fraction was purified using a CHEMCOSORB 3-ODS-H column (Chemco, Osaka, Japan) at a flow rate of 0.2 ml/min with a linear gradient from 10 to 60% CH₃CN/0.1% TFA for 40 min.

10. Approximately 10 pmol of the purified peptide was analyzed by a protein sequencer (model 494, Applied Biosystems, Foster City, CA). Five pmol of the purified peptide was redissolved in 5 μl of 50% (v/v) methanol containing 1% AcOH, and then determined the molecular weight using electrospray ionization mass spectrometry (ESI/MA) (SSQ 7000; Finnigan, San Jose, CA).

The three main ghrelin isoforms in frog have the following sequences: ghrelin-28, GLT(*O*-n-octanoyl)FLSPADMQKIAERQSQNKLRHGNMN; ghrelin-27, GLT(*O*-n-octanoyl)FLSPADMQKIAERQSQNKLRHGNM; and ghrelin-27-C10, GLT(*O*-n-decanoyl)FLSPADMQKIAERQSQNKLRHGNM. Northern blot analysis demonstrated that ghrelin mRNA is expressed predominantly in the stomach. Low levels of gene expression were observed in the heart, lungs, small intestine, gallbladder, pancreas, and testes, as revealed by RT-PCR analysis. Bullfrog ghrelin stimulated the secretion of both GH and prolactin in dispersed bullfrog pituitary cells with a potency two to three orders of magnitude greater than that of rat ghrelin. Bullfrog ghrelin, however, was only minimally effective in elevating plasma GH levels following intravenous injection into rats. These results indicate that although the mechanism by which ghrelin regulates GH secretion is evolutionarily conserved, the structural differences in the various ghrelins result in species-specific receptor binding properties.

3.5.2 Chicken Ghrelin

Chicken (*Gallus gallus*) ghrelin was purified by a combination of CM-ion-exchange, and anti-ghrelin IgG-affinity chromatographies and RP-HPLC.[37] Chicken ghrelin is composed of 26 amino acids (GSSFLSPTYKNIQQQKDTRKPTARLH) and possesses 54% sequence identity with human ghrelin. The serine residue at position 3 (Ser3) is conserved between the chicken and mammalian ghrelins, as is its acylation by either n-octanoic or n-decanoic acid. Chicken ghrelin mRNA is expressed predominantly in the stomach, where it is present in the proventriculus but absent in the gizzard. Using RT-PCR analysis, low levels of expression were also detectable in brain, lung, and intestine. Administration of chicken ghrelin increases plasma GH levels in both rats and chicks, with a potency similar to that of rat or human ghrelin. In addition, chicken ghrelin increases plasma corticosterone levels in growing chicks at a lower dose than in mammals.

3.5.3 FISH GHRELINS

We purified ghrelin from stomach extracts of a teleost fish, the Japanese eel (*Anguilla japonica*) after several steps of chromatographies: gel-filtration, CM-ion-exchange, and anti-ghrelin IgG-affinity chromatographies, and RP-HPLC.[38] Eel ghrelin contains an amide structure at its C-terminal end. Two forms of ghrelin, each comprising 21 amino acids, were identified: eel ghrelin-21, with an n-octanoyl modification, and eel-ghrelin-21-C10, with an n-decanoyl modification.

In conclusion, ghrelin exists not only in mammalian species, but also in non-mammalian species, such as in frog, chicken, and fish. Ghrelin, thus, may be an essential hormone for maintaining growth hormone release and energy homeostasis in vertebrates.

REFERENCES

1. Carter-Su, C. et al., Signalling pathway of GH, *Endocrinol. J.*, 43 Suppl, S65, 1996.
2. Muller, E.E., Locatelli, V., and Cocchi, D., Neuroendocrine control of growth hormone secretion, *Physiol. Rev.*, 79, 511, 1999.
3. Smith, R.G. et al., Peptidomimetic regulation of growth hormone secretion, *Endocrinol. Rev.*, 18, 621, 1997.
4. Bowers, C.Y., Growth hormone-releasing peptide (GHRP), *Cell. Mol. Life Sci.*, 54, 1316, 1998.
5. Kojima, M. et al., Ghrelin is a growth-hormone-releasing acylated peptide from stomach, *Nature*, 402, 656, 1999.
6. Vassilatis, D.K. et al., The G protein-coupled receptor repertoires of human and mouse, *Proc. Natl. Acad. Sci. USA*, 100, 4903, 2003.
7. Civelli, O., Functional genomics: the search for novel neurotransmitters and neuropeptides, *FEBS Lett.*, 430, 55, 1998.
8. Civelli, O. et al., Novel neurotransmitters as natural ligands of orphan G-protein-coupled receptors, *Trends Neurosci.*, 24, 230, 2001.
9. Reinscheid, R.K. et al., Orphanin FQ: a neuropeptide that activates an opioidlike G protein-coupled receptor, *Science*, 270, 792, 1995.
10. Meunier, J.C. et al., Isolation and structure of the endogenous agonist of opioid receptor-like ORL1 receptor, *Nature*, 377, 532, 1995.
11. Sakurai, T. et al., Orexins and orexin receptors: a family of hypothalamic neuropeptides and G protein-coupled receptors that regulate feeding behavior, *Cell*, 92, 573, 1998.
12. Hinuma, S. et al., A prolactin-releasing peptide in the brain, *Nature*, 393, 272, 1998.
13. Tatemoto, K. et al., Isolation and characterization of a novel endogenous peptide ligand for the human APJ receptor, *Biochem. Biophys. Res. Commun.*, 251, 471, 1998.
14. Ohtaki, T. et al., Metastasis suppressor gene KiSS-1 encodes peptide ligand of a G-protein-coupled receptor, *Nature*, 411, 613, 2001.

15. Fujii, R. et al., Identification of a neuropeptide modified with bromine as an endogenous ligand for GPR7, *J. Biol. Chem.*, 277, 34,010, 2002.
16. Tanaka, H. et al., Characterization of a family of endogenous neuropeptide ligands for the G protein-coupled receptors GPR7 and GPR8, *Proc. Natl. Acad. Sci. USA*, 100, 6251, 2003.
17. Shimomura, Y. et al., Identification of neuropeptide W as the endogenous ligand for orphan G-protein-coupled receptors GPR7 and GPR8, *J. Biol. Chem.*, 277, 35,826, 2002.
18. Thorner, M.O. et al., Growth hormone-releasing hormone and growth hormone-releasing peptide as therapeutic agents to enhance growth hormone secretion in disease and aging, *Recent Prog. Horm. Res.*, 52, 215; discussion 244, 1997.
19. Bennett, P.A. et al., Hypothalamic growth hormone secretagogue-receptor (GHS-R) expression is regulated by growth hormone in the rat, *Endocrinology*, 138, 4552, 1997.
20. Guan, X.M. et al., Distribution of mRNA encoding the growth hormone secretagogue receptor in brain and peripheral tissues, *Brain Res. Mol. Brain Res.*, 48, 23, 1997.
21. Howard, A.D. et al., A receptor in pituitary and hypothalamus that functions in growth hormone release, *Science*, 273, 974, 1996.
22. Bowers, C.Y. et al., Structure–activity relationships of a synthetic pentapeptide that specifically releases growth hormone in vitro, *Endocrinology*, 106, 663, 1980.
23. Bowers, C.Y. et al., On the *in vitro* and *in vivo* activity of a new synthetic hexapeptide that acts on the pituitary to specifically release growth hormone, *Endocrinology*, 114, 1537, 1984.
24. Smith, R.G. et al., A nonpeptidyl growth hormone secretagogue, *Science*, 260, 1640, 1993.
25. Akman, M.S. et al., Mechanisms of action of a second generation growth hormone-releasing peptide (Ala-His-D-beta Nal-Ala-Trp-D-Phe-Lys-NH2) in rat anterior pituitary cells, *Endocrinology*, 132, 1286, 1993.
26. Feighner, S.D. et al., Receptor for motilin identified in the human gastrointestinal system, *Science*, 284, 2184, 1999.
27. McKee, K.K. et al., Cloning and characterization of two human G protein-coupled receptor genes (GPR38 and GPR39) related to the growth hormone secretagogue and neurotensin receptors, *Genomics*, 46, 426, 1997.
28. Howard, A.D. et al., Identification of receptors for neuromedin U and its role in feeding, *Nature*, 406, 70, 2000.
29. Kojima, M. et al., Purification and identification of neuromedin U as an endogenous ligand for an orphan receptor GPR66 (FM3), *Biochem. Biophys. Res. Commun.*, 276, 435, 2000.
30. Fujii, R. et al., Identification of neuromedin U as the cognate ligand of the orphan G protein-coupled receptor FM-3, *J. Biol. Chem.*, 275, 21,068, 2000.
31. Hosoya, M. et al., Identification and functional characterization of a novel subtype of neuromedin U receptor, *J. Biol. Chem.*, 275, 29,528, 2000.
32. Kangawa, K. and Matsuo, H., Purification and complete amino acid sequence of alpha-human atrial natriuretic polypeptide (alpha-hANP), *Biochem. Biophys. Res. Commun.*, 118, 131, 1984.
33. Sudoh, T. et al., A new natriuretic peptide in porcine brain, *Nature*, 332, 78, 1988.
34. Hosoda, H. et al., Structural divergence of human ghrelin. Identification of multiple ghrelin-derived molecules produced by post-translational processing, *J. Biol. Chem.*, 278, 64, 2003.

35. Hosoda, H. et al., Purification and characterization of rat des-Gln14-Ghrelin, a second endogenous ligand for the growth hormone secretagogue receptor, *J. Biol. Chem.*, 275, 21,995, 2000.
36. Kaiya, H. et al., Bullfrog ghrelin is modified by n-octanoic acid at its third threonine residue, *J. Biol. Chem.*, 276, 40,441, 2001.
37. Kaiya, H. et al., Chicken ghrelin: purification, cDNA cloning, and biological activity, *Endocrinology*, 143, 3454, 2002.
38. Kaiya, H. et al., Amidated fish ghrelin: purification, cDNA cloning in the Japanese eel and its biological activity, *J. Endocrinol.*, 176, 415, 2003.

4 Ligand Screening of Olfactory Receptors

Kazushige Touhara, Sayako Katada,
Takao Nakagawa, and Yuki Oka

CONTENTS

4.1 INTRODUCTION

4.1.1 THE G PROTEIN-COUPLED OLFACTORY RECEPTOR SUPERFAMILY

Olfactory receptors (ORs), which comprise the largest G protein-coupled receptor (GPCR) family, play a pivotal role in recognizing a variety of odorants in the vertebrate olfactory system.[1-4] Genome-wide repertoires of the OR family have become available in a variety of vertebrate species, such as human (~350 intact ORs)[5-7] and mouse (1300 to 1500 intact ORs),[8,9] as well as in invertebrates, including fruit fly (62 ORs),[10-12] nematode (~500 chemosensory receptors),[13,14] and mosquito (79 ORs)[15] (0.5 to 4% of the predicted proteomes). The vertebrate OR sequences can be separated into two classes: the Class I family, which was first found in fish and was therefore predicted to recognize water-soluble odorant molecules, such as amino acids and carboxylic acids; and the Class II family, which was not found in fish, and was therefore thought to have appeared during the transition from an aquatic to a terrestrial lifestyle.[16,17] Conserved motifs of five to 10 amino acids have been found that can identify ORs and distinguish them from other GPCRs. In addition, variable regions are found within transmembrane domains of ORs. As in other GPCRs, these might function as the ligand-binding site.[18-20] Across the OR family, the amino acid similarity is 37% on average, with an identity ranging from less than 20% to over 90%.[4]

Variable numbers of pseudogenes have been identified in OR sequences across the evolutionary tree.[4] For example, approximately 60% of the total human OR genes are pseudogenes (~550 pseudogenes out of ~900 total OR genes), whereas 28 to 36% of the total OR genes in lower primate species are pseudogenes.[21] Also, of the 1300 to 1500 OR sequences in the mouse genome, approximately 20% are pseudogenes. An initial survey of the dog genome suggests that it contains only approximately 18% pseudogenes.[22] All of the fly ORs have an intact open reading frame, but more than 50% of the chemoreceptor genes in the nematode genome are pseudogenes. Thus, the numbers of intact ORs do not necessarily correlate with the evolutionary tree, raising the question of how diversity of the OR family across different phyla has been generated during evolution.[23]

Yeast and plant appear to be exceptional, because yeast has only three GPCRs, including two pheromone receptors and a glucose receptor, and plant has no apparent ORs.[24,25] Nonetheless, the OR family makes up a significant portion of the GPCR superfamily in many species. For example, approximately 350 of the approximately 900 human GPCRs are ORs. In *Drosophila*, 62 of the 270 GPCRs are ORs, and, in the mosquito, ORs account for 79 out of 276 GPCRs. This indicates that the OR family could be an excellent model for GPCR structure, pharmacology, and physiology.

4.1.2 ODORANT SIGNALING PATHWAYS

Odorant stimuli elicit a series of signal transduction events that are mediated by ORs expressed on the surface of olfactory neuronal cilia. Specifically, odorant binding to ORs activates G protein-mediated signal transduction cascades. These

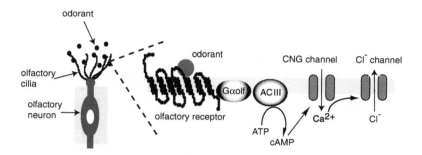

FIGURE 4.1 Odorant-mediated signal transduction cascade in olfactory neurons. AC: adenylyl cyclase; CNG channel: cyclic nucleotide-gated channel.

cascades lead to cation influx via cyclic nucleotide-gated channels[26–28] and Ca²⁺-activated Cl-channels,[29,30] which, in turn, causes neuronal depolarization (Figure 4.1).

Gene-knockout studies suggest that the cyclic adenosine monophosphate (cAMP) cascade, including the stimulatory G-protein α subunits ($G_{\alpha olf}$), adenylyl cyclase type III, and cyclic nucleotide-gated channels, plays a dominant role in transmitting the odorant signal in the olfactory neurons.[31–33] However, although some odorants clearly lead to an increase in cAMP, other odorants activate a signaling pathway that leads to an increase in inositol trisphosphate (IP_3).[27] Furthermore, recent studies in mice deficient in the cyclic nucleotide-gated channel suggested that cAMP-independent pathways participate in the response to a subset of odorants.[34] Thus, whether the cAMP cascade mediates signal transduction for all odorants in vertebrate olfactory neurons has not been established.

Regardless of the specific signaling cascade, the key function of the signal is to transmit binding of an odorant to production of an action potential. Electrophysiological techniques have enabled the recording of the elicitation of action potentials by various odorants in single olfactory neurons.[35,36] Measurement of the temporal and special properties of intracellular Ca²⁺ concentration with fluorescent dyes has also been used to detect physiological responses of single olfactory neurons to odorants.[37–39] The results from these two techniques suggest that individual olfactory neurons have a broad specificity that is determined by the structural features of the odorant molecules, including differences in chain length, terminal groups, and positions of functional groups.[40–44]

Despite the growing body of information on odorant response mechanisms and the signaling components in olfactory neurons, understanding of the OR function has progressed slowly due to a lack of appropriate heterologous systems for expressing and assaying odorant responses.[45] Until now, cognate ligands for only two rat, nine mouse, and two human ORs have been identified, so that the vast majority of the vertebrate ORs are orphan receptors.[4] Issues emerging in OR research following complete genome sequencing include clarification of the mechanism of membrane targeting and development of efficient and reproducible odorant-response assays. As with ORs, there are many problems in understanding the function of orphan GPCRs.

The purpose of this chapter is to introduce recent progress in functional expression of ORs in homologous and heterologous systems. These systems should help solve problems in the expression of many other GPCRs. This is of great importance, because more than half of the pharmaceuticals currently on the market or in development are targeted at GPCRs.

4.2 FUNCTIONAL EXPRESSION OF OLFACTORY RECEPTORS IN OLFACTORY NEURONS

4.2.1 ADENOVIRUS-MEDIATED GENE TRANSFER

Due to the difficulty in functionally expressing ORs in heterologous cell systems, olfactory neurons were first targeted as an OR expression system, because they should possess the appropriate cellular machinery for OR expression and transmission of odorant signals.[40,41] Because each OR is expressed in only a subset of olfactory neurons, overexpression of a particular OR in the entire olfactory epithelium would lead to an average increase in the ligand response. Adenovirus-mediated gene transfer has been successfully used for overexpression of ectopic ORs in many olfactory neurons.[40,41] This has been coupled with fluorescent Ca^{2+} imaging[41] or electro-olfactography[40] (EOG; see Section 4.2.3). This section provides a protocol for expression of an ectopic OR in olfactory neurons using adenovirus-mediated gene transfer.

PROTOCOL 4-1:   GENERATION OF ADENOVIRUS VECTORS AND GENE TRANSFER
INTO OLFACTORY NEURONS

1. Construct recombinant adenovirus transfer vectors containing a bicistronic expression unit for an OR and for green fluorescent protein (GFP) (Figure 4.2a). Note that the internal ribosome entry site (IRES) is inserted to allow for co-expression of an OR and GFP in single olfactory neurons.
2. Generate recombinant adenoviral plasmids by homologous recombination in *Escherichia coli* following a paper by Vogelstein.[46]
3. Transfect HEK293 cells with a linearized recombinant adenoviral plasmid.
4. Monitor transfected cells for GFP expression, and collect the cell culture supernatant.
5. Freeze the supernatant in a methanol/dry ice bath and rapidly thaw at 37°C. Repeat three times.
6. Apply the viral lysate to HEK293 cells, and harvest the viruses 3 to 4 days later.
7. Prepare an adenovirus solution at a titer of 10^8 to 10^9 pfu/ml.
8. Inject a total of 2.5 to 5 µl of the solution at 0.25 to 0.5 µl/min into the nostrils of 3- to 4-week-old mice.
9. Two to four days after injection, dissect out the olfactory epithelial tissue for use in an odorant response assay (Protocols 4-3, 4-4, and 4-5). For *in situ* hybridization or immunohistochemistry, fix the tissue in 4% paraformaldehyde for making cryostat sections.

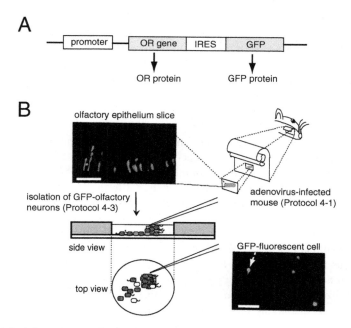

FIGURE 4.2 Adenovirus-mediated olfactory receptor gene transfer to olfactory neurons. (a) The bicistronic expression unit of a recombinant adenovirus vector. Co-expression of olfactory receptor (OR) and green fluorescent protein (GFP) is established by virtue of internal ribosomal entry site (IRES). (b) Isolation of adenovirus-infected GFP-positive olfactory neurons. Bars: 100 μm

4.2.2 Generation of GFP-Tagged Gene-Targeted Mice

Upon determination of ligand specificity of single olfactory neurons expressing an ectopic OR, co-expression with an endogenous OR is always a concern in the adenovirus-mediated homologous expression system. Further, identification of ligands by EOG (see Section 4.2.3.2) requires larger responses of infected animals to an exogenously expressed OR than of uninfected animals to endogenous ORs.[40] To overcome these problems, gene-targeted mice in which GFP is co-expressed with a defined OR have been used for studying the ligand response properties of an OR.[47] Although construction of knock-in mice is expensive, time-consuming, and technically challenging, it appears to be the most reliable strategy for characterizing endogenous OR responses to odorants.

Protocol 4-2: Generation of GFP-Tagged Gene-Targeted Mice

1. Isolate a genomic clone containing a targeted OR from a mouse (129/SvJ) lambda genomic library (Stratagene, La Jolla, CA). Subclone the fragment encompassing the OR gene into pBS-SK (Stratagene).
2. Insert a cassette containing IRES-*tauEGFP*-loxP-PGK-*neo*-loxP[48] downstream of the stop codon of the OR gene into the pBS-SK-OR vector.

3. Introduce 30 µg of the linearized targeting vector into 3×10^7 E14 embryonic stem cells by electroporation at 0.8 kV (3 µF) using a Gene Pulser (Bio-Rad, Hercules, CA). Culture the electroporated cells with mitomycin C-inactivated primary embryonic fibroblasts derived from neomycin-resistant transgenic mice.

4. At 24 h after electroporation, add 125 µg/ml of G418. After 7 to 8 days, transfer approximately 500 G418-resistant colonies to individual wells of 96-well plates.

5. Analyze genomic DNA from the G418-resistant embryonic stem cell clones by polymerase chain reaction (PCR) to define chromosomal rearrangements in the mouse genome. Further analyze PCR-positive clones by Southern blotting using probes external to the targeting vectors.

6. Inject the selected clone into C57BL/6 blastocysts to produce germline chimeras.

7. Cross the chimera mice with CRE transgenic mice to generate site-specific recombination, and obtain heterozygous mice. Interbreed the heterozygous mice and produce a line that is homozygous for the targeting gene and negative for the Cre transgene.

8. Utilize the gene-targeted mice for single-cell Ca^{2+} recording in olfactory neurons displaying GFP fluorescence (see Section 4.2.3.1).

4.2.3 Ca^{2+} Imaging and EOG Recording

4.2.3.1 Ca^{2+} Imaging Assay in Dissociated Olfactory Neurons

Activation of ORs leads to an increase in intracellular Ca^{2+} via the cyclic nucleotide-gated Ca^{2+} channel. Thus, Ca^{2+} imaging is very useful for detecting responses to odorants. Olfactory neurons showing GFP fluorescence are isolated from adenovirus-treated (Section 4.2.1) or gene-targeted (Section 4.2.2) mice[41,49] (Figure 4.2b). Imaging of Ca^{2+} in neurons co-expressing GFP and a given OR enables the identification of OR ligands. Changes in intracellular Ca^{2+} levels can be routinely monitored using a Ca^{2+}-sensitive fluorescent dye, such as Fura-2 or Fluo-4 (Figure 4.3). Olfactory neurons are usually incubated with a cell-permeable acetoxymethyl (AM) ester form of these dyes. Once in the cell, an intracellular esterase converts the AM form into the free acid, which cannot pass back across the membrane. Therefore, the dye is trapped in the cell so that it can be used to assess the intracellular Ca^{2+} level.

PROTOCOL 4-3:   ISOLATION OF OLFACTORY NEURONS

1. Anesthetize a mouse (16 to 19 g, 3- to 4-weeks-old) by injection with 0.15 ml Mioblock containing 20 µg/ml of bromopancronium (Sankyo Pharmaceuticals, Tokyo, Japan).

2. Dissociate the olfactory epithelium in Ringer's solution (140 mM NaCl, 5.6 mM KCl, 2 mM $CaCl_2$, 2 mM $MgCl_2$, 2 mM sodium pyruvate, 9.4 mM glucose, and 10 mM HEPES, pH 7.4). Using surgical blades, cut the epithelium into pieces approximately 300 µm square.

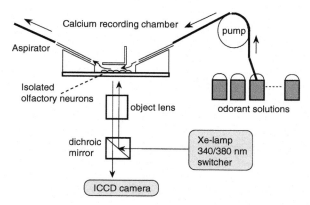

FIGURE 4.3 Ca^{2+} imaging set up to record odorant-induced Ca^{2+} increases in isolated olfactory neurons using Fura-2. Odorants are sequentially applied to cells, and the cells are washed continuously between odorant applications.

3. Incubate the pieces in 0.025% trypsin (Sigma, St. Louis, MO) at 37°C for 8 to 10 min. Terminate the trypsinization by adding 0.025% trypsin inhibitor (Sigma), and incubate for 10 min.

4. Incubate the digested pieces of epithelium for 2.5 min at room temperature in 0.1 mg/ml DNase in Ringer's solution (Sigma).

5. Transfer treated pieces to a 3.5 cm glass-bottomed dish (Iwaki) coated with Cell-Tak (Collaborative Research, Bedford, MA). Using an elastic glass capillary pipette and observing under a phase-contrast microscope, dissociate the olfactory neurons from the digested piece of olfactory epithelium by rolling the piece carefully so that the lateral side is kept in gentle contact with the glass dish.[41,50]

PROTOCOL 4-4: MEASURING ODORANT RESPONSES IN OLFACTORY NEURONS BY CA^{2+} IMAGING

1. Incubate dissociated olfactory neuron cells at room temperature for 15 min in Ringer's solution containing 5 μM fura-2-AM (diluted from a 1 mg/ml stock in DMSO; Molecular Probes, Eugene, OR).

2. At 2.5 min intervals, apply odorant solutions to the cells for 10 s at a flow rate of 1.5 ml/min (Figure 4.3). Between intervals, wash the chamber solution with Ringer's solution.

3. Monitor the transient increases in cytosolic Ca^{2+} of GFP-positive cells by ratiometric recording (excitation at 340 and 380 nm) using an AQUA COSMOS Ca^{2+}-imaging system (Hamamatsu Photonics, Shizuoka Pref., Japan). For GFP, the excitation wavelength is 475 nm, and the emission wavelength is 535 nm; for fura-2, the excitation is at 340 nm/380 nm and the emission is 510 nm. Thus, the GFP and fura-2 fluorescent signals can be separated using appropriate filters.

4.2.3.2 EOG Recording

Odorant-induced responses can be measured by EOG recording, in which an extracellular probe is used to measure the transepithelial potential.[51,52] The recorded potential is the sum of the activation of many olfactory neurons. The amplitude of the signal is determined by both the intensity of responses in individual neurons and the number of neurons responding. Overexpression of a given OR by adenovirus-mediated gene transfer (Section 4.2.1) results in an average increase in the agonist-induced EOG signal. Thus, ligands for a given OR ectopically overexpressed in the olfactory epithelium are reliably identified by the average increase in the transepithelial potential.[40] One of the advantages of EOG recording is that odorant stimulation can be performed via air phase, which allows for monitoring of OR activation under physiological conditions.

PROTOCOL 4-5: EOG RECORDING

1. Decapitate a rat overdosed with anesthetics (90 mg/kg ketamine and 5 mg/kg xylazine). Cut open the head sagitally, and remove the septum to expose the medial surface of the olfactory turbinates.
2. Mount the right half of the head in a wax dish filled with Ringer's solution. Next, expose the medial surface of turbinates to the air.
3. Blow humidified air gently and continuously on the turbinates through a glass tube (8 mm diameter) approximately 10 mm from the turbinate surface. The recording electrode is an Ag-AgCl wire in a capillary glass pipette filled with Ringer's solution containing 0.6% agarose gel. Adjust the electrode resistance to 0.5 to 1.0 MΩ. Connect the recording electrode to a DP-301 differential amplifier (Warner Instruments, Hamden, CT).
4. Use fluorescence microscopy to identify GFP-positive cells and determine the placement of the electrode. Observe the EOG signal on a chart recorder, and record the data with a digital audiotape recorder. Later, transfer the data to a computer. All experiments should be performed at 22 to 25°C.[40,52]

4.3 METHODS FOR FUNCTIONAL RECEPTOR EXPRESSION IN HETEROLOGOUS CELLS

4.3.1 GENERAL CONSIDERATIONS

Ligands for the rat I7 receptor have been identified by virtue of adenovirus-mediated expression in olfactory neurons.[40] Also, GFP tagging of neurons that expressed the mouse M71 receptor has led to identification of the ligands for M71.[47] These homologous *in vivo* strategies appear to be reliable, but they are low-throughput ligand screening approaches that are time-consuming and technically difficult. An alternative approach to matching ORs with their cognate ligands is to clone the OR gene from individual odorant-responsive olfactory neurons.[41,44,53] However, proof that the cloned OR is responsible for the observed response must be obtained either in homologous or heterologous systems. Functional expression of the cloned ORs has been successful in recapitulating odorant responses for some ORs, including MOR23, mOR-EG, and mOR-EV,[44] but it is not done routinely because of the difficulty in expressing ORs in heterologous cells. The main problem in this regard is that the mechanisms of OR transcription, translation, posttranslational modification, and plasma membrane targeting have not been elucidated.[45] Molecular dissection and reconstitution of the subcellular trafficking of an OR are required for the development of a high-throughput screening system, and they allow for the construction of complete odorant-activity matrices. In the next two sections, we introduce how some ORs have been functionally expressed in a heterologous expression system.

4.3.2 N-Terminal Tagging and Glycosylation

Currently, it is widely accepted that ORs are difficult to functionally express in heterologous cells due to a lack of proper membrane trafficking machinery in nonolfactory cells. Improper receptor targeting to the plasma membrane and retention in the endoplasmic reticulum appear to constitute the main reason for the failure of these functional expression studies.[54,55] In some cases, adding an N-terminal leader sequence from rhodopsin or the serotonin receptor results in a limited expression of functional receptors in the plasma membrane and in successful odorant-response recording in a heterologous system.[42,44,56]

Extra glycosylation at the N-terminus by virtue of the leader sequence may help increase cell surface expression. The OR family possesses a highly conserved potential N-glycosylation site (NXS/T) at the N-terminus that appears to be important for membrane localization in heterologous cells, such as HEK293. Thus, amino acid replacement of the glycosylation site or tunicamycin treatment of the cells abolishes the odorant responses.[57] Together with the evidence that OR expression is dramatically enhanced by addition of a sequence with extra glycosylation sites, this suggests that the glycosylation at the conserved N-terminal site is critical for proper translocation of the OR to the plasma membrane in mammalian cells. Finally, it is also apparently important to maintain the cells in good condition to obtain optimal responses.[58] Here we introduce protocols for constructing an epitope-tagged OR and for maintaining mammalian cell lines.

Protocol 4-6: Generation of an Epitope-Tagged OR and Maintenance of Mammalian Cells

1. Fuse a sequence encoding the N-terminal 20 amino acids of bovine rhodopsin to the N-terminus of an OR in a pME18S vector. The 20 amino acid leader sequence of bovine rhodopsin is MNGTEGPNFYVPFSNK TGVV (ATGAACGGGACCGAGGGCCCAAACTTCTACGTGCCTTT CTCCAACAAGACGGGCGTGGTG).

2. Cell culture conditions: Grow HEK293 and COS-7 cells in Dulbecco's modified Eagle's medium (DMEM) (Nacalai Tesque Inc., Kyoto, Japan) supplemented with 10% fetal bovine serum (FBS; JRH Biosciences, Lenexa, KS); CHO-K1 cells in F-12 Nutrient Mixture (Invitrogen, Carlsbad, CA) with 10% FBS; and PC12h cells on collagen-precoated dishes (Iwaki) in DMEM containing 10% FBS and 5% horse serum (Invitrogen). Culture all cells at 37°C in a humidified atmosphere containing 5% CO_2. To obtain reproducible Ca^{2+} responses, split cells every 2 days before becoming confluent. After 2 or 3 months of passaging, discard the cells, and prepare new ones from a frozen stock. Prepare frozen stocks in CELL-BANKER (Jyuuji-Field, Japan).

3. Cell passaging: Wash subconfluent (70 to 80% confluent) cells in a 10 cm dish with 1 ml of prewarmed phosphate-buffered saline (PBS), and then treat for 30 s with 1 ml of 0.05% Trypsin-0.53 mM EDTA (Invitrogen) to detach them from the dish. Collect the cells with a pipette, and then gently pipette them up and down several times to generate a single-cell suspension. Place the suspended cells in a centrifuge tube, mix them with 1 ml of prewarmed DMEM to inhibit trypsin, and centrifuge them for 30 s at 1000 rpm. Discard the supernatant, and resuspend the cell pellet in 400 µl of PBS. Place 80 µl of this cell suspension in a new dish containing 10 ml of DMEM.

4.3.3 Chaperones for Membrane Trafficking

Visualization of receptor trafficking within cells using epitope-tagged ORs has suggested that they stay in the intracellular network and are not efficiently transported to the surface of heterologous cells.[59] As described above, functional expression of limited amounts of ORs in the plasma membrane has been achieved by adding a membrane-targeting sequence from rhodopsin or a membrane-import sequence from the 5-HT3 receptor. Recently, we found that an untagged OR was also functionally expressed in the plasma membrane.[57] However, each OR appears to behave differently. These observations have led us to suspect that there is some unique machinery for functional expression of ORs in olfactory neurons.

In *Caenorhabditis elegans*, the *odr-4* gene is necessary for the trafficking of ORs to the cilia of sensory neurons.[60] Some GPCRs require a chaperone-like protein for functionality in the plasma membrane. For example, RAMP is needed for functional expression of the adrenomedullin receptor/calcitonin-gene-related peptide receptor.[61] The structure of the N-glycosylation site may also be critical, because a mutation at the N-terminal potential glycosylation site impairs the ability of ORs to travel to the plasma membrane in HEK293 cells.[57] Thus, we speculated that molecular chaperone(s) specific to ORs facilitate trafficking of ORs in olfactory neurons. Most recently, three genes that enhance cell surface expression of ORs in heterologous cells have been cloned.[62] These genes encode one-transmembrane proteins that can be referred to as OR chaperones. Although these OR chaperone genes should be powerful tools for developing a high-throughput ligand screening system for ORs, it appears that, for unknown reasons, they are not helpful for all ORs. Furthermore,

ORs are now known to be expressed in nonolfactory tissues, such as testis, developing heart, spleen, liver, and brain, suggesting that there might not be a specific regulatory mechanism for OR sorting. Further studies are necessary to clarify the mechanisms underlying proper maturation and trafficking of ORs in olfactory neurons.

4.4 METHODS FOR ODORANT RESPONSE ASSAYS OF OLFACTORY RECEPTORS

4.4.1 Ca²⁺ IMAGING ASSAY

Ca^{2+} imaging has been one of the most commonly used and reliable methods for detecting activation of orphan GPCRs in high-throughput ligand screening (Figure 4.3). However, the Ca^{2+} imaging assay is limited to GPCRs that couple to $G_{\alpha q}$-type G proteins. In the case of GPCRs that couple to unknown G proteins, the G protein α subunit, $G_{\alpha 15/16}$, is often utilized to force the signal to an inositol phosphate-mediated signaling cascade because of its promiscuity in coupling to GPCRs. This strategy has been successfully applied to ORs that most likely couple to the olfactory $G_{\alpha s}$-type G proteins, $G_{\alpha olf}$, which mediates increases in cAMP. For example, in HEK293 cells co-expressing $G_{\alpha 15}$ and mOR-EG, which is a mouse OR that recognizes eugenol (EG), Ca^{2+} responses are observed when stimulated with EG[44,58] (Figure 4.4). Using sequential application of odorants with a peristaltic pump in this system, we developed a high-throughput screen for mOR-EG agonists and antagonists.[63]

PROTOCOL 4-7:   AGONIST SCREENING OF AN OR IN HEK293 CELLS

1. Maintain HEK293 cells as described in Protocol 4-6.
2. Twenty-four hours prior to transfection, seed HEK293 cells at 6×10^5 cells/dish in 2 ml of DMEM into poly-L-lysine (Sigma)-coated 3.5 cm glass-based dishes (Iwaki).
3. Transfect 60 to 70% confluent HEK293 cells with 2.0 μg of an epitope tagged OR cDNA and 1.5 μg of $G_{\alpha 15}$ cDNA using Lipofectamine 2000 (Invitrogen; 2.5 μl per μg DNA).
4. At 24 to 28 h after transfection, incubate the transfected cells for 30 min at 37°C with DMEM containing 5 μM Fura 2-AM (diluted from a 1 mg/ml stock in DMSO; Molecular Probes, Eugene, OR).
5. Aspirate the medium and wash the recording dish for 5 min with Ringer's solution to stabilize the fluorescence ratio (340 nm/380 nm) using a peristaltic pump.
6. Apply odorant solutions sequentially to the cells using a peristaltic pump for 15 s at a flow rate of 1.5 ml/min. Wash the cells for 2.5 min with Ringer's solution between each application of odorant. Monitor the intracellular Ca^{2+} levels using an AQUA COSMOS Ca^{2+}-imaging system (Hamamatsu Photonics). For high-throughput agonist screening, in the

initial screen, sequentially apply odorant cocktails, each containing several different odorants, to the OR-expressing HEK293 cells.
7. Examine the responsiveness of the individual components in the odorant cocktails that elicit a response.

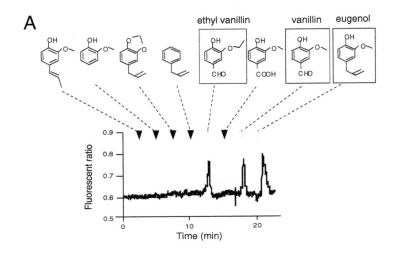

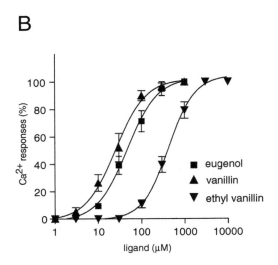

FIGURE 4.4 (a) Ca^{2+} response profile of HEK93 cells expressing a mouse olfactory receptor mOR-EG and $G_{\alpha15}$. Various odorants are applied at the times indicated by arrow heads. (b) Dose–response curves of mOR-EG to eugenol, vanillin, and ethyl vanillin.

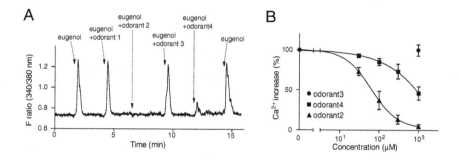

FIGURE 4.5 (a) Inhibition of eugenol-induced Ca^{2+} increases in mOR-EG-expressing HEK293 cells by odorants 2 and 4, but not by odorants 1 and 3. (b) Dose-dependent inhibition curves of mOR-EG by odorants 2 and 4. Odorant 3 did not show inhibition.

PROTOCOL 4-8: ANTAGONIST SCREENING OF AN OR IN HEK293 CELLS

1. Prepare OR-transfected HEK293 cells as described in Protocol 4-6.
2. Perform Ca^{2+}-imaging as described in Protocol 4-7.
3. For high-throughput antagonist screening, prepare an odorant cocktail containing an agonist and a threefold concentration of the odorant to be tested for antagonistic activity.
4. Apply the cocktail to OR-expressing HEK293 cells for screening of antagonistic odorants[63] (Figure 4.5).

4.4.2 cAMP ASSAY

Although the Ca^{2+} imaging assay appears to be useful for high-throughput ligand screening, it requires high level of expression of mature ORs. In contrast, the cAMP assay has been adapted to various mammalian cell lines that express a modest amount of OR. This is probably possible, because ORs couple to $G_{\alpha s}$-type G proteins more efficiently than to $G_{\alpha 15}$. Thus, without co-expression of $G_{\alpha 15}$, ORs activate endogenous stimulatory G proteins ($G_{\alpha s}$) upon ligand stimulation in various mammalian cell lines (e.g., HEK293, COS-7, and CHO-K1 cells)[58] (Figure 4.6). Odorant-induced cAMP increases have been measured using an enzyme-linked immunoassay (ELISA) kit (Applied Biosystems, Foster City, CA). Because this assay uses chemiluminescent detection, which is highly sensitive, it can be employed not only for high-throughput ligand screening but also for testing of constitutive activity or receptor desensitization.

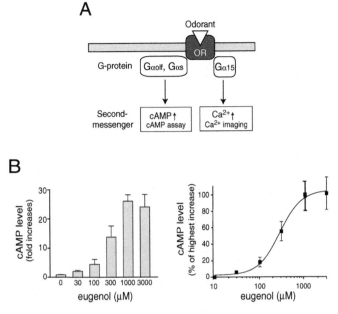

FIGURE 4.6 Eugenol-induced cAMP increases in HEK293 cells expressing mOR-EG. (a) Odorant responses are measured by cAMP increases via $G_{\alpha s/olf}$ coupling or by Ca^{2+} increases via $G_{\alpha 15}$. (b) Dose-dependent increases in cAMP by eugenol in HEK293 cells expressing mOR-EG.

PROTOCOL 4-9:   DETECTION OF ODORANT-INDUCED CAMP PRODUCTION USING AN ELISA SYSTEM

1. Forty-eight hours prior to the assay, seed cells (HEK293, COS-7, or CHO-K1) at 2×10^5 cells/well in 1 ml of growth medium into 24-well plates, and incubate overnight.

2. Transfect 60 to 70% confluent cells with 1 μg of a tagged OR cDNA using 2.5 μl of Lipofectamine 2000.

3. Culture the transfected cells for 24 h, and then incubate them for 30 min prior to the cAMP assay with 500 μl of serum-free growth medium containing 1 mM 3-isobutyl-1-methylxanthine (IBMX).

4. Add 500 μl of 2X odorant solution containing 1 mM IBMX to each well, and incubate for 10 min.

5. Wash the cells once with PBS, and add 150 μl of Assay/Lysis Buffer (kit component) to each well. Incubate at 37°C until the cells are lysed. (After this step, utilize the solutions from the cAMP ELISA kit according to the manufacturer's protocol.)

6. Add 60 μl/well of standard solution or sample and 30 μl/well of alkaline phosphatase-labeled cAMP conjugate to the wells of an assay plate. Mix the plates on a plate shaker for 5 min.

7. Add 60 μl/well of the anti-cAMP Antibody Solution (kit component), and mix the plates on a plate shaker for 1 h.
8. Remove the solution from the wells, and wash six times with 200 μl/well of Wash Buffer.
9. Add 100 μl/well of CSPD Substrate Solution (kit component) to each well, and incubate the plates for 30 min at 37°C.
10. Measure the luminescence for 1 s/well using a microplate luminometer. The luminescent signal is inversely proportional to the cAMP level in the sample.

4.4.3 LUCIFERASE REPORTER GENE ASSAY

Zif268 is an immediate early gene encoding a transcription factor that may play a role in differentiation and neuronal plasticity.[64] The Zif268 promoter contains serum response elements and cAMP-response elements (CREs) (Figure 4.7a). In a variety of cells, stimulation with the proper agonist induces an increase in intracellular cAMP, which initiates a cascade leading to CRE-mediated gene expression. A luciferase-reporter assay system has been developed using a reporter gene that expresses luciferase under control of the zif268 promoter. This system allows luminescent detection of cAMP increases upon stimulation with an odorant. For example, in PC12h cells cotransfected with an OR and the luciferase-reporter construct, odorant stimulation produces dose-dependent increases in luciferase activity (Figure 4.7b). Deletion of the CRE-sequences from the zif268 promoter significantly reduces the response, suggesting that the increase in the luciferase activity is mainly dependent on cAMP-mediated activation via CRE-sequences. Hence, this technique can be applied not only to ligand screening but also to functional characterization of receptors and their signal transduction pathways.[58] Herein, we introduce an assay using "dual reporters" (firefly and *Renilla* luciferases), which is used to improve experimental accuracy. Thymidine kinase promoter-linked *Renilla* luciferase is cotransfected as an internal control for the baseline response. Experimental variability caused by differences in cell viability or transfection efficiency is minimized by normalizing the data with this internal control.

PROTOCOL 4-10: LUCIFERASE REPORTER GENE ASSAY FOR ORs IN PC12H CELLS

1. Seed PC12h cells in 24-well plates at a density of 1×10^5 cells/well in 1 ml of growth medium, and incubate overnight.
2. Transfect 50 to 60% confluent cells in a 24-well plate with 0.3 μg of the receptor cDNA, 0.25 μg of a zif268 promoter-linked firefly luciferase cDNA, and 0.25 μg of a thymidine kinase promoter-linked *Renilla* luciferase cDNA (Promega, Madison, WI) using 3 μl of TransFast (Promega).
3. Forty-eight hours after transfection, incubate the cells for 1 h with 500 μl of serum-free culture medium.

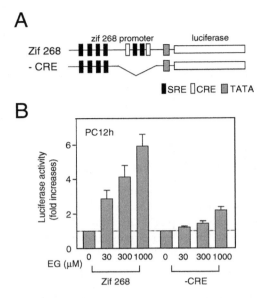

FIGURE 4.7 Eugenol responses to mOR-EG detected by zif268 promoter-mediated expression of luciferase activity in PC12h cells. (From Katada, S. et al., *Biochem. Biophys. Res. Communi.*, 305, 964, 2003. With permission.) (a) Luciferase reporter genes. (b) Dose-dependent increases in luciferase activities in eugenol-stimulated PC12h cells transfected with reporter genes and mOR-EG.

4. Stimulate the cells for 6 h with 500 μl of 2X odorant solutions, and measure luciferase activity using a dual reporter assay system (Promega).
5. Wash the cells once with PBS, and add 150 μl of Passive Lysis Buffer (kit component) to each well. Incubate the cells at 37°C until they are lysed.
6. Transfer 20 μl of the cell lysate to a 96-well plate, and measure the luciferase activities with a dual-injector luminometer.

4.4.4 *Xenopus* Oocyte Assay

4.4.4.1 Assays for Multiple G Protein-Mediated Signals

An expression system using *Xenopus laevis* oocytes was developed by Gurdon et al. in 1971.[65] Injection of exogenous mRNA into oocytes leads to expression of functional proteins translated from the injected mRNA. By taking advantage of signaling components and ion channels that were endogenously expressed in an oocyte, it was possible to functionally clone genes encoding G-protein coupled receptors.[66] These studies found that the binding of ligands to GPCRs activates various G protein-mediated cascades, including an adenylyl cyclase-cAMP pathway, and a phospholipase C-inositol polyphosphate pathway (Figure 4.8a). In *Xenopus* oocytes, an IP$_3$ pathway leads to an increase in cytosolic Ca^{2+} that, in turn, activates an endogenous Ca^{2+}-dependent Cl$^-$ channel. The cAMP pathway can be detected by electrophysiological techniques, if the GPCR is co-expressed with a cAMP-dependent Cl$^-$ channel,

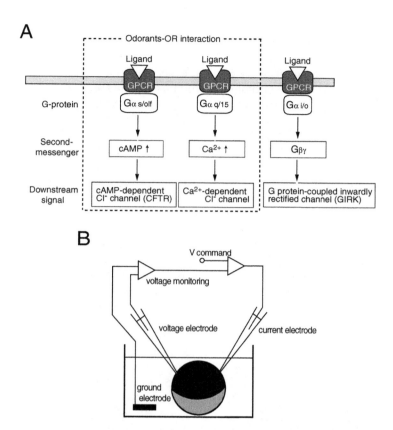

FIGURE 4.8 (a) Signaling cascades utilized to detect ligand responses of G protein-coupled receptors in *X. laevis* oocytes. (b) Two-electrode voltage-clamp recording setup.

such as the cystic fibrosis transmembrane regulator (CFTR).[58] For detection of signaling by $G_{\alpha i}$-coupled receptors, the G protein-coupled inwardly rectified K^+-channel is co-expressed with the GPCR. Alternatively, coinjection of $G_{\alpha 15}$ cRNA allows for the detection of signals that are mediated by various GPCRs.[58] Electrophysiological recording of odorant responses in *Xenopus* oocytes has been demonstrated for ORs from mouse, rat, goldfish, and fruit fly.[42,58,67,68] In this section, we present an overview of an experimental protocol for expression of an ectopic OR in *Xenopus* oocytes.

PROTOCOL 4-11: cRNA SYNTHESIS

1. Subclone the coding region of an OR into a vector for expression in *Xenopus* oocytes that is under the control of the bacteriophage promoter SP6, T7, or T3. This vector contains *Xenopus* β-globin 5′ and 3′ untranslated regions, and a 3′ poly A tail to increase the stability and ribosomal binding of the cRNA.

2. Linearize the vector with a restriction enzyme that does not cut in the coding region. Synthesize the cRNA at 37°C for 1.5 h using RNA polymerases in the presence of the capping analog $m^7G(5')ppp(5')G$.
3. Add DNase, and incubate at room temperature for 5 min.
4. Extract twice with phenol-chloroform-isoamylalchol.
5. Precipitate the cRNA with ethanol, and wash with 70% ethanol.
6. Check the purity of the cRNA by agarose electrophoresis.

PROTOCOL 4-12: INJECTION OF cRNA IN *XENOPUS* OOCYTES

1. Anesthetize *Xenopus laevis* for 20 min on ice in 0.1% MS-222/0.3% $NaHCO_3$.
2. Place a frog on its back in the ice. Make a 1 cm incision with a surgical blade to remove the lobes of the ovary. Place the isolated oocytes in Ca^{2+}-free solution (82.5 mM NaCl, 2 mM KCl, 1 mM $MgCl_2$, 5 mM HEPES, pH 7.5).
3. To remove follicle cells around the oocytes, treat the oocytes for 1 to 2 h at room temperature with 2 mg/ml collagenase Type I in Ca^{2+}-free solution. Wash out the enzyme with Ca^{2+}-free solution. Collect Stages V to VII oocytes for injection with cRNA.
4. Healthy oocytes are selected for injection of cRNAs. Coinject oocytes with 25 ng of an OR cRNA and either 25 ng of CFTR cRNA or 250 pg of $G_{\alpha15}$ cRNA. For $G_{\alpha q}$-coupled receptors, inject 50 ng of receptor cRNA alone.
5. Culture the injected oocytes for 2 to 5 days at 18°C in Barth's solution (88 mM NaCl, 1 mM KCl, 0.3 mM $Ca(NO_3)_2$, 0.4 mM $CaCl_2$, 0.8 mM $MgSO_4$, 2.4 mM $NaHCO_3$, and 15 mM HEPES, pH 7.4) supplemented with 50 μg/ml streptomycin and 100 unit penicillin.

4.4.4.2 Measurement of Ion Currents with a Voltage Clamp Method

Voltage clamp is one of the most powerful methods for characterizing ion channels, especially voltage-dependent channels (Figure 4.8b). The whole cell voltage clamp method allows for measurement of whole-cell ion currents flowing through the membrane via ion channels. This method has been adapted to the study of an OR that mediates G protein signals. Odorant responses are monitored as an inward Cl⁻ current via endogenous Ca^{2+}-dependent Cl⁻ channels or via heterologously expressed CFTRs[58] (Figure 4.9).

PROTOCOL 4-13: WHOLE-CELL VOLTAGE-CLAMP METHOD

1. Perform whole-cell voltage-clamp experiments with two microelectrodes made from borosilicate glass capillaries. When the microelectrodes are filled with 3M KCl, the electrode has a resistance of 0.1 to 0.5 MΩ.

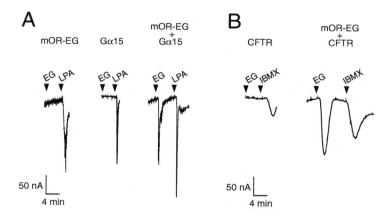

FIGURE 4.9 Electrophysiological recording of eugenol-induced currents in *X. laevis* oocytes. (From Katada, S. et al., *Biochem. Biophys. Res. Commun.*, 305, 964, 2003. With permission.) Cl⁻ conductance induced by eugenol in oocytes injected with the indicated cRNA(s). (Panel A) Current recording of Ca^{2+} dependent Cl⁻ channel, (Panel B) Current recording of cAMP dependent Cl⁻ channel (CFTR: cystic fibrosis transmembrane regulator).

2. Place the oocyte in a chamber filled with a bath solution (88 mM NaCl, 1 mM KCl, 0.3 mM $Ca(NO_3)_2$, 0.4 mM $CaCl_2$, 0.8 mM $MgSO_4$, 2.4 mM $NaHCO_3$, and 15 mM HEPES, pH7.4), and impale it with the two electrodes (Figure 4.8b).

3. Apply odorant to the bath solution through a silicon tube connected to a peristaltic pump, as described in Protocol 2-4.

4. Perform a two-electrode voltage-clamp recording at a holding potential of −80 mV using an Oocyte Clamp OC-725-B amplifier (Warner Instruments). Digitize the output of the amplifier using a Digidata 1322A (Axon Instruments, Union City, CA) that is connected to a computer. Perform data acquisition and analysis using pClamp8 software (Axon Instruments).

4.4.4.3   Measurement with Depolarization Step Pulses

A Ca^{2+}-activated Cl⁻ channel carrying a current activated via $G_{\alpha q}$ acts as an outward rectifying channel. The conductance of this type of channel at +50 mV is higher than that at −80 mV, and, thus, the amplitude of a current induced by an odorant stimulus increases more at a holding potential of +60 mV than at −80 mV. For this reason, measurement of the conductance at +60 mV with depolarization step pulses allows for amplification of a small response[69] (Figure 4.10).

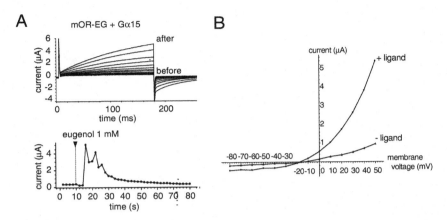

FIGURE 4.10 (a) Current traces (upper panel) and time course (lower panel) of eugenol response of *X. laevis* oocyte expressing mOR-EG and $G_{\alpha 15}$ plotted as the current amplitude at the end of each depolarizing pulse. (b) A current–voltage plot of oocyte responses in the presence or absence of a ligand (eugenol).

PROTOCOL 4-14: STEP PULSE TO +60 MV FROM A HOLDING POTENTIAL OF –80 MV

1. Perform a two-electrode voltage-clamp method with a depolarizing step pulse using an Oocyte Clamp OC-725-B amplifier (Warner Instruments) and pClamp8 software (Axon Instruments).
2. Monitor the Ca^{2+}-activated Cl^- current by applying 200 ms of a depolarizing step pulse to +60 mV every 2 s from a holding potential of –80 mV.
3. Note that a slow-activating outward current can be observed upon odorant application. Plot the amplitude at the end of the 200 ms depolarizing pulse.

4.5 OTHER FUNCTIONAL ASSAY SYSTEMS

In addition to mammalian cell lines and *Xenopus* oocytes, cell lines derived from olfactory neurons can be used as an expression system. For example, *odora* cells, which express neuronal and olfactory markers, have been utilized to characterize OR function.[70] Further studies, however, appear to be necessary to determine whether the *odora* cells mimic native olfactory neurons and are suitable for analysis of OR function. Freshly dissociated olfactory neurons have also been tested for the ability to express exogenous ORs. Odorant responses were successfully recorded by Ca^{2+} imaging in primary culture cells infected with recombinant I7-expressing adenovirus.[49] These results suggest that optimizing the conditions for primary olfactory neurons or developing a methodology for establishing olfactory neuronal cell lines will be helpful.

A reasonable approach for reconstituting odorant responses could be to construct cells that express the necessary signaling components. Stable expression of $G_{\alpha olf}$, adenylyl cyclase Type-III, and cyclic nucleotide-gated channel a2 in HeLa cells has allowed for identification of cognate ligands for some ORs.[71] Expression of OR chaperones in HEK293 cells is also a promising avenue for constructing a high-throughput screening system for ORs.[62] However, there are some discrepancies between the ligands identified in various homologous and heterologous assays.[4] Odorant responses should ideally be measured in cells that normally express ORs. Because the expression level of an OR on the cell surface appears to affect the dose response of ligands, care must be taken when comparing ligand specificities between ORs in heterologous systems.

4.6 STRUCTURE–ACTIVITY MATRICES OF OLFACTORY RECEPTORS

Ligand specificity studies of several ORs have shown that an OR recognizes various odorants with a high specificity for some functional groups or molecular features but also has a high tolerance for some other aspects of the odorant structure.[40,41,43,44] ORs are able to discriminate structurally related odorants and can recognize a wide variety of odorants. The two most extensively characterized ORs to date are the rat I7[43] and mouse mOR-EG receptors.[44] The structure–function relationships for mOR-EG are summarized in Figure 4.11. Among 21 identified mOR-EG agonists, the most potent odorant is 4-hydroxy-3-methyl benzaldehyde, with a threshold concentration of 30 to 300 nM. These studies have also shown that an OR does not necessarily recognize odorants with the same odor. For example, mOR-EG recognizes both EG and vanillin with similar EC_{50} values, although EG smells different than vanillin[44] (Figure 4.4). ORs are sensitive to odorants in the nanomolar to millimolar range, which appears to be higher than the range that can actually be detected. This may be due to the fact that the olfactory epithelium is covered with mucus that contains odorant-binding proteins and other unknown components that increase the local concentration of odorants around the olfactory neuron cilia. Unfortunately, direct binding studies have not yet been reported for ORs due to the difficulty of handling volatile radioactive compounds.

Because ORs belong to the GPCR superfamily, which is the most common target of therapeutic drugs, it is possible that odorants could not only activate ORs as agonists but also antagonize odorant responses to ORs. In fact, compounds that inhibit the response of the mOR-EG to EG have been recently identified using the HEK293 assay system[63] (Figure 4.5). These OR antagonists were found to be structurally related to the agonists, as is often the case for other GPCRs. Antagonism between components in an odorant mixture suggests that coding of odorant response is not simply the sum of responses to the individual components. Interestingly, an antagonist for human OR17-40, which has been identified in both in the olfactory epithelium and testis, was shown to block agonist-induced sperm chemotaxis.[72] Such OR antagonists could be used to control perceived odor quality or sperm chemotaxis.

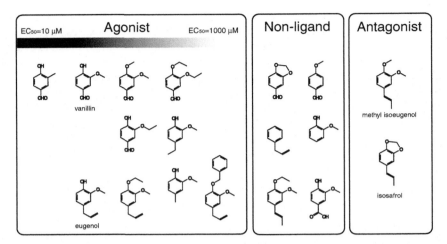

FIGURE 4.11 Structure of agonists and antagonists of mOR-EG. The compounds on the left-most side exhibit the lowest EC_{50} values.

4.7 CONCLUSIONS

ORs and other GPCRs share the ability to sense small molecules. Structure–function analysis of ORs will surely stimulate pharmacological screening and design of GPCR agonists and antagonists. Clarification of the atomic-level structure of bovine rhodopsin has provided an opportunity to elucidate the structural basis for odorant recognition by ORs. Recent studies of mOR-EG and its cognate ligand EG using a combination of computer modeling and ligand-docking analysis with site-directed mutagenesis revealed the ligand binding site of an OR for the first time.[73] The molecular basis for the binding of agonists and antagonists to ORs will help in the design of novel ligands and will help elucidate the mechanisms underlying G protein activation by GPCRs. The functional odorant response assays described in this chapter should help understand not only our sense of smell but also how GPCRs recognize small compounds with high specificity and affinity.

ACKNOWLEDGMENTS

We thank K. Inaki, K. Kajiya, and M. Tanaka for the initial stage of studies on olfactory receptor expression, and all other Touhara lab members for helpful discussion. The support by T. Haga, U. Kikkawa, and H. Kataoka is appreciated. This work is supported in part by grants from the MEXT and from the PROBRAIN in Japan. K.T. is recipient of grants from Uehara Memorial Foundation and Kato Memorial Bioscience Foundation.

REFERENCES

1. Buck, L. and Axel, R., A novel multigene family may encode odorant receptors: a molecular basis for odor recognition, *Cell*, 65, 175, 1991.
2. Firestein, S., How the olfactory system makes sense of scents, *Nature*, 413, 211, 2001.
3. Touhara, K., Odor discrimination by G protein-coupled olfactory receptors, *Microsc. Res. Tech.*, 58, 135, 2002.
4. Mombaerts, P., Genes and ligands for odorant, vomeronasal and taste receptors, *Nat. Rev. Neurosci.*, 5, 263, 2004.
5. Zozulya, S. et al., The human olfactory receptor repertoire, *Genome Biol.*, 2, RESEARCH0018, 2001.
6. Glusman, G. et al., The complete human olfactory subgenome, *Genome Res.*, 11, 685, 2001.
7. Young, J.M. and Trask, B.J., The sense of smell: genomics of vertebrate odorant receptors, *Hum. Mol. Genet.*, 11, 1153, 2002.
8. Zhang, X. and Firestein, S., The olfactory receptor gene superfamily of the mouse, *Nat. Neurosci.*, 5, 124, 2002.
9. Zhang, X. et al., Odorant and vomeronasal receptor genes in two mouse genome assemblies, *Genomics*, 83, 802, 2004.
10. Vosshall, L.B. et al., A spatial map of olfactory receptor expression in the *Drosophila* antenna, *Cell*, 96, 725, 1999.
11. Clyne, P.J. et al., A novel family of divergent seven-transmembrane proteins: candidate odorant receptors in *Drosophila*, *Neuron*, 22, 327, 1999.
12. Gao, Q. and Chess, A., Identification of candidate *Drosophila* olfactory receptors from genomic DNA sequence, *Genomics*, 60, 31, 1999.
13. Bargmann, C.I., Neurobiology of the *Caenorhabditis elegans* genome, *Science*, 282, 2028, 1998.
14. Robertson, H., Updating the str and srj (stl) families of chemoreceptors in *Caenorhabditis* nematodes reveals frequent gene movement within and between chromosomes, *Chem. Senses.*, 26, 151, 2001.
15. Hill, C.A. et al., G protein-coupled receptors in *Anopheles gambiae*, *Science*, 298, 176, 2002.
16. Freitag, J. et al., Two classes of olfactory receptors in *Xenopus laevis*, *Neuron*, 15, 1383, 1995.
17. Mezler, M. et al., Characteristic features and ligand specificity of the two olfactory receptor classes from *Xenopus laevis*, *J. Exp. Biol.*, 204, 2987, 2001.
18. Singer, M.S. et al., Potential ligand-binding residues in rat olfactory receptors identified by correlated mutation analysis, *Receptors Channels*, 3, 89, 1995.
19. Pilpel, Y. and Lancet, D., The variable and conserved interfaces of modeled olfactory receptor proteins, *Protein Sci.*, 8, 969, 1999.
20. Singer, M.S., Analysis of the molecular basis for octanal interactions in the expressed rat 17 olfactory receptor, *Chem. Senses*, 25, 155, 2000.
21. Gilad, Y. et al., Human specific loss of olfactory receptor genes, *Proc. Natl. Acad. Sci. USA*, 100, 3324, 2003.
22. Quignon, P. et al., Comparison of the canine and human olfactory receptor gene repertoires, *Genome Biol.*, 4, R80, 2003.
23. Dryer, L., Evolution of odorant receptors, *Bioessays*, 22, 803, 2000.
24. Versele, M. et al., Sex and sugar in yeast: two distinct GPCR systems, *EMBO Rep.*, 2, 574, 2001.

25. Jones, A.M., G-protein-coupled signaling in *Arabidopsis*, *Curr. Opin. Plant Biol.*, 5, 402, 2002.

26. Reed, R.R. et al., The molecular basis of signal transduction in olfactory sensory neurons, *Soc. Gen. Physiol. Ser.*, 47, 53, 1992.

27. Breer, H. et al., Molecular mechanisms of olfactory signal transduction, *Soc. Gen. Physiol. Ser.*, 47, 93, 1992.

28. Schild, D. and Restrepo, D., Transduction mechanisms in vertebrate olfactory receptor cells, *Physiol. Rev.*, 78, 429, 1998.

29. Kurahashi, T. and Kaneko, A., Gating properties of the cAMP-gated channel in toad olfactory receptor cells, *J. Physiol.*, 466, 287, 1993.

30. Frings, S. et al., Neuronal Ca^{2+}-activated Cl channels — homing in on an elusive channel species, *Prog. Neurobiol.*, 60, 247, 2000.

31. Belluscio, L. et al., Mice deficient in G(olf) are anosmic, *Neuron*, 20, 69, 1998.

32. Brunet, L.J. et al., General anosmia caused by a targeted disruption of the mouse olfactory cyclic nucleotide-gated cation channel, *Neuron*, 17, 681, 1996.

33. Wong, S.T. et al., Disruption of the type III adenylyl cyclase gene leads to peripheral and behavioral anosmia in transgenic mice, *Neuron*, 27, 487, 2000.

34. Lin, W. et al., Odors detected by mice deficient in cyclic nucleotide-gated channel subunit A2 stimulate the main olfactory system, *J. Neurosci.*, 24, 3703, 2004.

35. Firestein, S., Scentsational ion channels, *Neuron*, 17, 803, 1996.

36. Kurahashi, T. et al., Suppression of odorant responses by odorants in olfactory receptor cells, *Science*, 265, 118, 1994.

37. Restrepo, D. and Boyle, A.G., Stimulation of olfactory receptors alters regulation of Ca^{2+} in olfactory neurons of the catfish (*Ictalurus punctatus*), *J. Membr. Biol.*, 120, 223, 1991.

38. Restrepo, D. et al., Second messenger signaling in olfactory transduction, *J. Neurobiol.*, 30, 37, 1996.

39. Tareilus, E. et al., Calcium signals in olfactory neurons, *Biochim. Biophys. Acta*, 1269, 129, 1995.

40. Zhao, H. et al., Functional expression of a mammalian odorant receptor, *Science*, 279, 237, 1998.

41. Touhara, K. et al., Functional identification and reconstitution of an odorant receptor in single olfactory neurons, *Proc. Natl. Acad. Sci. USA*, 96, 4040, 1999.

42. Wetzel, C.H. et al., Specificity and sensitivity of a human olfactory receptor functionally expressed in human embryonic kidney 293 cells and *Xenopus Laevis* oocytes, *J. Neurosci.*, 19, 7426, 1999.

43. Araneda, R.C. et al., The molecular receptive range of an odorant receptor, *Nat. Neurosci.*, 3, 1248, 2000.

44. Kajiya, K. et al., Molecular bases of odor discrimination: reconstitution of olfactory receptors that recognize overlapping sets of odorants, *J. Neurosci.*, 21, 6018, 2001.

45. McClintock, T.S. and Sammeta, N., Trafficking prerogatives of olfactory receptors, *Neuroreport*, 14, 1547, 2003.

46. He, T.C. et al., A simplified system for generating recombinant adenoviruses, *Proc. Natl. Acad. Sci. USA*, 95, 2509, 1998.

47. Bozza, T. et al., Odorant receptor expression defines functional units in the mouse olfactory system, *J. Neurosci.*, 22, 3033, 2002.

48. Mombaerts, P. et al., Visualizing an olfactory sensory map, *Cell*, 87, 675, 1996.

49. Ivic, L. et al., Intracellular trafficking of a tagged and functional mammalian olfactory receptor, *J. Neurobiol.*, 50, 56, 2002.

50. Sato, T. et al., Tuning specificities to aliphatic odorants in mouse olfactory receptor neurons and their local distribution, *J. Neurophysiol.*, 72, 2980, 1994.

51. Silver, W.L. et al., The underwater electro-olfactogram: a tool for the study of the sense of smell of marine fishes, *Experientia*, 32, 1216, 1976.

52. Scott, J.W. et al., Spatially organized response zones in rat olfactory epithelium, *J. Neurophysiol.*, 77, 1950, 1997.

53. Malnic, B. et al., Combinatorial receptor codes for odors, *Cell*, 96, 713, 1999.

54. Gimelbrant, A.A. et al., Olfactory receptor trafficking involves conserved regulatory steps, *J. Biol. Chem.*, 276, 7285, 2001.

55. Lu, M. et al., Endoplasmic reticulum retention, degradation, and aggregation of olfactory G-protein coupled receptors, *Traffic*, 4, 416, 2003.

56. Krautwurst, D. et al., Identification of ligands for olfactory receptors by functional expression of a receptor library, *Cell*, 95, 917, 1998.

57. Katada, S. et al., Structural determinants for membrane trafficking and G protein selectivity of a mouse olfactory receptor, *J. Neurochem.*, 90, 1453, 2004.

58. Katada, S. et al., Odorant response assays for a heterologously expressed olfactory receptor, *Biochem. Biophys. Res. Commun.*, 305, 964, 2003.

59. Gimelbrant, A.A. et al., Truncation releases olfactory receptors from the endoplasmic reticulum of heterologous cells, *J. Neurochem.*, 72, 2301, 1999.

60. Dwyer, N.D. et al., Odorant receptor localization to olfactory cilia is mediated by ODR-4, a novel membrane-associated protein, *Cell*, 93, 455, 1998.

61. Foord, S.M. and Marshall, F.H., RAMPs: accessory proteins for seven transmembrane domain receptors, *Trends Pharmacol. Sci.*, 20, 184, 1999.

62. Saito, H. et al., RTP family members induce functional expression of mammalian odorant receptors, *Cell*, 119, 679, 2004.

63. Oka, Y. et al., Olfactory receptor antagonism between odorants, *EMBO J.*, 23, 120, 2004.

64. Gashler, A. and Sukhatme, V.P., Early growth response protein 1 (Egr-1): prototype of a zinc-finger family of transcription factors, *Prog. Nucleic Acid Res. Mol. Biol.*, 50, 191, 1995.

65. Gurdon, J.B. et al., Use of frog eggs and oocytes for the study of messenger RNA and its translation in living cells, *Nature*, 233, 177, 1971.

66. Masu, Y. et al., cDNA cloning of bovine substance-K receptor through oocyte expression system, *Nature*, 329, 836, 1987.

67. Speca, D.J. et al., Functional identification of a goldfish odorant receptor, *Neuron*, 23, 487, 1999.

68. Wetzel, C.H. et al., Functional expression and characterization of a *Drosophila* odorant receptor in a heterologous cell system, *Proc. Natl. Acad. Sci. USA*, 98, 9377, 2001.

69. Kubo, Y. et al., Structural basis for a Ca^{2+}-sensing function of the metabotropic glutamate receptors, *Science*, 279, 1722, 1998.

70. Murrell, J.R. and Hunter, D.D., An olfactory sensory neuron line, odora, properly targets olfactory proteins and responds to odorants, *J. Neurosci.*, 19, 8260, 1999.

71. Shirokova, E. et al., Hela cells designed for functional genomics of odorant receptors and pheromone receptors, *AchemS Abstr.*, 362, 2004.

72. Spehr, M. et al., Identification of a testicular odorant receptor mediating human sperm chemotaxis, *Science*, 299, 2054, 2003.

73. Katada, S. et al., Structural basis for a broad but selective ligand spectrum of a mouse olfactory receptor: mapping the odorant-binding site, *J. Neurosci.*, 25, 1806, 2005.

Part II

Functions and Regulations of GPCRs

5 Generation and Phenotypical Analysis of Muscarinic Acetylcholine Receptor Knockout Mice

*Jürgen Wess, Dinesh Gautam,
Sung-Jun Han, Jongrye Jeon, Cuiling Li,
and Chuxia Deng*

CONTENTS

5.1   INTRODUCTION

The individual members of the muscarinic acetylcholine receptor (mAChR) family are prototypical members of the class A subfamily of G protein-coupled receptors (rhodopsin-like GPCRs).[1,2] Molecular cloning studies have shown that mAChRs form a family of five molecularly distinct members (M_1 to M_5). At a molecular level, the M_1, M_3, and M_5 receptor subtypes preferentially couple to G proteins of the G_q family, whereas the M_2 and M_4 receptors are primarily linked to G proteins of the G_i class.[1,2]

The individual mAChRs are widely expressed in nearly all parts of the central nervous system and in many peripheral tissues.[1-5] Characteristically, each tissue or organ expresses multiple mAChR subtypes, with only very few exceptions.[1-5]

mAChRs modulate the activity of many fundamental central and peripheral functions. In the central nervous system, mAChRs play important roles in learning and memory and many key motor, autonomous, sensory, and behavioral processes.[6,8-11] Importantly, disturbances in the central muscarinic cholinergic system have been implicated in the pathophysiology of Alzheimer's and Parkinson's diseases, depression, epilepsy, and schizophrenia.[6,8-11] In the body periphery, mAChRs mediate the well-known functions of acetylcholine (ACh) at parasympathetically innervated effector organs.[6,7]

An important question is which specific mAChR subtypes are involved in mediating the various muscarinic functions of ACh? Work in this area has been complicated by the fact that most organs, tissues, and cell types express multiple mAChRs, as already indicated above. Moreover, except for some recently identified snake toxins,[12] muscarinic ligands with a high degree of receptor subtype selectivity are not available at present.[1,2] For these reasons, the precise physiological and pathophysiological roles of the individual mAChR subtypes are not well understood at present, specifically, as far as the central muscarinic actions of ACh are concerned. To shed light on this issue, we and other investigators decided to use gene targeting technology to generate mutant mouse strains deficient in each of the five mAChR genes. (For a recent review, see Wess.[13])

5.2 GENERAL STRATEGY USED TO GENERATE MICE LACKING SPECIFIC MACHRS

During the past few years, we have used gene ablation techniques to generate mutant mouse strains lacking functional M_1,[14,15] M_2,[16] M_3,[17] M_4,[18] or M_5[19] mAChRs. All mutant mouse strains were obtained by using the general strategy summarized in Figure 5.1.

Initially, genomic fragments containing the murine M_1 to M_5 mAChR coding sequences were isolated from a 129SvJ mouse genomic library (Genome Systems, Palo Alto, CA) by hybridization with polymerase chain reaction (PCR) fragments coding for the central nonconserved portions of the third intracellular loops of the mouse M_1 to M_5 mAChRs. It is generally recommended that the mouse strain from which the genomic library is obtained is identical to or similar to the mouse strain from which the mouse embryonic stem cells (ES cells) to be used are derived. The isolated genomic fragments contained the M_1 to M_5 mAChR coding sequences (note that the M_1 to M_5 mAChR coding sequences are intronless), and several kilobits of sequence upstream and downstream of the coding sequences. These fragments were then subjected to restriction mapping and partial sequencing. The mouse M_1[20] and

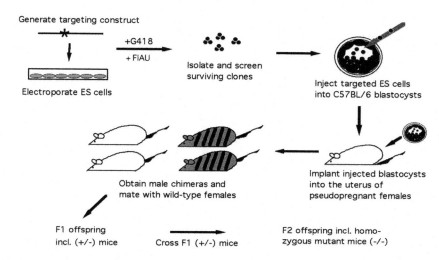

FIGURE 5.1 General scheme depicting the different steps involved in the generation of homozygous M_1 to M_5 mAChR mutant mice. To inactivate a specific mAChR gene in the mouse genome, we first generated a pPN2T-based DNA construct containing an inactivated version of the receptor gene of interest. Following electroporation of the targeting construct into mouse ES cells, the ES cells are subjected to a G418/FIAU double selection procedure. The surviving ES cell clones are then screened, most commonly by Southern blot analysis, for the occurrence of the proper targeting event. Properly targeted ES cell clones are then microinjected into mouse blastocysts derived from C57BL/6 mice, and the injected blastocysts are implanted into the uterus of pseudopregnant foster mothers. The resulting pups are chimeric in nature, consisting of cells derived from the injected ES cells and of cells derived from the blastocyst host. Because the ES cells are usually derived from different 129 substrains that have an agouti (brownish-yellow) coat color, and the blastocysts usually come from a mouse strain with a black coat color (typically C57BL/6), the degree of chimerism can be judged by the color pattern of the mouse fur. To check for germline transmission of the mutant allele, chimeric male mice are then crossed with WT females that do not have an agouti coat color. Because the agouti coat color is dominant, the presence of agouti offspring indicates that the targeted ES cells have populated the germline. F1 heterozygotes are then intermated to produce F2 offspring, 1/4 of which is predicted to be homozygous for the desired receptor mutation.

M_4[21] receptor coding sequences were previously published. Recently, we deposited the mouse M_2, M_3, and M_5 mAChR coding sequences at GenBank (Bethesda, MD; accession numbers: M_2, AF264049; M_3, AF264050; M_5, AF264051). It should be noted that the recent availability of the sequence of the entire mouse genome will greatly facilitate the construction of gene targeting vectors (see below).

All targeting vectors were derived from pPN2T,[22] which is characterized by the presence of two copies of the herpes simplex virus thymidine kinase gene (HSV-TK) and a PGK-neomycin resistance cassette (*neo*) that replaced functionally essential segments of the mAChR coding sequences.[14–19] The structure of a representative targeting vector is shown in Figure 5.2. In all cases, the targeting constructs contained at least 5 kb of homologous sequence, and each "arm" of homology was at least

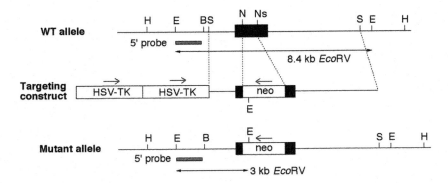

FIGURE 5.2 Targeted disruption of the mouse M_2 mAChR gene. Restriction maps of WT receptor locus, targeting vector, and targeted allele are shown (H, HindIII; E, EcoRV; B, BamHI; S, StuI; N, NheI; Ns, NsiI). The M_2 receptor coding region is represented by a filled bar. The M_2 targeting vector was derived from pPN2T (see Paszty, C. et al., *Nat. Genet.*, 11, 33, 1995), which contains a PGK-neomycin resistance gene (*neo*) and two copies of the herpes simplex virus thymidine kinase gene (HSV-TK). Note that an 0.67 kb NheI–NsiI genomic fragment (which corresponds to the M_2 mAChR sequence coding for the region between the third transmembrane domain and the C-terminus of the third intracellular loop) with the PGK-*neo* cassette was obtained. The targeting vector was linearized with NotI (not shown) and introduced into mouse ES cells by electroporation. G418- and FIAU-resistant clones were isolated and screened by Southern blotting for homologous recombination. The 5' probe used for Southern analysis and the sizes of the restriction fragments detected with this probe are indicated. Note that this probe is located outside the region of homology (indicated by the stippled lines). When genomic DNA is digested with EcoRV, the 5' probe detects an 8.4 kb band in the presence of the WT receptor allele and a 3 kb band in the presence of the mutant M_2 receptor allele, respectively.

1 kb long. A detailed discussion of the design of gene targeting vectors is beyond the scope of this chapter.[23]

The targeting vectors were linearized (usually at a unique NotI site contained in the pPN2T vector) and introduced into mouse ES cells by electroporation (see Protocol 5-1). The M_1, M_3, M_4, and M_5 receptor constructs were electroporated into TC1 (129SvEv) cells,[24] and the M_2 receptor construct was introduced into "J1" cells[25] derived from a different 129 mouse substrain. Clones resistant to G418 and FIAU were isolated (see Protocol 5-1) and tested for the occurrence of homologous recombination via Southern hybridization (see, for example, Figure 5.2). Using the protocols and reagents described below, about 2 to 5% of the ES cell clones screened showed the desired targeting event. Properly targeted ES cell clones were microinjected into C57BL/6J blastocysts (see Protocols 5-2 and 5-3) to generate male chimeric offspring that, in turn, were mated with female CF-1 (Charles River Laboratories, Wilmington, MA), C57BL/6J (The Jackson Laboratories, Bar Harbor, ME), or 129SvEV mice (Taconic, Germantown, NY; this 129 substrain is now referred to as "129S6") to generate F1 offspring. F1 animals heterozygous for the desired mAChR mutation were then intermated to produce homozygous mAChR mutant mice (F2; Figure 5.1). Several issues critically related to the genetic background of the analyzed mAChR mutant mice are discussed in Section 5.3.3.

Initial genotyping studies were done via Southern blotting analysis. Subsequently, for increased convenience, mouse genotyping was carried out via PCR analysis of mouse tail DNA, using primers specific for the individual wild-type (WT) and mutant M_1 to M_5 mAChR alleles. Successful gene disruption was confirmed by the use of receptor subtype-selective antisera or antibodies, radioligand biding studies, and reverse transmission (RT)-PCR studies.

Protocol 5-1: Isolation of ES Cell Clones Containing Targeted Disruptions of Individual mAChR Genes

1. Materials required:
   a. TC1 ES cell line[24]; these cells are derived from blastocysts of 129SvEv mice (Taconic)
   b. 129SvJ mouse genomic library (Genome Systems)
   c. pPN2T vector[22]
   d. G418 (#11811031; Invitrogen, Carlsbad, CA; Life Technologies, Gaithersburg, MD) and 1-(2-deoxy-2-fluoro--arabinofuranosyl)-5-iodouracil (FIAU; MC251, Moravek Biochemicals, Brea, CA)
   e. Composition of the ES cell medium: 500 ml of DMEM (high glucose, without L-glutamine and sodium pyruvate; #11960044, Invitrogen, Life Technologies), 15% fetal calf serum (heat inactivated), 10^{-4} M 2-mercaptoethanol (M-7522; Sigma-Aldrich, St. Louis, MO), MEM none essential amino acids (#11140050, 100X; Invitrogen, Life Technologies), penicillin/streptomycin (#15140122, 100X; Invitrogen, Life Technologies), 2% L-glutamine (#25030081; Invitrogen, Life Technologies), trypsin (#25200056, 0.25%; Invitrogen, Life Technologies), mitomycin C (M-4287; Sigma), and leukemia inhibitory factor (LIF) at a concentration of 1000 Unit/ml (Chemicon, Temecula, CA) (note that LIF is essential for maintaining ES cells in a pluripotent state)
2. Preparation of the targeting vector for electroporation:
   a. Linearize 100 μg of Qiagen column-purified DNA by an appropriate restriction enzyme (in the case of pPN2T-based constructs, NotI can usually be used; linearize the targeting construct at a site in the backbone of the plasmid, not within the homologous DNA region or area of interest; perform a large-scale, overnight digestion for linearization)
   b. Extract linearized DNA with phenol/chloroform followed by ethanol precipitation
   c. Resuspend linearized DNA in sterile TE at a concentration of 1 to 1.5 μg/μl
3. Preparation of ES cells for electroporation:
   a. Quickly thaw one vial of frozen TC1 ES cells in a 37°C water bath
   b. Transfer the cells to a 12 ml tube containing 5 ml ES cell medium

    c. Centrifuge at 270 × g for 2 min, aspirate the supernatant, resuspend the cell pellet in 10 ml ES cell medium, and plate the cells on a 10 cm dish containing a mitomycin C-treated embryonic fibroblast feeder layer

    d. Change the medium on the second day after plating the ES cells

    e. On the third day, aspirate the medium, wash the cells with 5 ml of PBS, add 1 ml trypsin/EDTA, incubate for 5 min at 37°C, add 5 ml ES cell medium, and then pipette up and down to break up cell clumps

    f. Collect the cells into a 12 ml tube, and repellet the cells via centrifugation (see above)

    g. Aspirate the supernatant, gently resuspend the cells in ES cell medium, and add ~2 × 10^6 cells to a fresh 10 cm plate (with feeder layer)

    h. Change the medium daily; on the third day, 2 to 4 h after changing the medium, trypsinize the cells (see above), wash once with PBS, and resuspend the cells in PBS at a concentration of 1 to 2 × 10^7 cells/ml

    i. Add DNA (up to 50 μg) to 1 ml of the cell suspension, and electroporate cells at 25 μF and 600 mV using a Gene Pulser (Bio-Rad, Hercules, CA) electroporator (under these conditions, ~50% of the cells will survive after electroporation)

    j. Following electroporation, dilute the cells with 4 ml of growth medium; 5 to 10 min later, plate the cells on five 10-cm dishes containing feeder cells

    k. Change the ES cell medium (15 to 25 ml) daily (grow cells at 37°C in a 5% CO_2 incubator).

    l. Supplement the medium with G418 (300 μg/ml) 24 h after electroporation, and add FIAU (0.2 μM) 48 h after electroporation, respectively; maintain the cells under G418/FIAU selection for 6 to 7 days

4. Isolation of ES cell clones:

    a. Eight days after transfection, replace ES cell medium with PBS (10 ml)

    b. Pick ES cell clones with a yellow pipette tip under a microscope using a pipetman adjusted to a 20 μl volume

    c. Transfer each colony to an individual well of a 96-well plate

    d. Once the plate is filled with ES cell clones, add 50 μl of trypsin (use a 1:1 mixture of PBS:trypsin) to each well, and incubate at room temperature (RT) for 3 to 5 min

    e. Add 100 μl of ES cell medium to each well, and pipette up and down (five to 10 times), using an eight-channel pipetman adjusted to a 80 μl volume

    f. Label two 96-well plates containing feeder cells with A and B

    g. Transfer 80 μl of cell suspension to the A plate and 80 μl to the B plate

    h. Change medium daily using 150 μl of ES cell medium on the second day and 200 μl of medium on the third day

5. Freezing (storage) of ES cell clones (96-well Plate A):

    a. Four days after plating the ES cells, aspirate the ES cell medium from 96-well Plate A, and rinse cells with 100 μl of PBS

    b. Add 50 µl of trypsin:PBS (1:1), and incubate the plate for 5 min at room temperature

    c. Add 100 µl of freezing medium (15% DMSO, 85% ES cell medium without LIF)

    d. Pipette up and down repeatedly (5 to 10 times) to dissociate the ES cell colonies

    e. Wrap the plate with one piece of heavy-duty cleaning paper (Kimberly Clark #39215), label the plate properly (name of investigator, date, construct used, etc.), place the plate into a polyethylene bag (Fisher #01-812-10C), seal the bag, and place the sealed plate into a –70°C freezer

6. Expansion of ES cell clones (96-well Plate B):

    a. Four days after plating the ES cells, aspirate the ES cell medium from 96-well Plate B, and rinse cells with 100 µl of PBS

    b. Add 50 µl of trypsin:PBS (1:1), and incubate the plate for 5 min at room temperature

    c. Add 100 µl of growth medium, and repeatedly pipette up and down (5 to 10 times) to dissociate the ES cell colonies

    d. Transfer the cell clones to four 24-well plates containing a reduced density of feeder cells (i.e., 50%; label the 24-well plates properly so that the corresponding well in Plate A can be identified at a later point)

    e. On the first day, add 1.5 ml of ES cell medium (there is no need to change the medium during the first 3 to 4 days)

    f. After 3 to 4 days of cell growth, the medium in most wells should turn yellow; these cells can be lysed for the isolation of genomic DNA; after the fourth day, trypsinize the ES cells in the remaining wells using 150 to 200 µl (about three drops) of trypsin:PBS (1:1), and resuspend the cells in 1.5 ml of ES cell medium (cell lysates can be prepared from these cells when the medium turns yellow [1 to 3 days after replating]; this procedure can be repeated one more time until all clones have been harvested)

The use of the G418/FIAU double-selection procedure (see above) is thought to increase the yield of ES cell clones displaying the proper genomic targeting event. However, nonhomologous recombination events (random integration of the targeting construct) usually still occur at a far higher frequency than do homologous recombination events. Initially, we, like most other investigators, used Southern blot analysis to identify properly targeted ES cells. The probes to be used for this analysis should be located outside the region of homology and should have a corresponding restriction site located beyond the probe sequence (see, for example, Figure 5.2; note that the use of restriction sites within the arms of homology of the targeting construct will not differentiate between homologous and nonhomologous recombinants). In some cases, it is also possible to engineer PCR primers that will give a product specific for the homologous recombination event. Although this approach can save time during the initial round of screening, it is highly recommended that the desired recombination event be confirmed via Southern blotting analysis.

PROTOCOL 5-2: PREPARATION OF TARGETED ES CELLS FOR BLASTOCYST INJECTIONS

1. Identify the properly targeted ES cell clones on 96-well Plate A (stored in a −70°C freezer), and thaw the cells in a 37°C water bath (remove the cover of the plate before placing the plate into the water bath)
2. Three to five min later, transfer the plate into a tissue culture hood, and add 100 µl ES cell medium to the wells containing properly targeted ES cells
3. Transfer the targeted ES cell clones into 12 ml tubes, and centrifuge at 270 × g for 2 min
4. Resuspend the cells in ES cell medium, and transfer them into a 24-well plate containing feeder cells
5. Change the ES cell medium daily
6. Three to four days after plating, transfer the cells from the 24-well plate to 12-well plates containing feeder cells
7. After having reached ~50% confluency, passage the cells to multiple wells, and freeze back two vials for storage at −70°C (at this point, DNA can also be isolated from the cells to confirm their genotypes)
8. Change the medium 3 h before blastocyst injections, trypsinize the ES cells, and resuspend them in ES cell medium at the desired concentration (based on the injector's preference; at this stage, ES cells are ready to be injected)

PROTOCOL 5-3: BLASTOCYST INJECTIONS AND IMPLANTATION OF INJECTED BLASTOCYSTS

1. Prepare blastocysts from C57BL/6 mice according to standard procedures[23]
2. Treat [via intraperitoneal (i.p.) injection] 10 female C57BL/6 mice (3.5 weeks old) with five units of pregnant mare serum (PMS; Sigma G4877; dissolved in PBS) at around 1:00 p.m. 6 days prior to injection of blastocysts
3. Two days after PMS injection, treat (via i.p. injection) the mice with five units of human chorionic gonadotropin (hCG; Sigma C1063; dissolved in PBS) at around 11:00 a.m. (i.e., 46 h after PMS injection), then set up the injected female mice for matings with male mice of the same strain
4. Three days later, sacrifice the 10 female mice, and harvest the blastocysts in Brinster's BMOC-3 medium (Invitrogen, Catalog #11126-034), according to standard procedures[23] (choose only those blastocysts that are well developed)
5. Inject 10 to 15 ES cells into each blastocyst, and culture the injected blastocysts for 1 h prior to implantation
6. Use pseudopregnant female FVB/n mice as foster mothers for the implantation of injected blastocysts

7. Mate female FVB/n mice (at least 8 weeks old) with vasectomized male FVB/n mice 3 days prior to blastocyst injection; on the next morning, check the female mice for the presence of vaginal plugs, and keep the mice displaying a vaginal plug as potential foster mothers (pseudopregnant mice)

8. Anesthetize a pseudopregnant female mouse, and collect five to seven injected blastocysts in a transfer pipette marked with an air bubble on each side (set the pipette aside while preparing the mouse for the implantation of blastocysts)

9. Cut the fur of the pseudopregnant mouse on the lower back, clean the skin with alcohol, make a 1-cm transverse skin incision on both sides at the level of the first lumbar vertebra, make the ovarian fat pad visible through the muscle layer, make a 3 to 5 mm incision through the peritoneum, grasp the fat pad and pull it out toward the midline, and stabilize the ovaries outside the surgical area with clamps

10. Use a 3/8 inch 26G hypodermic needle to punch a hole into the uterus, ~2 mm away from the utero-tubal junction

11. Insert the pipette tip into the opening generated by the needle, and blow the blastocysts into the uterus while monitoring the movements of the air bubbles (note that the first air bubble is small and enters into the uterus together with the blastocysts, while the second air bubble is big and is located on the other side of the blastocysts; when the second air bubble reaches the pipette tip, all blastocysts should be in the uterus; check the pipette after transfer to ensure that all blastocysts have been expelled)

12. Place all tissues back into the proper position in the peritoneum, seal the muscle layer with suture, and close the skin incision with a small wound clip (use the same procedure for both sides), and leave the animals on a heating plate to keep them warm until recovery

5.3 PHENOTYPICAL ANALYSIS OF MACHR KNOCKOUT (KO) MICE

5.3.1 SUMMARY OF MUTANT PHENOTYPES

In all cases, homozygous M_1 to M_5 mAChR mutant mice (referred to as M1R$^{-/-}$mice, M2R$^{-/-}$mice, etc.) were obtained at the expected Mendelian frequency.[14–19] The M_1 to M_5 mAChR mutant mice are viable, fertile, appear generally healthy, and seem to have similar life spans as their corresponding WT mice.[13] They also do not show any obvious morphological or anatomical abnormalities. However, we, in collaboration with many collaborators inside and outside of the National Institutes of Health (Bethesda, MD), demonstrated that each of the five mAChR mutant mouse strains displayed characteristic pharmacological, physiological, behavioral, electrophysiological, biochemical, neurochemical, and behavioral deficits or changes. A summary of the major phenotypes that have been observed so far is given in Table 5.1. This table not only summarizes work carried out with mutant mice generated at NIH/National Institute of Diabetes and Digestive and Kidney Diseases (NIDDK) but also includes phenotyping studies performed by other investigators using independently generated mAChR mutant mice.

TABLE 5.1
Summary of Phenotyping Studies Carried Out with M$_1$ to M$_5$ mAChR-Deficient Mice

Experimental Parameter Studied	Observed Phenotypical Changes	Ref.
Pharmacological/Physiological Phenotypes (CNS)		
Pilocarpine-mediated seizure activity	Abolished in M1R$^{-/-}$ mice	26
Oxotremorine-induced tremor	Absent in M2R$^{-/-}$ mice	16
Oxotremorine-mediated hypothermia	Greatly reduced in M2R$^{-/-}$ mice	16
Oxotremorine-mediated analgesic responses following systemic, intrathecal, or intracerebroventricular administration	Greatly reduced in M2R$^{-/-}$ mice; little impairment in M4R$^{-/-}$ mice; abolished in M2R$^{-/-}$/M4R$^{-/-}$ double KO mice	16, 31
Behavioral Phenotypes		
Spontaneous locomotor activity	Strongly increased in M1R$^{-/-}$ mice	14, 32
	Modestly increased in M4R$^{-/-}$ mice	18
Social discrimination and win-shift spatial working memory	Impaired in M1R$^{-/-}$ mice	56
Passive avoidance learning	Impaired in M2R$^{-/-}$ mice	57
Locomotor stimulation after administration of a D1 dopamine receptor agonist	Potentiated in M4R$^{-/-}$ mice	18
Sensitivity to phencyclidine-mediated disruptions in prepulse inhibition	Increased in M4R$^{-/-}$ mice	44
Ability of scopolamine to antagonize haloperidol-mediated cataleptic effects	Absent in M4R$^{-/-}$ mice	46
Rewarding effects of morphine studied in a conditioned place preference (CPP) test	Reduced in M5R$^{-/-}$ mice	30
Rewarding effects of cocaine studied in CPP and self-administration tests	Reduced in M5R$^{-/-}$ mice	54
Morphine and cocaine withdrawal symptoms	Reduced in M5R$^{-/-}$ mice	30, 54
Water intake after an extended period of food and water deprivation	Increased in M5R$^{-/-}$ mice	58
Scopolamine (given i.p.)-mediated increase in locomotor activity	Slightly enhanced in M5R$^{-/-}$ mice	59
Electrophysiological Phenotypes		
Muscarinic agonist-mediated M current (I$_m$) inhibition in sympathetic ganglion neurons	Absent in M1R$^{-/-}$ mice	26
Slow, voltage-independent muscarinic inhibition of N- and P/Q-type Ca^{2+} channels in sympathetic ganglion neurons	Absent in M1R$^{-/-}$ mice	60
Muscarine-mediated γ oscillations in area CA3 of the hippocampus	Abolished in M1R$^{-/-}$ mice	15

(continued)

TABLE 5.1 (CONTINUED)
Summary of Phenotyping Studies Carried Out with M_1 to M_5 mAChR-Deficient Mice

Experimental Parameter Studied	Observed Phenotypical Changes	Ref.
Electrophysiological Phenotypes (continued)		
Hippocampal LTP at the Schaffer collateral-CA1 synapse (induced by theta burst stimulation)	Slightly reduced in M1R[-/-] mice	56
Fast, voltage-dependent muscarinic inhibition of N- and P/Q-type Ca^{2+} channels in sympathetic ganglion neurons	Absent in M2R[-/-] mice	60
Muscarinic enhancement of depolarization-induced suppression of inhibitory postsynaptic currents (IPSCs) in cultured hippocampal neurons	Absent in M1R[-/-]/M3R[-/-] double KO mice	61
Cannabinoid receptor type 1-independent muscarinic suppression of IPSCs in cultured hippocampal neurons	Abolished in M2R[-/-] mice	62
Cannabinoid receptor type 1-dependent muscarinic suppression of IPSCs in cultured hippocampal neurons	Absent in M1R[-/-]/M3R[-/-] double KO mice	62
Biochemical Phenotypes		
Carbachol-mediated mitogen-activated protein kinase (MAPK) activation in CA1 hippocampal pyramidal neurons	Abolished in M1R[-/-] mice	63
Muscarinic agonist-mediated phosphoinositide (PI) hydrolysis in primary cortical cultures	Strongly reduced in M1R[-/-] mice	64
Pilocarpine-stimulated *in vivo* PI hydrolysis in hippocampus and cerebral cortex	Abolished in M1R[-/-] mice	65
Muscarinic agonist-induced GTPγS binding to G proteins of the G_q family in hippocampus and cerebral cortex	Absent in M1R[-/-] mice	66
Neurochemical Phenotypes		
Extracellular dopamine levels in the striatum studied by *in vivo* microdialysis	Increased in M1R[-/-] mice	32
Oxotremorine-mediated inhibition of [^{3}H]ACh release from K[+]-depolarized hippocampal and cortical slices	Absent in M2R[-/-] mice	67
Oxotremorine-mediated inhibition of [^{3}H]ACh release from K[+]-depolarized striatal slices	Absent in M4R[-/-] mice	67
Muscarine-mediated inhibition of ACh release from electrically stimulated phrenic diaphragm neuromuscular preparations (*in vitro*)	Absent in M2R[-/-] mice	68

(continued)

TABLE 5.1 (CONTINUED)
**Summary of Phenotyping Studies Carried Out with M₁ to
M₅ mAChR-Deficient Mice**

Experimental Parameter Studied	Observed Phenotypical Changes	Ref.
Neurochemical Phenotypes (continued)		
Kinetics of ACh release at the phrenic diaphragm neuromuscular junction (*in vitro*)	Altered in M2R[-/-] mice	68
Carbachol-mediated inhibition of electrically stimulated [³H]norepinephrine release from heart atria, urinary bladder, and vas deferens (*in vitro*)	Reduced in M2R[-/-] mice	69
Oxotremorine-stimulated [³H]dopamine outflow from K⁺-depolarized striatal slices	Increased in M3R[-/-] mice	70
	Abolished in M4R[-/-] mice	70
	Impaired in M5R[-/-] mice	19, 70
Muscarinic-agonist mediated autoinhibiton of[³H]ACh release in heart atria and urinary bladder	Reduced in M4R[-/-] mice	71
Basal ACh efflux in the hippocampus studied by *in vivo* microdialysis	Increased in M4R[-/-] mice	57
Enhancement of ACh efflux in the hippocampus induced by local administration of scopolamine	Reduced in M2R[-/-] mice; absent in M2R[-/-]/M4R[-/-] double KO mice	57
Enhancement of ACh efflux in the hippocampus after exposure to a novel environment	Increased in M2R[-/-] mice	57
Basal dopamine levels in the nucleus accumbens studied by *in vivo* microdialysis	Increased in M4R[-/-] mice	72
Increase in dopamine efflux in the nucleus accumbens following i.p. administration of d-amphetamine and phencyclidine studied by *in vivo* microdialysis	Enhanced in M4R[-/-] mice	72
Basal ACh levels in the midbrain studied by *in vivo* microdialysis	Increased in M4R[-/-] mice	72
Sustained increase in dopamine levels in the nucleus accumbens triggered by electrical stimulation of the laterodorsal tegmental nucleus	Absent in M5R[-/-] mice	55
Cardiovascular Phenotypes		
Cardiovascular stimulation following systemic administration of McN-A-343	Abolished in M1R[-/-] mice	73
Carbachol-mediated bradycardia in isolated spontaneously beating heart atria	Abolished in M2R[-/-] mice	74
Bradycardia induced by vagal stimulation or systemic administration of methacholine *in vivo*	Abolished in M2R[-/-] mice	75
Decrease in mean arterial pressure following administration of methacholine *in vivo*	Greatly reduced in M2R[-/-] mice; slightly reduced in M3R[-/-] mice	75

(continued)

TABLE 5.1 (CONTINUED)
Summary of Phenotyping Studies Carried Out with M$_1$ to M$_5$ mAChR-Deficient Mice

Experimental Parameter Studied	Observed Phenotypical Changes	Ref.
Smooth Muscle Phenotypes		
Carbachol-mediated contractions of different smooth preparations (*in vitro*)	Pronounced reduction in E_{max} in M3R$^{-/-}$ mice	37, 38, 39
	Small reduction in carbachol potency in M2R$^{-/-}$ mice	35, 74
	Abolished in M2R$^{-/-}$/M3R$^{-/-}$ double KO mice	33, 35, 36
Pupil size (*in vivo*)	Increased in M3R$^{-/-}$ mice	37, 65
Appearance of the urinary bladder (*in vivo*)	Distended in male M3R$^{-/-}$ mice	37
Relaxant effects of forskolin and isoproterenol on oxotremorine-M-mediated contractions of different smooth muscle tissues (*in vitro*)	Increased in M2R$^{-/-}$ mice	76
Muscarine-induced constriction of peripheral airways (*in vitro*)	Increased in M1R-/- mice; greatly reduced in M3R$^{-/-}$ mice; abolished in M2R$^{-/-}$/M3R$^{-/-}$ double KO mice	33
Bronchoconstriction in response to vagal stimulation or systemic administration of methacholine *in vivo*	Increased in M2R$^{-/-}$ mice; absent in M3R$^{-/-}$ mice	75
ACh-induced heterologous desensitization in ileal preparations (*in vitro*)	Absent in M2R$^{-/-}$ and M3R$^{-/-}$ mice	77
ACh-mediated dilation of peripheral blood vessels (aorta and coronary arteries)	Greatly reduced in M3R$^{-/-}$ mice	78, 79
ACh-mediated dilation of a cerebral artery (basilar artery *in vivo*)	Greatly reduced in M5R$^{-/-}$ mice	19
ACh-mediated dilation of cerebral arterioles (pial vessels *in vivo*)	Abolished in M5R$^{-/-}$ mice	19
Glandular Phenotypes		
Muscarinic agonist-induced salivation	Impaired in M3R$^{-/-}$ mice	17, 27, 34, 37, 65
	Slightly impaired in M1R$^{-/-}$ mice	27, 65
	Abolished in M1R$^{-/-}$/M3R$^{-/-}$ double KO mice	27, 34
Serum gastrin levels	Significantly increased in M3R$^{-/-}$ mice	80
Basal and carbachol-, histamine-, and gastrin-stimulated gastric acid secretion	Reduced in M3R$^{-/-}$ mice	80

(continued)

TABLE 5.1 (CONTINUED)
Summary of Phenotyping Studies Carried Out with M₁ to
M₅ mAChR-Deficient Mice

Experimental Parameter Studied	Observed Phenotypical Changes	Ref.
Metabolic Phenotypes		
Body weight, mass of peripheral fat deposits, and food intake	Reduced in M3R$^{-/-}$ mice	17
Plasma insulin, leptin, glucagon, and triglyceride levels	Reduced in M3R$^{-/-}$ mice	17, 42
Muscarinic agonist-mediated potentiation of glucose-dependent insulin release from isolated pancreatic islets	Absent in M3R$^{-/-}$ mice	42, 43
Muscarinic agonist-mediated release of glucagon from isolated pancreatic islets	Absent in M3R$^{-/-}$ mice	42
Insulin sensitivity and glucose tolerance	Increased in M3R$^{-/-}$ mice	17, 42
Skin Phenotypes		
Muscarine-mediated desensitization of skin nociceptors	Absent in M2R$^{-/-}$ mice	41
Keratinocyte adhesion	Impaired in M3R$^{-/-}$ mice	81
Migration of epidermal keratinocytes and wound reepithelialization	Increased in M3R$^{-/-}$ mice; impaired in M4R$^{-/-}$ mice	45
Miscellaneous Phenotypes		
Partial muscarinic agonist (BuTAC)-induced increases in serum corticosterone levels	Abolished in M2R$^{-/-}$ mice	82
Ventilation at rest	Increased in M3R$^{-/-}$ mice	83
Ventilatory responses to hypercapnia and hypoxia	Reduced in M3R$^{-/-}$ mice	83

5.3.2 POTENTIAL COMPENSATORY CHANGES OF MACHR EXPRESSION LEVELS IN MACHR KO MICE

As summarized in Table 5.1, the phenotypical analysis of mAChR-deficient mice has led to a wealth of interesting new information regarding the physiological roles of the individual mAChR subtypes. However, it should be pointed out that all mAChR mutant mice that have been analyzed so far lack one or more mAChR subtypes throughout development. The possibility therefore exists that compensatory changes may have led to altered expression levels of other mAChR subtypes (or other classes of receptors) and components of downstream signal transduction cascades. However, several studies examining mAChR expression levels in various tissues suggest that disruption of one specific mAChR gene does not seem to have major effects on the expression levels of the remaining four mAChRs.[14–18,26,27] For example, immunoprecipitation studies using mAChR-specific antisera have shown that inactivation of the M_1 mAChR gene has no significant effects on the expression levels of the M_2 to M_5 mAChR in hippocampus[15] and striatum.[14]

Analogously, quantitative RT-PCR (TaqMan) studies with salivary gland tissue (submandibular gland) showed that the inactivation of the M_1 or M_3 mAChR genes did not lead to significantly altered mRNA levels of the remaining mAChR sub-types.[27] However, these observations cannot completely rule out the possibility that functional inactivation of a specific mAChR gene leads to changes in the distribution of the remaining mAChRs at the ultrastructural level. Fortunately, new genetic tools (e.g., the CRE/loxP recombination system) are now available that will allow the inactivation of specific mAChR genes in a tissue-specific and time-dependent fashion.[28]

5.3.3 ISSUES RELATED TO THE MOUSE GENETIC BACKGROUND OF THE ANALYZED MAChR KO MICE

Because the 129 mouse strains from which most ES cells are derived are difficult to breed, the mutant mice used in most of the studies listed in Table 5.1 were hybrid mice with a mixed genetic background. Several cases have been reported in which phenotypical differences between mutant mice and WT littermates of hybrid genetic origin do not appear to be due to the null mutation but due to background genes linked to the target genes.[29] Such problems are of particular relevance when the studied phenotype (e.g., drug-seeking behavior) differs between the mouse strain from which the ES cells were derived and the mouse strain with which the chimeras were crossed. To address this issue, we carried out several studies in which we analyzed the two parental mouse strains in parallel with the mAChR mutant mice of mixed genetic background.[30,31]

Another possibility is to completely avoid the use of hybrid mice and to generate so-called isogenic mice in which the KO mutation is present on an isogenic mouse background.[29] For example, most of our gene-targeting studies were carried out with ES cells derived from the 129SvEv strain.[24] (This 129 substrain is now referred to as 129S6.) Mating of chimeras derived from the use of this type of ES cells with 129SvEv mice results in the generation of isogenic mAChR mutant mice. We used such isogenic mAChR mutant mice in several cases to confirm phenotypical changes observed with mutant mice of a mixed genetic background.[17,30] However, a disadvantage of these mice is that they are more difficult to breed (for example, their litter size is relatively small) than mice of mixed genetic background, and they are not preferred for behavioral studies.[29] Because the use of C57BL/6 mice is now preferred for most behavioral studies, ES cells have also been developed from this mouse strain. For example, in a recent study, Gerber et al.[32] generated M1R$^{-/-}$ mice by using ES cells derived from C57BL/6 mice and mating the resulting chimeras with C57BL/6 mice. The analysis of the resulting isogenic M1R$^{-/-}$ mice is, therefore, not confounded by the problems potentially associated with the use of mice of a mixed genetic background.

Alternatively, mice of a mixed genetic background can be backcrossed onto a mouse strain of choice (usually C57BL/6 mice) to generate so-called congenic mice that carry the KO mutation on a desired genetic background. After 10 backcrosses, for example, only ~0.1% of the genome of the resulting congenic mice will be derived from the mouse strain from which the ES cells were obtained. The use of such congenic mice, therefore, greatly reduces the likelihood that an observed phenotype is caused by genes linked to the KO locus rather than by the actual KO mutation.

5.3.4 Generation and Analysis of mAChR Double KO Mice

As already indicated in the introduction, the M_2 and M_4 mAChRs couple to a similar set of G proteins (G_i family) and are co-expressed in many tissues, particularly in the CNS.[1–5] Analogously, the M_1, M_3, and M_5 mAChRs are all coupled to G proteins of the G_q family.[1,2] Moreover, most peripheral and central tissues that express M_3 receptors also express M_1 receptors and vice versa.[1–5] We, therefore, speculated that the analysis of M_1/M_3 and M_2/M_4 mAChR double KO mice might reveal novel phenotypes that may not manifest themselves in the receptor single KO mice due to functional redundancy. In addition to the M_1/M_3[27] and M_2/M_4[31] mAChR double KO mice, we recently generated M_2/M_3[33] and M_1/M_4[27] mAChR double KO mice. These latter mouse strains were generated because M_2 and M_3 receptors are co-localized in many tissues, and M_1 and M_4 receptors are the most abundant mAChRs expressed in higher brain regions, including the hippocampus and cerebral cortex.[1–5]

To generate mice deficient in two different mAChRs, homozygous receptor single KO mice lacking functional Subtype X receptors were intermated with homozygous receptor single KO mice lacking functional Subtype Y receptors. The resulting F1 compound heterozygotes were then intercrossed to generate F2 mice. According to Mendelian inheritance, 1/16 of the F2 pups are predicted to be homozygous for both the X and Y receptor gene disruptions (homozygous double KO mice). Similarly, 1/16 of the F2 pups are predicted to carry two copies of the WT X and Y receptor genes.

Consistent with our hypothesis that the analysis of mAChR double KO mice might reveal novel phenotypes, we recently demonstrated that centrally active muscarinic agonists were unable to induce analgesic responses in mice deficient in both M_2 and M_4 receptor subtypes (M2R$^{-/-}$/M4R$^{-/-}$ mice).[31] In contrast, muscarinic agonist-induced analgesic responses were significantly reduced but not abolished in M2R$^{-/-}$ single KO mice.[16,31] These findings strongly suggest that the analgesic effects of muscarinic agonists, both at the spinal and supraspinal levels,[31] are mediated by a mixture of M_2 and M_4 receptors.

In another study, we showed that pilocarpine-induced salivation responses were abolished in M1R$^{-/-}$/M3R$^{-/-}$ mice.[27] In contrast, high doses of pilocarpine were able to induce robust salivation responses in both M_1 and M_3 receptor single KO mice.[27] Similar results were obtained by Nakamura et al.[34] using independently generated mutant mouse lines. Taken together, these studies support the concept that cholinergic stimulation of salivary flow is mediated by a mixture of M_1 and M_3 receptors.

Another investigation demonstrated that carbachol-mediated smooth muscle contractions were virtually abolished in ileal and urinary bladder preparations from mice deficient in both M_2 and M_3 mAChRs (M2R$^{-/-}$/M3R$^{-/-}$ mice).[35] Similar results were obtained with smooth muscle preparations from peripheral airways,[33] trachea,[36] and stomach fundus[36] obtained from M2R$^{-/-}$/M3R$^{-/-}$ mice. In contrast, carbachol-mediated contractile responses were greatly reduced, but not abolished, in smooth muscle preparations from M3R$^{-/-}$ single KO mice.[37–39] These findings indicated that muscarinic agonist-induced smooth muscle contractions involve the activation of both M_2 and M_3 mAChRs. (Note, however, that the M_3 receptor clearly predominates in mediating this activity.)

5.4 RELEVANCE FOR DRUG THERAPY

The phenotypical analysis of mAChR mutant mice revealed that each individual mAChR subtype is involved in mediating multiple central and peripheral physiological responses[13] (Table 5.1). This knowledge should be highly useful in predicting side effects of muscarinic drugs designed to activate or block specific mAChR subtypes. In addition, these studies revealed several potential new drug targets and suggested the use of novel animal models of human disease. Several examples for this are given below.

5.4.1 M₁ RECEPTORS

Interestingly, M1R$^{-/-}$ mice showed a pronounced increase in locomotor activity that was consistently observed in all tests that included locomotor activity measurements.[14,32]

The behavioral pattern displayed by the M1R$^{-/-}$ mice was somewhat reminiscent of human attention deficit/hyperactivity disorder (ADHD), in which hyperactivity is often accompanied by cognitive deficits.[40] The potential use of M1R$^{-/-}$ mice as an animal model of human ADHD clearly deserves further study. Gerber et al.[32] reported that the hyperactivity phenotype of the M1R$^{-/-}$ mice was associated with a significant increase (approximately twofold) in extracellular dopamine concentrations in the striatum, most probably due to an increase in dopamine release. Centrally active M_1 mAChR antagonists may therefore be useful in the treatment of Parkinson's disease, a brain disorder characterized by drastically reduced striatal dopamine levels. Moreover, because schizophrenia is thought to be associated with increased dopaminergic transmission in various forebrain areas, improper signaling through M_1 receptors may contribute to the pathophysiology of certain forms of schizophrenia.[32]

5.4.2 M₂ RECEPTORS

The therapeutic role of muscarinic antagonists able to block cardiac M_2 receptors is well established.[6] Perhaps one of the most interesting findings emerging from the analysis of M2R$^{-/-}$ mice is that activation of M_2 receptors leads to potent analgesic effects, both at the spinal and supraspinal levels.[16,31] Interestingly, activation of M_2 mAChRs located on peripheral nociceptors in the skin may also contribute to the analgesic effects observed after systemic administration of muscarinic agonists.[41] The development of selective M_2 receptor agonists may therefore lead to a new class of analgesic drugs. However, because M_2 receptors mediate the bradycardic effects of ACh on the heart,[6,7,16] such agents might be more suitable for local (e.g., spinal) rather than systemic administration.

5.4.3 M_3 Receptors

Food intake studies showed that M3R[-/-] mice consumed considerably less food than their WT littermates and showed a pronounced decrease in body fat.[17] Additional studies suggested that activation of central M_3 receptors located on a subset of hypothalamic neurons expressing melanin-concentrating hormone (MCH) can stimulate food intake.[17] We, therefore, speculated that centrally active M_3 receptor antagonists might have appetite-suppressing effects and may be potentially useful as antiobesity drugs.

Studies with isolated pancreatic islets prepared from WT and M3R[-/-] mice recently showed that activation of islet M_3 receptors leads to a pronounced potentiation of glucose-dependent insulin release.[42,43] This observation suggests that stimulation of β-cell M_3 receptors or M_3 receptor-regulated downstream signaling components may represent a useful approach to boost insulin output in type 2 diabetes.

As shown in Table 5.1, peripheral M_3 receptors also play important roles in stimulating salivary secretion and contracting smooth muscle tissues. These peripheral actions are predicted to complicate the development of M_3 receptor selective agents for the treatment of clinical disorders, such as obesity or type 2 diabetes.

5.4.4 M_4 Receptors

M4R[-/-] mice displayed a significant increase in sensitivity to the prepulse inhibition (PPI)-disrupting effect of the psychomimetic, phencyclidine, a noncompetitive N-methyl-D-aspartate (NMDA) receptor antagonist.[44] Because phencyclidine-mediated disruption of PPI is often used as an animal model of psychosis, central M_4 receptors may represent a novel drug target for the treatment of schizophrenia and related neurological disorders. Studies with M4R[-/-] mice also indicate that stimulation of M_4 receptors in the skin may facilitate keratinocyte migration and wound reepithelialization.[45] Moreover, several lines of evidence suggest that centrally active M_4 receptor antagonists may be of benefit in the treatment of Parkinson's disease[18] and drug-induced parkinsonism.[46]

Studies with M2R[-/-]/M4R[-/-] mice have shown that M4 receptors can also mediate pronounced analgesic effects, both at the spinal and supraspinal levels.[31] Because activation of peripheral M4 receptors, in contrast to peripheral M2 receptors, is unlikely to cause any serious side effects, the development of selective, centrally active M_4 receptor agonists as novel analgesic drugs appears particularly attractive.

5.4.5 M_5 Receptors

Until recently, very little was known about the physiological functions of the M_5 receptor, which was the last mAChR subtype to be cloned.[47,48] However, we recently demonstrated that ACh virtually lost its ability to dilate cerebral arteries and arterioles prepared from M5R[-/-] mice,[19] indicating that the vasorelaxing effects of ACh on these blood vessels are mediated by M_5 receptors that are probably present on vascular endothelial cells.[49] Several studies suggest that deficits in cortical cholinergic vaso-dilation may play a role in the pathophysiology of Alzheimer's disease[50,51] and that activation of cholinergic vasodilator fibers can reduce neuronal damage during certain forms of focal cerebral ischemia.[52,53] Vascular M_5 mAChRs may, therefore, represent an attractive novel therapeutic target for the treatment of a variety of cerebrovascular disorders.

Behavioral studies demonstrated that M5R[-/-] mice showed reduced sensitivity to the rewarding effects of morphine and cocaine.[30,54] In addition, the severity of morphine and cocaine withdrawal symptoms was shown to be reduced in M5R[-/-] mice.[30,54] Consistent with these observations, *in vivo* microdialysis studies demon-strated that activation of mesolimbic M_5 receptors causes a sustained release of dopamine in the nucleus accumbens.[55] Taken together, these findings suggest that centrally active M_5 receptor antagonists may become therapeutically useful for the treatment of drug addiction.

5.5 CONCLUSIONS AND FUTURE OUTLOOK

In conclusion, the phenotypical analyses of M_1 to M_5 mAChR single KO and various mAChR double KO strains shed new light on the physiological and pathophysio-logical roles of the individual mAChR subtypes. It is likely that the development of mutant mice in which specific mAChR subtypes can be inactivated in a conditional fashion will provide even more powerful research tools. For example, the use of the CRE/loxP recombination system[28] will allow the generation of mutant mouse strains that lack specific mAChRs only in certain tissues or cell types. These studies should pave the way toward the development of novel strategies for the treatment of a variety of important pathophysiological conditions, including pain, obesity, diabetes, and various disorders of the CNS.

ACKNOWLEDGMENTS

I would like to thank all my past and present co-workers and collaborators, especially, Drs. Jesus Gomeza and Masahisa Yamada who played key roles in generating the M_1 to M_5 mAChR single KO mouse lines. I am particularly grateful to Dr. Christian C. Felder and his colleagues at the Eli Lilly Research Laboratories (Indianapolis, IN) for their many major contributions to the phenotypical analysis of the different mAChR mutant mouse strains generated at the NIDDK.

REFERENCES

1. Wess, J., Molecular biology of muscarinic acetylcholine receptors, *Crit. Rev. Neuro-biol.*, 10, 69, 1996.
2. Caulfield, M.P. and Birdsall, N.J.M., International Union of Pharmacology. XVII. Classification of muscarinic acetylcholine receptors, *Pharmacol. Rev.*, 50, 279, 1998.
3. Levey, A.I., Immunological localization of M1–M5 muscarinic acetylcholine recep-tors in peripheral tissues and brain, *Life Sci.*, 52, 441, 1993.

4. Vilaro, M.T., Mengod, G., and Palacios, J.M., Advances and limitations of the molecular neuroanatomy of cholinergic receptors: the example of multiple muscarinic receptors, *Prog. Brain Res.*, 98, 95, 1993.

5. Wolfe, B.B. and Yasuda, R.P., Development of selective antisera for muscarinic cholinergic receptor subtypes, *Ann. N.Y. Acad. Sci.*, 757, 186, 1995.

6. Wess, J. et al., Cholinergic receptors, in *Comprehensive Medicinal Chemistry*, Vol. 3, Emmett, E.C., Ed., Pergamon Press, Oxford, 1990, chap. 12.6.

7. Caulfield, M.P., Muscarinic receptors — characterization, coupling and function, *Pharmacol. Ther.*, 58, 319, 1993.

8. *Proceedings of the Eight International Symposium on Subtypes of Muscarinic Receptors*, Levine, R.R., Birdsall, N.J.M., and Nathanson, N.M., Eds., *Life Sci.*, 64, 355–593, 1999.

9. *Proceedings of the Ninth International Symposium on Subtypes of Muscarinic Receptors*, Levine, R.R., Birdsall, N.J.M., and Nathanson, N.M., Eds., *Life Sci.*, 68, 2449–2642, 2001.

10. Eglen, R.M. et al., Muscarinic receptor ligands and their therapeutic potential, *Curr. Opin. Chem. Biol.*, 3, 426, 1999.

11. Felder, C.C. et al., Therapeutic opportunities for muscarinic receptors in the central nervous system, *J. Med. Chem.*, 43, 4333, 2000.

12. Potter, L.T., Snake toxins that bind specifically to individual subtypes of muscarinic receptors, *Life Sci.*, 68, 254, 2001.

13. Wess, J., Muscarinic acetylcholine receptor knockout mice: novel phenotypes and clinical implications, *Annu. Rev. Pharmacol. Toxicol.*, 44, 423, 2004.

14. Miyakawa, T. et al., Hyperactivity and intact hippocampus-dependent learning in mice lacking the M_1 muscarinic acetylcholine receptor, *J. Neurosci.*, 21, 5239, 2001.

15. Fisahn, A. et al., Muscarinic induction of hippocampal gamma oscillations requires coupling of the M_1 receptor to two mixed cation channels, *Neuron*, 33, 615, 2002.

16. Gomeza, J. et al., Pronounced pharmacologic deficits in M2 muscarinic acetylcholine receptor knockout mice, *Proc. Natl. Acad. Sci. USA*, 96, 1692, 1999.

17. Yamada, M. et al., Mice lacking the M3 muscarinic acetylcholine receptor are hypophagic and lean, *Nature*, 410, 207, 2001.

18. Gomeza, J. et al., Enhancement of D1 dopamine receptor-mediated locomotor stimulation in M4 muscarinic acetylcholine receptor knockout mice, *Proc. Natl. Acad. Sci. USA*, 96, 10,483, 1999.

19. Yamada, M. et al., Cholinergic dilation of cerebral blood vessels is abolished in M5 muscarinic acetylcholine receptor knockout mice, *Proc. Natl. Acad. Sci. USA*, 98, 14,096, 2001.

20. Shapiro, R.A. et al., Isolation, sequence, and functional expression of the mouse M1 muscarinic acetylcholine receptor gene, *J. Biol. Chem.*, 263, 18,397, 1988.

21. van Koppen, C.J., Lenz, W., and Nathanson, N.M., Isolation, sequence and functional expression of the mouse M4 muscarinic acetylcholine receptor gene, *Biochim. Biophys. Acta*, 1173, 342, 1993.

22. Paszty, C. et al., Lethal alpha-thalassaemia created by gene targeting in mice and its genetic rescue, *Nat. Genet.*, 11, 33, 1995.

23. Joyner, A.L., *Gene Targeting: A Practical Approach*, Oxford University Press, Oxford, 1993.

24. Deng, C. et al., Fibroblast growth factor receptor 3 is a negative regulator of bone growth, *Cell*, 84, 911, 1996.

25. Li, E., Bestor, T.H., and Jaenisch, R., Targeted mutation of the DNA methyltransferase gene results in embryonic lethality, *Cell*, 69, 915, 1992.
26. Hamilton, S.E. et al., Disruption of the M_1 receptor gene ablates muscarinic receptor-dependent M current regulation and seizure activity in mice, *Proc. Natl. Acad. Sci. USA*, 94, 13311, 1997.
27. Gautam, D. et al., Cholinergic stimulation of salivary secretion studied with M_1 and M_3 muscarinic receptor single- and double-knockout mice, *Mol. Pharmacol.*, 66, 260, 2004.
28. Lewandoski, M., Conditional control of gene expression in the mouse, *Nat. Rev. Genet.*, 2, 743, 2001.
29. Gerlai, R., Gene-targeting studies of mammalian behavior: is it the mutation or the background genotype? *Trends Neurosci.*, 19, 177, 1996.
30. Basile, A.S. et al., Deletion of the M_5 muscarinic acetylcholine receptor attenuates morphine reinforcement and withdrawal but not morphine analgesia, *Proc. Natl. Acad. Sci. USA*, 99, 11,452, 2002.
31. Duttaroy, A. et al., Evaluation of muscarinic agonist-induced analgesia in muscarinic acetylcholine receptor knockout mice, *Mol. Pharmacol.*, 62, 1084, 2002.
32. Gerber, D.J. et al., Hyperactivity, elevated dopaminergic transmission, and response to amphetamine in M1 muscarinic acetylcholine receptor-deficient mice, *Proc. Natl. Acad. Sci. USA*, 98, 15312, 2001.
33. Struckmann, N. et al., Role of muscarinic receptor subtypes in the constriction of peripheral airways: studies on receptor-deficient mice, *Mol. Pharmacol.*, 64, 1444, 2003.
34. Nakamura, T. et al., M_3 muscarinic acetylcholine receptor plays a critical role in parasympathetic control of salivation in mice, *J. Physiol.*, 558, 561, 2004.
35. Matsui, M. et al., Mice lacking M_2 and M_3 muscarinic acetylcholine receptors are devoid of cholinergic smooth muscle contractions but still viable, *J. Neurosci.*, 22, 10627, 2002.
36. Stengel, P.W. et al., unpublished data, 2002.
37. Matsui, M. et al., Multiple functional defects in peripheral autonomic organs in mice lacking muscarinic acetylcholine receptor gene for the M_3 subtype, *Proc. Natl. Acad. Sci. USA*, 97, 9579, 2000.
38. Stengel, P.W. et al., M_3-receptor knockout mice: muscarinic receptor function in atria, stomach fundus, urinary bladder, and trachea, *Am. J. Physiol. Regul. Integr. Comp. Physiol.*, 282, R1443, 2002.
39. Stengel, P.W. and Cohen, M.L., Muscarinic receptor knockout mice: role of muscarinic acetylcholine receptors M_2, M_3, and M_4 in carbamylcholine-induced gallbladder contractility, *J. Pharmacol. Exp. Ther.*, 301, 643, 2002.
40. Paule, M.G. et al., Attention deficit/hyperactivity disorder: characteristics, interventions and models, *Neurotoxicol. Teratol.*, 22, 631, 2000.
41. Bernardini, N. et al., Muscarinic M2 receptors on peripheral nerve endings: a molecular target of nociception, *J. Neurosci.*, 22, RC229, 1, 2002.
42. Duttaroy, A. et al., Muscarinic stimulation of pancreatic insulin and glucagon release is abolished in M_3 muscarinic acetylcholine receptor-deficient mice, *Diabetes*, 53, 1714, 2004.
43. Zawalich, W.S. et al., Effects of muscarinic receptor type 3 knockout on mouse islet secretory responses, *Biochem. Biophys. Res. Commun.*, 315, 872, 2004.
44. Felder, C.C. et al., Elucidating the role of muscarinic receptors in psychosis, *Life Sci.*, 68, 2605, 2001.

45. Chernyavsky, A.I. et al., Novel signaling pathways mediating reciprocal control of keratinocyte migration and wound epithelialization through M3 and M4 muscarinic receptors, *J. Cell. Biol.*, 166, 261, 2004.

46. Karasawa, H., Taketo, M.M., and Matsui, M., Loss of anti-cataleptic effect of scopolamine in mice lacking muscarinic acetylcholine receptor subtype 4, *Eur. J. Pharmacol.*, 468, 15, 2003.

47. Bonner, T.I. et al., Cloning and expression of the human and rat M5 muscarinic acetylcholine receptor genes, *Neuron*, 1, 403, 1988.

48. Eglen, R.M. and Nahorski, S.R., The muscarinic M5 receptor: a silent or emerging subtype? *Br. J. Pharmacol.*, 130, 13, 2000.

49. Elhusseiny, A. et al., Functional acetylcholine muscarinic receptor subtypes in human brain microcirculation: identification and cellular localization, *J. Cereb. Blood Flow Metab.*, 19, 794, 1999.

50. Geaney, D. et al., Effect of central cholinergic stimulation on regional cerebral blood flow in Alzheimer disease, *Lancet*, 335, 1484, 1990.

51. Tong, X.K. and Hamel, E., Regional cholinergic denervation of cortical microvessels and nitric oxide synthase-containing neurons in Alzheimer's disease, *Neuroscience*, 92, 163, 1999.

52. Scremin, O.U. and Jenden, D.J., Cholinergic control of cerebral blood flow in stroke, trauma and aging, *Life Sci.*, 5, 2011, 1996.

53. Kano, M., Moskowitz, M.A., and Yokota, M., Parasympathetic denervation of rat pial vessels significantly increases infarction volume following middle cerebral artery occlusion, *J. Cereb. Blood Flow Metab.*, 11, 628, 1991.

54. Fink-Jensen, A. et al., Role for M_5 muscarinic acetylcholine receptors in cocaine addiction, *J. Neurosci. Res.*, 74, 91, 2003.

55. Forster, G.L. et al., M5 muscarinic receptors are required for prolonged accumbal dopamine release after electrical stimulation of the pons in mice, *J. Neurosci.*, 22, RC190, 1, 2002.

56. Anagnostaras, S.G. et al., Selective cognitive dysfunction in acetylcholine M1 muscarinic receptor mutant mice, *Nat. Neurosci.*, 6, 51, 2003.

57. Tzavara, E.T. et al., Dysregulated hippocampal acetylcholine neurotransmission and impaired cognition in M2, M4 and M2/M4 muscarinic receptor knock-out mice, *Mol. Psychiat.*, 8, 673, 2003.

58. Takeuchi, J. et al., Increased drinking in mutant mice with truncated M5 muscarinic receptor genes, *Pharmacol. Biochem. Behav.*, 72, 117, 2002.

59. Chino, A. et al., Role of cholinergic receptors in locomotion induced by scopolamine and oxotremorine-M, *Pharmacol. Biochem. Behav.*, 76, 53, 2003.

60. Shapiro, M.S. et al., Assignment of muscarinic receptor subtypes mediating G-protein modulation of Ca^{2+} channels by using knockout mice, *Proc. Natl. Acad Sci. USA*, 96, 10,899, 1999.

61. Ohno-Shosaku, T. et al., Postsynaptic M1 and M3 receptors are responsible for the muscarinic enhancement of retrograde endocannabinoid signalling in the hippocampus, *Eur. J. Neurosci.*, 18, 109, 2003.

62. Fukudome, Y. et al., Two distinct classes of muscarinic action on hippocampal inhibitory synapses: M2-mediated direct suppression and M1/M3-mediated indirect suppression through endocannabinoid signalling, *Eur. J. Neurosci.*, 19, 2682, 2004.

63. Berkeley, J.L. et al., M1 Muscarinic acetylcholine receptors activate extracellular signal-regulated kinase in CA1 pyramidal neurons in mouse hippocampal slices, *Mol. Cell. Neurosci.*, 18, 512, 2001.

64. Hamilton, S.E. and Nathanson, N.M., The M_1 receptor is required for muscarinic activation of mitogen-activated protein (MAP) kinase in murine cerebral cortical neurons, *J. Biol. Chem.*, 276, 15,850, 2001.

65. Bymaster, F.P. et al., Role of specific muscarinic receptor subtypes in cholinergic parasympathomimetic responses, *in vivo* phosphoinositide hydrolysis, and pilocarpine-induced seizure activity, *Eur. J. Neurosci.*, 17, 1403, 2003.

66. Porter, A.C. et al., M1 muscarinic receptor signaling in mouse hippocampus and cortex, *Brain Res.*, 944, 82, 2002.

67. Zhang, W. et al., Characterization of central inhibitory muscarinic autoreceptors by the use of muscarinic acetylcholine receptor knock-out mice, *J. Neurosci.*, 22, 1709, 2002.

68. Slutsky, I. et al., Use of knockout mice reveals involvement of M_2-muscarinic receptors in control of the kinetics of acetylcholine release, *J. Neurophysiol.*, 89, 1954, 2003.

69. Trendelenburg, A.U. et al., Heterogeneity of presynaptic muscarinic receptors mediating inhibition of sympathetic transmitter release: a study with M_2- and M_4-receptor-deficient mice, *Br. J. Pharmacol.*, 138, 469, 2002.

70. Zhang, W. et al., Multiple muscarinic acetylcholine receptor subtypes modulate striatal dopamine release, as studied with M_1–M_5 muscarinic receptor knock-out mice, *J. Neurosci.*, 22, 6347, 2002.

71. Zhou, H. et al., Heterogeneity of release-inhibiting muscarinic autoreceptors in heart atria and urinary bladder: a study with M_2- and M_4-receptor-deficient mice, *Naunyn Schmiedebergs Arch. Pharmacol.*, 365, 112, 2002.

72. Tzavara, E.T. et al., M_4 muscarinic receptors regulate the dynamics of cholinergic and dopaminergic neurotransmission: relevance to the pathophysiology and treatment of related central nervous system pathologies, *FASEB J.*, 18, 410, 2004.

73. Hardouin, S.N. et al., Altered cardiovascular responses in mice lacking the M_1 muscarinic acetylcholine receptor, *J. Pharmacol. Exp. Ther.*, 301, 129, 2002.

74. Stengel, P.W. et al., M_2 and M_4 receptor knockout mice: muscarinic receptor function in cardiac and smooth muscle *in vitro*, *J. Pharmacol. Exp. Ther.*, 292, 877, 2000.

75. Fisher, J.T. et al., Loss of vagally mediated bradycardia and bronchoconstriction in mice lacking M_2 or M_3 muscarinic acetylcholine receptors, *FASEB J.*, Electronically published ahead of print, February 20, 2004.

76. Matsui, M. et al., Increased relaxant action of forskolin and isoproterenol against muscarinic agonist-induced contractions in smooth muscle from M_2 receptor knockout mice, *J. Pharmacol. Exp. Ther.*, 305, 106, 2003.

77. Griffin, M.T. et al., Muscarinic agonist-mediated heterologous desensitization in isolated ileum requires activation of both muscarinic M_2 and M_3 receptors, *J. Pharmacol. Exp. Ther.*, 308, 339, 2004.

78. Lamping, K.G. et al., Muscarinic (M) receptors in coronary circulation: gene-targeted mice define the role of M_2 and M_3 receptors in response to acetylcholine, *Arterioscler. Thromb. Vasc. Biol.*, 24, 1253, 2004.

79. Khurana, S. et al., Vasodilatory effects of cholinergic agonists are greatly diminished in aorta from $M_3R^{-/-}$ mice, *Eur. J. Pharmacol.*, 493, 127, 2004.

80. Aihara, T. et al., Impaired gastric secretion and lack of trophic responses to hypergastrinemia in M_3 muscarinic receptor knockout mice, *Gastroenterology*, 125, 1774, 2003.

81. Nguyen, V.T. et al., Synergistic control of keratinocyte adhesion through muscarinic and nicotinic acetylcholine receptor subtypes, *Exp. Cell Res.*, 294, 534, 2004.

82. Hemrick-Luecke, S.K. et al., Muscarinic agonist-mediated increases in serum corticosterone levels are abolished in M_2 muscarinic acetylcholine receptor knockout mice, *J. Pharmacol. Exp. Ther.*, 303, 99, 2002.

83. Boudinot, E. et al., Ventilatory pattern and chemosensitivity in M1 and M3 muscarinic receptor knockout mice, *Respir. Physiol. Neurobiol.*, 139, 237, 2004.

6 Systematic Mutagenesis of M₁ Muscarinic Acetylcholine Receptors

Edward C. Hulme

CONTENTS

6.1 INTRODUCTION

6.1.1 G PROTEIN-COUPLED RECEPTORS

The seven-transmembrane (7-TM) G protein-coupled receptors (GPCRs) form the largest superfamily of transmembrane signaling molecules in the mammalian genome. There are over 600 GPCRs in humans, and they provide the targets for 50% of all of the drugs in clinical use.[1–3]

The rhodopsin-like family (family A) is the biggest subgroup. It contains about 300 distinct receptors for endogenous neurotransmitters, hormones, and modulators as well as numerous receptors for exogenous odorants and tastants. It is supplemented by smaller receptor families that share the 7-TM fold but have distinct primary sequences: In mammals, these are family B exemplified by the secretin and glucagon receptors; family C, containing the metabotropic glutamate and $GABA_B$ receptors; and family D, the frizzled receptors.

The 7-TM receptors are transmembrane switches. Their fundamental property is to undergo a transition from an inactive state to an active state triggered by the binding of the cognate agonist from the extracellular milieu. This causes breakage of certain intramolecular bonds that characterize the inactive state, and their replacement by a new set of bonds, incorporating the agonist into the structure, that stabilize the activated state. The structural change is read out on the intracellular surface of the membrane, where the active conformation of the receptor acts as a dissociable allosteric subunit of a complex with a specific heterotrimeric combination of G protein subunits The agonist–receptor complex catalyzes guanosine diphosphate (GDP)–guanosine triphosphate (GTP) exchange at the α-subunit, so modifying or breaking its tight interaction with the $\beta\gamma$ subunit.[4,5] The GTP-bound α subunit, and the $\beta\gamma$ subunit typically activate different signaling pathways, until turned off by the GTP-ase activity of the α-subunit. This allows them to revert to the inactive state.

Rhodopsin is the only 7-TM receptor for which direct structural information has been obtained, initially by a pioneering sequence of cryoelectron microscopy and crystallography experiments on two-dimensional crystals of bovine and frog rhodopsin, which were interpreted in the context of multiple 7-TM receptor sequence alignments,[6,7] and subsequently by a landmark x-ray diffraction analysis of three-dimensional crystals of bovine rhodopsin (in the inactive form), leading to a structure at 2.8 Å.[8,9] This has recently been extended to 2.2 Å.[10]

At present, the structure of rhodopsin provides the only template for modeling the other 7-TM receptors. However, it should be cautioned that even within family A, the rhodopsins form a separate group, distinguishable from the mainstream both by sequence analysis[11] and by the presence of introns in the coding sequence (several neuropeptide and peptide hormone receptor genes also contain introns). This may imply that even the transmembrane domain shows functionally significant structural differences in different receptors. Furthermore, although they all bind and activate G proteins, which are highly conserved, the ligand binding modalities of the family A receptors show wide variations, ranging from the interaction of a small molecule with a site within the transmembrane helices to the binding of a large glycoprotein hormone to a specialized extracellular domain. In fact, rhodopsin is the only GPCR with its ligand, 11-cis retinal, covalently immobilized within the binding pocket, acting as an inverse agonist until isomerized by the absorption of a photon of light. In other receptors, the agonist must diffuse into the binding pocket, bind, and then dissociate during the activation and deactivation cycle.

Thus, while a canonical transmembrane switch mechanism may exist,[12] the members of family A are expected to exhibit particular adaptations related to their ligand binding properties and the transmission of conformational information from the binding site to the transmembrane region, as well as to G protein recognition and cellular regulation. These differences must be addressed experimentally in each case. In the absence of an x-ray structure, systematic site-directed mutagenesis experiments interpreted in the context of a good homology model enable progress to be made. In this chapter, a description of how Alanine-scanning mutagenesis experiments have been used in my laboratory to help us to understand the structure and function of the M_1 muscarinic acetylcholine receptor (mAChR) will be provided.

6.1.2 THE MUSCARINIC ACETYLCHOLINE RECEPTORS

The mAChRs were among the first neurotransmitter receptors to be defined pharmacologically. They have been investigated systematically for over 100 years. Cloning and sequencing of the five muscarinic receptor genes followed shortly after the cloning of the beta-adrenergic receptor[13] and showed that the receptors shared essential sequence motifs with rhodopsin, which had been sequenced earlier by protein chemical techniques.[14]

The five mAChR subtypes fall into two groups.[15] M_1, M_3, and M_5 mAChRs activate G proteins of the G_q/G_{11} class. Activated Gq_α stimulates phospholipase C-$\beta 1$ isoforms, breaking down PIP_2 to diacylglycerol (DAG) and inositol trisphosphate (IP_3). DAG stimulates protein kinase C, IP_3 mobilizes intracellular calcium, while depletion of membrane PIP_2 modulates slow ion conductances, for instance, the M potassium current, and certain calcium conductances. M_2 and M_4 mAChRs couple primarily to G proteins of the G_i and G_o class. Activated Gi_α inhibits adenylyl cyclase, while the $\beta\gamma$ subunits activate fast inward rectifier potassium channels, as well as inhibit voltage-sensitive, calcium channels. All of the mAChR subtypes can activate ERK kinase cascades.

Gene knockout studies have clarified the physiological roles of the mAChR subtypes.[16] In the periphery, M_2 mAChRs inhibit cardiac contraction, while M_3

mAChRs promote smooth muscle contraction and salivation. In the CNS, M_2 and M_4 inhibit presynaptic ACh release, interact with dopamine to control locomotor behavior, and mediate cholinergic analgesia. M_5 dilates cerebral arterioles and modulates the rewarding effects of dopamine release in the midbrain. M_1 receptors mediate virtually all of the ACh-stimulated PIP_2 breakdown and MAP kinase activation in the hippocampus and cerebral cortex[17] and activate ion channels that mediate prolonged oscillations in hippocampal circuits. M_1 mAChR stimulation can promote the cell-surface ectodomain shedding of growth factors and amyloid-precursor protein. These phenomena provide a link between muscarinic receptors, synaptic plasticity, cognition, pathogenesis, and psychosis. Their rich physiology renders mAChRs a rewarding target for drug development.

The mAChR subtypes can be distinguished pharmacologically, but the selectivity of ligands interacting with the acetylcholine binding site is limited,[18] because sequence conservation restricts the potential for diverse binding interactions. Higher selectivity may be achieved by allosteric ligands and muscarinic toxins acting outside the primary binding site, although the molecular details of their binding remain poorly defined.[19]

The free energy of the activating conformational change appears to differ slightly between the mAChR subtypes.[20] The origin of these differences has yet to be investigated systematically, but they would be consistent with small sequence differences in the vicinity of the important interhelical contacts.

6.2 MUTAGENESIS STUDIES

At the outset of any mutagenesis study, a decision needs to be made about the path to follow. If several genetic subtypes of the protein are known that diverge in a particular property, then chimeric proteins may be constructed to try to narrow down the amino acid sequences responsible for the differences. If a three-dimensional structure of the target protein is available, then a selected set of point mutants may be constructed to test a specific hypothesis, for instance, whether particular amino acids participate in ligand interactions. These may be characterized as hypothesis-based experimental strategies.

In contrast, there are more inclusive and exploratory experimental approaches that do not assume prior knowledge. They include the various forms of scanning and random mutagenesis. Here the aim is either systematically to replace all of the amino acids in a target sequence by a particular residue that was preselected for its properties, or by amino acids drawn from a library of sequences encoded by oligonucleotides that were synthesized with random substitutions. The mutant proteins are expressed and characterized by a panel of functional measures. In the case of variants generated by random mutagenesis, the sequences are determined after phenotypic selection.[21–23]

Different scanning mutagenesis strategies have been used. The earliest to be introduced, and still arguably the most informative, was Alanine-scanning. This technique was pioneered in the receptor field in studies of the human growth hormone receptor.[24,25] Mutation to Ala deletes the side chain of the amino acid beyond the beta carbon atom, abolishing the interactions these atoms may make, whether intra- or intermolecular. The substitution may leave a small cavity in the three-dimensional structure of the protein, which, in most cases (although not always), does not propagate significant structural disturbances beyond the local second-shell side chains.[26] In our work on the M₁ muscarinic receptor, Ala-substitution has been well-tolerated, even when the substituted residue is buried in a functionally critical position in the structure. A drawback is that Ala is not a particularly good replacement for the structurally important residues Gly and Pro. In our experiments, Ala was replaced by Gly, although this runs the risk of introducing extra flexibility into the protein backbone.

Mapping the effects of Ala-substitution mutants onto the protein sequence can give information about the secondary structure of the target sequence, or even the tertiary fold of the protein. Ala-scanning has been particularly valuable in mapping binding interfaces in proteins, where it identifies "hot spots" of binding energy,[27] amino acids whose mutation leads to a substantial (e.g., 10-fold) drop in binding affinity. These tend to be clustered in the center of the structural binding interface and may be surrounded by a ring of residues with the function of excluding solvent. Hot-spot residues at ligand binding interfaces are often structurally conserved and were recently identified with some success by computational Ala-scanning.[28] In an extension of these studies, we found interactions between key residues in different elements of the transmembrane domain of the M₁ mAChR, which may help to transmit the conformational changes induced by agonist binding.

6.3 THE MECHANISTIC FRAMEWORK FOR THE INTERPRETATION OF MUTAGENESIS STUDIES ON GPCRs

The activation of a G protein by a 7-TM receptor is a multistep process. The parameters that can be measured experimentally are combinations of the underlying intrinsic rate or affinity constants characterizing the mechanism. So, to understand the significance of a mutation-induced change, it is necessary to attempt to relate the effects observed to the underlying molecular events using a mechanistic model.

To be useful, a mechanistic model should provide a basis for the quantitation of the effects observed. Even the simplest model of receptor activation of a G protein, the ternary complex model, continues to prove useful in this respect.

6.3.1 THE TERNARY COMPLEX MODEL OF RECEPTOR–G PROTEIN INTERACTION

In the ternary complex mechanism, the binding of the agonist to the receptor is distinct from the process of receptor binding to the G protein and its subsequent activation (Figure 6.1a).[29] The first assumption made is that there is a membrane-delimited reversible association between the receptor and the inactive GDP-ligated G protein, characterized by an apparent affinity constant K_G. The second is that the RG complex is the form that undergoes activation, by catalyzed dissociation of GDP, and rapid binding of GTP from the high relative concentration available in the intracellular milieu, a process crudely characterized by a catalytic rate constant K_{cat}. Ligand binding to the receptor, described by an affinity constant K_{bin}, influences this process by changing K_G. The ligand effect is expressed by a cooperativity factor α, greater than one for agonists, and less than one for inverse agonists. Thermodynamic reversibility demands that the same cooperativity factor be applied to the affinity of the ligand for the RG complex relative to the free receptor. Thus, the binding of an agonist promotes the activation of the G protein by enhancing the formation of receptor–G protein complexes, and agonist molecules bind to pre-existing RG complexes with a higher affinity than to free receptor molecules. The converse applies to inverse agonists, while the binding of neutral antagonists is neither influenced by nor influences the binding of the G protein. Activation is considered to be accompanied by the dissociation of the receptor from the GTP-ligated G protein. Inactivation of the G protein proceeds by hydrolysis of the bound GTP as a result of the intrinsic GTPase activity of the α-subunit, governed by a rate constant K_h. This may be accelerated greatly (>100-fold) by association with effector molecules, such as phospholipase C-β1, or accessory proteins such as RGS proteins.[30,31]

Even this simplest mechanism makes useful predictions. The first is that a combination of receptor and G protein is expected to exhibit basal signaling activity with a level governed by the concentrations of the two proteins in their membrane compartment, and by the value of K_G. The second is that while the binding affinity of a ligand is governed by K_{bin}, its potency in signaling and the magnitude of the maximum signal achieved will also be influenced by αK_G, as well as the relative and absolute concentrations of receptor and G protein.

The equilibrium concentrations of active G protein complexes predicted by the ternary complex model can be expressed mathematically. This is most simply achieved when the total concentration of functional receptor, R_T is substantially (>threefold) greater than the total concentration of functional G protein, G_T. Initially, the G protein is assumed to be in the inactive GDP-bound form. Catalytically active forms are generated by binding to R or AR.

$$\frac{RG_{GDP} + ARG_{GDP}}{G_{TGDP}} = \varphi\left(R_T, A\right) = \frac{K_G.R_T + \alpha K_G.R_T.K_{bin}.A}{1 + K_G.R_T + K_{bin}.A + \alpha K_G.R_T.K_{bin}.A} \quad (6.1)$$

Here, G_{TGDP} represents the total concentration of GDP-ligated G protein at the steady state. This equation is similar to Equation 22 of Black and Leff.[32]

To calculate the functional response of the system, the balance between the rates of activation and inactivation of the G protein must be considered. At the steady state, these rates are equal.

$$K_{cat}\varphi\left(R_T, A\right).\left(G_T\text{-}G_{GTP}\right) = K_h.G_{GTP}$$

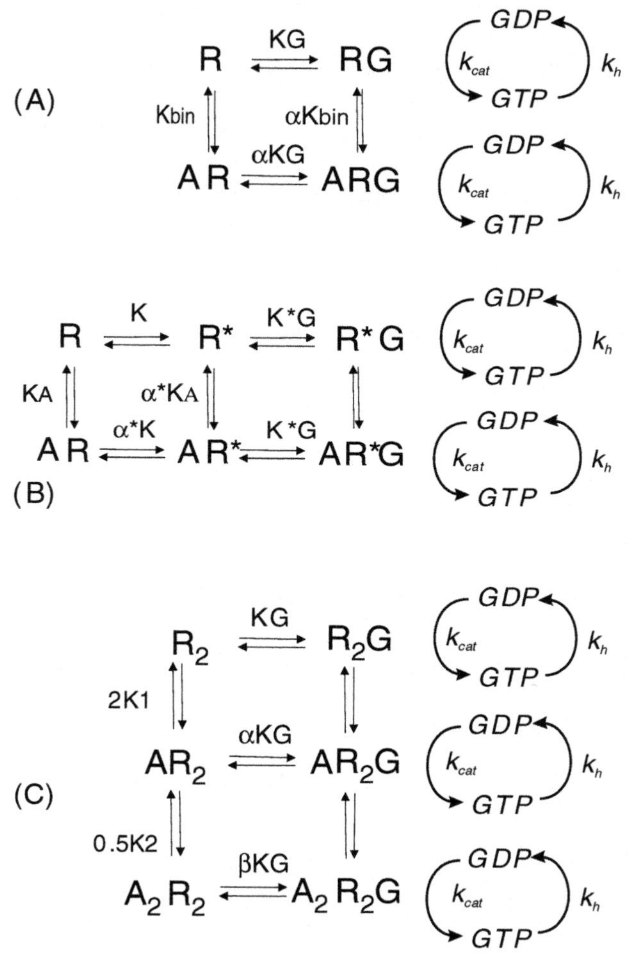

FIGURE 6.1 G protein activation by GPCRs. (a) The simple ternary complex model. (b) The extended ternary complex model. (c) The dimeric receptor ternary complex model.

The fraction of the total G protein existing in an active state is given by

$$G_{GTP}/G_T = K_{cat}\varphi/\left(K_{cat}\varphi + K_h\right))$$ (6.2)

Substituting from Equation (6.1) into Equation (6.2), we obtain

$$\underline{G}_{GTP} = \frac{K_{cat}/K_h\left(\underline{K}_G.\underline{R}_T + \alpha\underline{K}_G.\underline{R}_T.K_{bin}.A\right)}{1+\left(1+K_{cat}/K_h\right).\underline{K}_G.\underline{R}_T + K_{bin}.A+\left(1+K_{cat}/K_h\right).\alpha\underline{K}_G.\underline{R}_T.K_{bin}.A}$$ (6.3)

In this equation, we introduce a convention whereby an underscore represents a *ratio* of the concentration of a complex to the total concentration of G protein ($\underline{G}_{GTP}$, $\underline{R}_T$), or a *product* of an affinity constant with the total concentration of G protein ($\underline{K}_G$). This convention is applied to species and the affinity constants governing their reactions, in the membrane compartment where absolute values cannot be estimated but relative values are easily understood. Thus, $\underline{R}_T$ can be regarded as the ratio of functional receptor to functional G protein in the cell membrane, while $\underline{K}_G$ represents the avidity of the G protein for the receptor in the membrane compartment; doubling the concentration of G protein will increase $\underline{K}_G$ by a factor of 2.

By setting $A = 0$, the basal activity of the system is

$$B_{asal} = \frac{\left(K_{cat}/K_h\right).\underline{K}_G.\underline{R}_T}{1+\left(1+K_{cat}/K_h\right).\underline{K}_G.\underline{R}_T}$$

and by setting A at a receptor-saturating concentration, the maximum response is

$$E_{max} = \frac{\left(K_{cat}/K_h\right).\alpha\underline{K}_G.\underline{R}_T}{1+\left(1+K_{cat}/K_h\right).\alpha\underline{K}_G.\underline{R}_T}$$

The reciprocal of the concentration of ligand at which the half-maximal response is obtained is

$$K_{act} = 1/EC_{50} = \frac{K_{bin}.\left(1+\left(1+K_{cat}/K_h\right).\alpha\underline{K}_G.\underline{R}_T\right)}{1+\left(1+K_{cat}/K_h\right).\underline{K}_G.\underline{R}_T} \tag{6.4}$$

Using these definitions, the response of the system can be expressed as

$$E = \underline{G}_{GTP} = \frac{B_{asal}+E_{max}.K_{act}.A}{1+K_{act}.A}$$

Activation of the whole of the G protein pool by an agonist acting at a particular receptor can only be achieved if $K_{cat} \gg K_h$. If this is not the case, then the maximum activation attainable by a full agonist is $K_{cat}/(K_{cat} + K_h)$. Reconstitution experiments carried out with purified M_1 mAChRs, G_q and phospholipase C- β1 have shown that the phospholipase has strong GTPase activating activity.[30,31] In these experiments, K_h was about 10 s^{-1}, and K_{cat} about 1.5 s^{-1} at 30°C. Assuming that these relative values are preserved in the cellular environment, the value of $K_{cat}/(K_{cat} + K_h)$ is 0.15, suggesting that the wild-type receptor may not be capable of fully activating the G protein population. This leaves room for discovering mutants that increase, as well as decrease, the E_{max} of the receptor. It may also mean that the value of $\underline{K}_G$ should be considered to relate to the G_q-PLC-β1 complex during the steady-state reaction.

In the case of the wild-type M_1 mAChR, expressed in COS-7 cells, K_{act}/K_{bin}, the ratio of the agonist concentration necessary to give 50% occupancy of the receptor binding sites to that needed to give 50% of the maximum functional response is about 100. Because $(1 + K_{cat}/K_h) \sim 1.15$, the expression for K_{act} requires that $\alpha \underline{K}_G.\underline{R}_T \sim 100$. This being so, the wild-type maximum response, $E_{max,wt}$, is directly proportional to $K_{cat,wt}/(K_{cat,wt} + K_h)$.

A natural way to express functional dose–response data for a single agonist acting at a series of mutant receptors (or for different agonists acting at the same receptor) is by normalization by division by $E_{max,wt}$. Then, Equation (6.3) becomes

$$E/E_{max.wt} = \frac{\left(1 + K_h/K_{cat.wt}\right).\left(K_{cat}/K_h\right).\left(\underline{K}_G.\underline{R}_T + \alpha \underline{K}_G.\underline{R}_T.K_{bin}.A\right)}{1 + \left(1 + K_{cat}/K_h\right).\underline{K}_G.\underline{R}_T + K_{bin}.A + \left(1 + K_{cat}/K_h\right).\alpha \underline{K}_G.\underline{R}_T.K_{bin}.A} \quad (6.5)$$

If $K_{cat} = K_{cat,wt}$, this equation simplifies to

$$E/E_{max.wt} = \frac{\left(1 + K_{cat}/K_h\right).\left(\underline{K}_G.\underline{R}_T + \alpha \underline{K}_G.\underline{R}_T.K_{bin}.A\right)}{1 + \left(1 + K_{cat}/K_h\right).\underline{K}_G.\underline{R}_T + K_{bin}.A + \left(1 + K_{cat}/K_h\right).\alpha \underline{K}_G.\underline{R}_T.K_{bin}.A} \quad (6.6)$$

which gives

$$B_{asal}/E_{max.wt} = \frac{\left(1 + K_{cat}/K_h\right).\underline{K}_G.\underline{R}_T}{1 + \left(1 + K_{cat}/K_h\right).\underline{K}_G.\underline{R}_T} \quad (6.7)$$

and

$$E_{max}/E_{max.wt} = \frac{\left(1 + K_{cat}/K_h\right).\alpha \underline{K}_G.\underline{R}_T}{1 + \left(1 + K_{cat}/K_h\right).\alpha \underline{K}_G.\underline{R}_T} \quad (6.8)$$

while the expression for K_{act} remains as before.

It is natural to define the basal signaling efficacy of the unoccupied receptor as

$$e_0 = \left(1 + K_{cat}/K_h\right).\underline{K}_G \quad (6.9)$$

and the signaling efficacy of the agonist-occupied receptor as

$$e_A = \left(1 + K_{cat}/K_h\right).\alpha \underline{K}_G \quad (6.10)$$

The reconstitution experiments discussed above suggest that $(1 + K_{cat}/K_h)$ for the wild-type receptor is 1.15 (approximately equal to 1) in Equations (6.4) through

(6.10). This implies that for the M_1 mAChR, the dominant factors determining basal and agonist-stimulated signaling efficacy and the value of K_{act} should be those associated with the agonist–receptor and receptor–G protein binding interactions.

With K_{act} estimated from the EC_{50} and B_{asal} and E_{max} *defined relative to the wild-type maximum signal*, as in Equations (6.5) and (6.6), e can be estimated, provided that we also have an estimate of the concentration of functional receptor, either by the expression

$$e_A = \left(K_{act} \Big/ \left(K_{bin} \cdot \left(1 - B_{asal} \right) \right) - 1 \right) \Big/ \underline{R}_T \tag{6.11}$$

or by the expression

$$e_A = \left(E_{max} \Big/ \left(1 - E_{max} \right) \right) \Big/ \underline{R}_T \tag{6.12}$$

while

$$e_0 = \left(B_{asal} \Big/ \left(1 - B_{asal} \right) \right) \Big/ \underline{R}_T \tag{6.13}$$

In practice, Equation (6.11) is useful when the functional dose–response curve lies well to the left of the binding curve, and the relative E_{max} is close to 1.0, while Equation (6.12) is useful when E_{max} is significantly less than 1, which usually means that the functional dose–response curve lies close to the binding curve. This follows from the relationship, derived by equating Equations (6.11) and (6.12):

$$E_{max} = 1 - \left(K_{bin} \left(1 - B_{asal} \right) \Big/ K_{act} \right) \tag{6.14}$$

As will be seen, this relationship is well obeyed by nearly all M_1 mAChR mutants. When exceptions are found, the implication is that K_{cat} for the mutant cannot be assumed to be the same as K_{cat} for the wild-type receptor. These exceptional cases will be considered later. Equations (6.11) and (6.12) are related to those derived for the β-adrenergic receptor adenylyl cyclase interaction by Whaley et al.[33]

Using these experimentally derived parameters, the dose–response curve can be expressed in the following form:

$$E = \frac{B_{asal} + E_{max} \left(1 - B_{asal} \right) \left(1 + e_A \cdot \underline{R}_T \right) \cdot K_{bin} \cdot A}{1 + \left(1 - B_{asal} \right) \left(1 + e_A \cdot \underline{R}_T \right) \cdot K_{bin} \cdot A}$$

Note that from Equation (6.12),

$$E_{max}/E_{max,wt} = e_A.\underline{R}_T/\left(1 + e_A.\underline{R}_T\right)$$

The above analysis is subject to the assumption that $\underline{R}_T$, the ratio of functional receptor to functional G protein, is significantly greater than one. When this ceases to be the case ($\underline{R}_T \le 2$), the expression given in Equation (6.1) for the fraction of GDP-ligated G protein in the form of catalytically competent receptor-bound complexes takes a quadratic form:

$$\frac{\left(RG_{GDP} + ARG_{GDP}\right)}{G_{TGDP}} = \varphi\left(R_T, A\right) = \frac{-T3 + \sqrt{\left(T3^2 + 4T1.T2.\underline{K}_G.\underline{R}_T\right)}}{2T1 - T3 + \sqrt{\left(T3^2 + 4T1.T2.\underline{K}_G.\underline{R}_T\right)}} \quad (6.15)$$

where $T1 = 1 + K.A$; $T2 = 1 + \alpha K.A$; $T3 = T1 + \underline{K}_G.(1 - \underline{R}_T).T2$.

This expression can be inserted into Equation (6.2) to provide a general description of the catalytic process analogous to Equation (6.3).

6.3.2 THE EXTENDED TERNARY COMPLEX MODEL

The simple ternary complex model provides no insight into the mechanistic connection between ligand binding to the receptor and binding of the G protein. The link is provided by the multistate model of receptor activation.[34] In the simplest case, the receptor is predisposed to exist in two states: an active (R*) state, and an inactive (R) state (Figure 6.1b). These are assumed to be in equilibrium with one another, governed by an activation constant K. As before, ligands that bind to the receptor may favor the activated state, the inactive state, or manifest neutral properties. Again, this is expressed quantitatively by a cooperativity factor $\alpha*$ (>1 for agonists, equal to 1 for neutral antagonists, <1 for inverse agonists).

In the context of the two-state model, the binding constant of a ligand for the receptor *that we can measure experimentally* is given by an expression that reflects both the underlying ligand interaction and the energetics of the induced conformational change:

$$K_{bin} = K_A\left(1 + \alpha* K\right)/\left(1 + K\right)$$

K_{bin} depends not only on K_A, the affinity of the agonist for the inactive state of the receptor, but also on the conformational constant K, and the cooperativity factor $\alpha*$.

In the case of the wild-type M_1 mAChR, we conclude that $K \ll 1$ because there is very little basal activity, so

$$K_{bin} \approx K_A\left(1 + \alpha* K\right)$$

Incorporating the two-state model of receptor activation into the G protein activation cycle leads to the extended ternary complex model of agonist–receptor–G protein interaction (Figure 6.1b). This was first proposed to account for the phenomenon of agonist-independent signaling induced by particular mutations.[35] Its value is that it allows us to assign the mutant phenotypes that we observe to functional categories.

It is proposed that only the activated states of the receptor, R* and AR* bind the G protein, with affinity K^*_G, and catalyze GDP–GTP exchange, leading to downstream signaling. We make the simplifying assumption that these two states have equal catalytic activities; in reality, there is now evidence that multiple activated states of 7-TM receptors are possible.[36] Assuming that $K \ll 1$, the equivalent of Equation (6.1) is

$$\underline{R*G}_{GDP} + \underline{AR*G}_{GDP} = \frac{K.\underline{K}^*_G.\underline{R}_T + \alpha * K.\underline{K}^*_G.\underline{R}_T.K_A.A}{\left(1 + K.\underline{K}^*_G.\underline{R}_T + K_{bin}.A + \alpha * K.\underline{K}^*_G.\underline{R}_T.K_A.A\right)} \quad (6.16)$$

The form of Equations (6.11 to 6.13) for B_{asal}, E_{max} and signaling efficacy remain unaltered, but as in the case of the ligand binding constant, the unitary parameters used in the simple ternary complex model are revealed to be composite parameters dependent on the underlying molecular processes in the extended ternary complex model. Thus,

$$\underline{K}_G \quad \text{becomes} \quad \underline{K}^*_G.K/(1+K) \approx K\underline{K}^*_G$$

$$\alpha \quad \text{becomes} \quad \alpha */(1+\alpha * K)$$

$$\alpha\underline{K}_G \quad \text{becomes} \quad \underline{K}^*_G.\alpha K/(1+\alpha * K)$$

so, for instance,

$$e_A = \frac{\left(1 + K_{cat}/K_h\right)\alpha * K.\underline{K}^*_G}{\left(1+\alpha * K\right)}$$

Interpreting this expression, the signaling efficacy of the agonist-occupied receptor is the effective affinity of the G protein for the activated state of the receptor weighted by the fraction of receptors that are in the activated state when the receptors are fully occupied by the agonist and multiplied by $(1 + K_{cat}/K_h)$.

6.3.3 THE DIMERIC RECEPTOR TERNARY COMPLEX MODEL

Family C GPCRs, such as the metabotropic glutamate, and GABA B receptors are constitutive dimers in which both the receptor subunits have a function in signal transduction.[37] Increasingly, evidence suggests that dimerization may also be a feature of family A receptors, possibly acting as a quality control mechanism regulating their export from the endoplasmic reticulum to the plasma membrane.[38,39]

There is little evidence that dimer formation is a dynamic process regulated by receptor ligands. The simplest modification of the ternary complex scheme in which a preexisting receptor dimer binds to and activates a single G protein is shown in Figure 6.1c.

In principle, the binding of agonist molecules can occur in two sequential steps, which can exhibit either positive or negative cooperativity, leading to steep or flat binding curves. This is modeled by allowing different affinity constants K_1 and K_2 for the successive binding steps. Flat curves, showing high- and low-affinity binding sites, are a well-known phenomenon associated with agonist binding to 7-TM GPCRs. Traditionally, high-affinity binding has been attributed to the association of a fraction of the receptors with an effector molecule, such as a G protein, but in fact, high-affinity agonist binding is often partly or completely insensitive to the presence of GTP analogues; this is true in our studies of wild-type and mutant M_1 and M_2 mAChRs expressed in COS-7 cells. If analyzed using a two-site model of binding, a binding isotherm arising from negative cooperativity in a homodimer should give equal fractions of high- and low-affinity sites, with apparent affinity constants K_H and K_L. It is easily shown that $K_H + K_L = 2K_1$, while $K_H.K_L = K_1.K_2$.

In principle, a dimer in which one agonist binding site is occupied may have a different affinity for the G protein than the doubly occupied dimer. This is expressed by allowing different cooperativity factors α and β for the singly and doubly occupied receptor. This can lead to a complex dose–response curve, for instance, a bell-shaped curve if $\beta < \alpha$, even if the different forms of the RG complex have equal catalytic activities. Such behavior has been reported, but we have not seen it in our studies of the PI response of the M_1 mAChR.

The general expression for the relative fraction of GDP-ligated G protein in the form of catalytically competent receptor-bound complexes R_2G, AR_2G, and A_2R_2G is the same as Equation (6.15), with different definitions of T1 and T2 to reflect the presence of two ligand binding sites within the functional unit; $T1 = 1 + 2K1.A + K1.K2.A^2$; $T2 = 1 + 2\alpha K1.A + \beta K1.K2.A^2$. Again, this expression can be inserted into Equation (6.2). It should be noted that there is not necessarily any unitary measure of agonist signaling efficacy in this case, because the singly and doubly agonist-occupied forms of the receptor, in general, have different efficacies. The effect of the more complex binding isotherm is that the slope of the functional dose–response curve follows that of the binding curve, for instance, it is flat if the underlying binding curve is flat.

6.4 EXPERIMENTAL TECHNIQUES

6.4.1 CHOICE OF TARGET SEQUENCES

In our studies, we tried to target entire domains by Ala-scanning rather than "cherry picking" selected residues. In this way, we strove to avoid imposing our prejudices on the sequence. Once the significant residues have been identified, and their phenotypes established and interpreted in a structural context, we often attempt to follow up by making additional point or combinatorial mutations to test specific hypotheses.

In this context, it should be remembered that it is the overall pattern of a mutational scan that is informative, and that null mutations are an important part of that pattern.

The ternary complex analysis tells us that in order to quantitate the effects of mutations on muscarinic receptor binding and signaling, we need to measure the following basic set of parameters for each mutant that is expressed:

1. The affinity of the mutant receptor for a high-affinity antagonist: In the case of the M_1 mAChRs, this is the classical atropine-like antagonist (-)-[^{3}H]-N-methyl scopolamine ([^{3}H]NMS; $K_{bin} = 10^{10}$ M^{-1} for the wild-type receptor); if this does not bind with adequate affinity, we will try another radioligand, typically (-)-[^{3}H]-3-quinuclidinyl benzilate ([^{3}H]QNB; $K_{bin} = 3 \times 10^{10}$ M^{-1}).

2. The expression level of the mutant receptor, relative to the wild type, calculated from the B_{max} for [^{3}H]NMS, or [^{3}H]QNB: From this measurement, we estimate $\underline{R_T}$.

3. The affinity, K_{bin}, of the mutant for the agonist acetylcholine by competition with [^{3}H]NMS or [^{3}H]QNB.

4. The EC_{50} of ACh in eliciting the phosphoinositide (PI) response: This provides a measure of functional potency, defined as $K_{act} = 1/EC_{50}$, which we can use to estimate ACh signaling efficacy.

5. The maximum PI response to ACh, E_{max}: This is also used in the calculation of agonist signaling efficacy. An inconsistent value may indicate a change in the catalytic rate constant of the agonist–receptor–G protein complex.

6. The level of basal PI signaling activity, expressed as a percentage of the wild-type E_{max}: This may indicate whether the underlying conformational constant of the receptor has changed.

6.4.2 EXPRESSION VECTOR

In our experiments, the rat M_1 mAChR coding sequence (1380 bp) was expressed under the control of an SV40 promoter that allows high copy number replication in mammalian cells such as COS-7 that express the SV40 large T antigen. It also contains a PBR322 origin of replication and an ampicillin-resistance gene for propagation in *Escherichia coli*.[40] The DNA sequence was modified to ensure the presence of well-spaced unique restriction sites at ca. 400 base pair intervals, without alteration of the amino acid sequence, to facilitate the subcloning of mutated fragments.

6.4.3 SITE-DIRECTED MUTAGENESIS

For the introduction of point mutants, we originally used the PCR-based method of splicing by overlap extension.[41] This requires two sets of primers, first, sense and antisense "outer" primers, lying respectively 5' and 3' to a convenient pair of restriction sites bracketing the target sequence, and a pair of complementary "inner" primers encoding the point mutant in a central position. These oligos do not need to be 5′-phosphorylated. Separate PCR reactions are carried out using *Pfu* polymerase with the sense outer/antisense mutant, and antisense outer/sense mutant primers using the wild-type sequence as template. The products of this first pair of PCR reactions are gel-isolated, mixed, and amplified using the outer primers. The mutant sequence is cut out and cloned back into the expression plasmid from which the wild-type sequence was excised.

More recently, we switched to the Quickchange™ double-stranded, site-directed mutagenesis kit (Stratagene, La Jolla, CA). Two complementary oligonucleotides are designed that contain the desired codon for each mutation in the middle of the sequence. For work on the M$_1$ mAChR, the primers are between 25 and 45 bases in length, and their melting temperatures are equal to or greater than 78°C. Primers usually have a minimum GC content of 40% and terminate in either a G or a C. The equation used to calculate the melting temperature is as follows:

$$T_m = 81.5 + 0.41(\%GC\ content) - 675/number\ of\ bases - \%mismatch.$$

They are 5′ phosphorylated during synthesis and HPLC purified.

The forward and reverse oligonucleotides are annealed to the double stranded wild-type plasmid and then extended by means of *PfuTurbo* DNA polymerase. A mutant plasmid containing staggered nicks results. Temperature cycling results in the linear amplification of the population of mutant plasmids. Selection against the parent sequence is accomplished by digestion with *Dpn* I endonuclease. This is specific for methylated and hemimethylated DNA strands that originate from the parent sequence, while the *in vitro* synthesized DNA is not methylated. The process is completed by transformation of the mutant nicked DNA into XL1-Blue supercompetent cells that repair the nicks. The mutated target sequence can then be excised using appropriate restriction sites, and subcloned into the parent plasmid from which the wild-type sequence has been excised, as before. The presence of the mutation must be confirmed by di-deoxy sequencing, so this maneuver cuts down on the sequencing burden.

6.4.4 PREPARATION OF PLASMID DNA

Good-quality, endotoxin-free plasmid DNA preparations are essential for the reproducible transfection of mammalian cells. In our work, we preferred to use the Qiagen Maxi preparation kit, with chloramphenicol amplification of the culture to maximize the plasmid DNA:protein ratio. The quality and quantity of plasmid DNA is checked spectroscopically at 260 and 280 nm.

6.4.5 TRANSFECTION OF CELLS

The transient expression of mutant M$_1$ mAChRs in COS-7 cells allows the binding and signaling properties of a series of mutants to be evaluated rapidly and quantitatively.

PROTOCOL 6-1:   TRANSFECTION OF COS-7 CELLS

1. Materials required: COS-7 cells maintained in α-minimal essential media (α-MEM), supplemented with antibiotics: penicillin and streptomycin, 10% heat inactivated (60 min, 56°C) newborn calf serum and 2 mM L-glutamine. Cells are incubated at 37°C, 95% O_2, 5% CO_2 in a humidified incubator and are subcultured every 3 to 4 days.
2. Grow cells in 175 cm^2 tissue culture flasks until almost confluent.
3. Wash cells twice with warm phosphate-buffered saline (PBS).
4. Dislodge cells from the surface of the flask by adding 1 to 2 ml of trypsin versene, and take up in 15 ml of warm supplemented media (as above), increasing volumes and pooling as necessary.
5. Harvest suspended cells at 2000 rpm, 5°C for 3 min. Resuspend in ice-cold *serum-free* α-MEM, and centrifuge again.
6. Resuspend cells in ice-cold *serum-free* α-MEM using 0.8 ml for every 1.5 flasks.
7. Place 10 to 15 μg of DNA in TE buffer in cooled 0.4 cm gap cuvettes. Add 0.8 ml of cell suspension (approximately 4×10^7 cells per cuvette).
8. Electroporate using a Bio-Rad Gene Pulser at optimized settings. These need to be determined experimentally. In our lab, they are 260 volts, 960 microfarads, with capacitance on the "external" setting.
9. Leave cuvettes to recover at room temperature for 10 min.
10. Carefully resuspend cells, and pipette them into warm supplemented media.
11. Plate into 10 cm diameter tissue culture dishes for harvesting for binding studies (30 ml per cuvette, 10 ml aliquots) or 12-well plates for phospho-inositide (PI) assays (12 ml/cuvette, 1 ml/well).
12. Grow for 72 h at 37°C, 95% O_2, 5% CO_2 in a humidified incubator.
13. Note that atropine can be used as a chemical chaperone for mutants that are not well expressed in COS-7 cells. At 24 h after transfection of the cells, the medium is changed to α-MEM containing 1×10^{-6} M atropine. The cells are incubated for a further 48 h at 37°C. After incubation, the cells are washed an extra four times with warm PBS over a period of 20 min prior to either membrane harvesting or use in phosphoinositide turnover assays.

6.4.6 MEMBRANE PREPARATIONS FOR BINDING ASSAYS

PROTOCOL 6-2:   MEMBRANE PREPARATIONS FROM TRANSFECTED COS-7 CELLS

1. Wash tissue culture dishes containing transfected COS-7 cells twice with PBS.
2. Add 1 ml of ice-cold harvesting buffer (20 mM HEPES, 10 mM EDTA, pH 7.5) to each dish, and incubate at 5°C for 15 min.

3. Scrape cells from the surface of the dishes using a Teflon cell scraper, harvest, and wash plate with further 1 ml of harvesting buffer to collect residue.

4. Homogenize the cells with 20 strokes in a glass on glass homogenizer on ice, and centrifuge at 28,000 rpm 5°C for 30 min.

5. Resuspend pellets in storage buffer (20 mM HEPES, pH 7.5, 1 mM EDTA), 1 ml per dish.

6. Rehomogenize suspension using a Polytron PT3100 homogenizer at setting 12 for 2 × 15 s with cooling on ice.

7. Snap-freeze 1 ml aliquots on dry ice, and store at −70°C until use.

8. Measure protein concentration on 50 μl aliquots (a) by micro-BCA or (b) by the Lowry method after precipitation with 1 ml of 10% (w/v) trichloroacetic acid followed by centrifugation for 10 min at 14,000 g to pellet the protein, while removing interfering substances, such as HEPES. BSA standards (between 0 and 100 μg) are processed in parallel. The protein concentration is usually about 1 mg/ml.

6.4.7 Binding Assays on Wild-Type and Mutant M$_1$ mAChRs

Radioligand binding assays can be carried out either on membrane preparations or on whole cells in culture plates. It is often advisable, or necessary, to use more than one radioligand to fully characterize the properties of a mutant receptor.

Protocol 6-3: Membrane Binding Assays

1. Thaw aliquots of membrane preparations at 0°C.

2. Centrifuge for 5 min at 14,000 × g.

3. Resuspend the membranes to a final concentration of 5 to 50 μg protein/ml (depending on the expression level and radioligand affinity of the mutant) in binding buffer (20 mM Na-HEPES, 0.1 M NaCl, 1 mM MgCl$_2$, pH 7.5). Homogenize using a polytron PT3100 at setting 12 for 30 s in 15 s bursts, with cooling on ice, to ensure an even suspension.

4. Prepare radioligand dilutions at 100 times their final assay concentrations. For radioligand saturation binding assays, use eight concentrations of [³H]NMS (78 Ci/mmol, Amersham Pharmacia Biotech, Piscataway, NJ) or [³H]QNB (50 Ci/mmol, Amersham) in binding buffer to yield final concentrations in the range 3×10^{-12} to 3×10^{-9} M. For competition assays, use a fixed concentration of 3×10^{-10} M [³H]NMS or 10^{-10} M [³H]QNB, increasing or decreasing this in accordance with the properties (affinity, expression level) of the mutant receptor. It is advisable to keep the level of radioligand depletion caused by receptor-specific binding to below 10%.

5. Working in triplicate or quadruplicate, pipette 10 μl aliquots of the radioligand dilutions into 5 ml polystyrene tubes or 96-well microtiter plates.

6. If performing a competition assay, add 10 µl aliquots of a solution of the unlabeled competing ligand, again freshly made up at 100 times the final assay concentration. For ACh, typically, 10 concentrations are used in the range 10^{-8} to 10^{-2} M, final assay concentration.

7. To tubes designated for total binding, or nonspecific binding, add 10 µl aliquots of binding buffer, or 10^{-4} M atropine (final assay concentration 10^{-6} M), respectively. In saturation assays, each radioligand concentration is assayed without and with atropine. In competition assays, a set of nonspecific binding measurements is included with each set of assays. In setting up the assays, take care to avoid cross-contamination.

8. Pipette 1 ml of membrane suspension into each tube (0.9 ml in microtiter plates), taking care to maintain an even suspension by frequent mixing. Mix by shaking.

9. Incubate the assays at 30°C for a minimum of 60 min ([^{3}H]NMS), or 180 min ([^{3}H]QNB) to allow for equilibrium to be achieved.

10. Terminate the assays by rapid vacuum filtration using either a Brandel cell harvester onto Whatman GF/B or a Tomtec 96-well cell harvester onto Wallac filtermat A glass-fiber paper. The filters are pretreated for 30 min with 0.15% polyethylenimine to reduce the amount of nonspecific binding. After harvesting, wash the filters three times with ice-cold H_2O.

11. Determine radioactivity either by liquid scintillation counting for the Brandel harvester method or solid β-plate counting for the Tomtec method. For liquid scintillation counting, the individual filters are punched into scintillation vials and mixed with 4.5 ml of liquid scintillant. After standing overnight, and mixing, the radioactivity is determined using a liquid scintillation counter. For β-plate counting, the filters are dried, and Meltilex scintillant is melted onto them using a Wallac 1495 microsealer. Each filter is fitted into a rack, and its radioactivity is determined using a Wallac 1450 Microbeta Trilux beta-plate counter. For the microbeta counter, counting efficiencies (compared to liquid scintillation) may be calculated by comparing 2 µl of 10^{-9} M [^{3}H]NMS spotted onto a filtermat or deposited into liquid scintillant. Counting efficiencies are generally calculated to be between 40% and 52%.

PROTOCOL 6-4: MEASURING THE CONCENTRATION OF M$_1$ MACHRS ON THE SURFACE OF INTACT CELLS

1. Grow transfected COS-7 cells in complete α-MEM in 12-well plates at 37°C, 5% CO_2 in a humidified incubator for 72 h.

2. Wash the COS-7 cells three times with cold serum-free α-MEM media.

3. Measure the concentration of cell-surface receptor by adding 1 ml of 10^{-9} M [^{3}H]NMS in α-MEM. Use 10^{-6} M atropine to measure nonspecific binding.

4. Incubate the cells for 4 h at 4°C.

5. Wash the cells with α-MEM media, then twice with ice-cold PBS.
6. Harvest the bound radioactivity in 500 µl 1% Triton X-100, and determine its level by liquid scintillation counting.

6.4.8 FUNCTIONAL ASSAYS ON MUTANT RECEPTORS

We found that electroporated cells function well in PI assays. Other methods of transfection may also be used, for instance, DEAE dextran, or Transfast™. To obtain good reproducibility, it is important to achieve uniform coverage of the wells of the 12-well plate by having an excess of cells. It is also essential to handle the cells gently during the medium changes during the assay, to avoid unpredictable losses. It is advisable to do test transfections with different amounts of DNA to optimize receptor expression and the PI response.

PROTOCOL 6-5: PHOSPHOINOSITIDE ASSAYS ON TRANSFECTED CELLS

1. Electroporate COS-7 cells that are in rapid growth with 15 µg DNA as described in Protocol 6-1, and grow them in 12-well plates at 37°C, 95% O_2, 5% CO_2. It is important to have an initial excess of transfected cells to completely cover the bottom of the well.
2. After 24 h, replace the medium with medium containing 1 µCi/ml myo-D-[³H]inositol (80 Ci/mmol, Amersham). Grow the cells for a further 48 h to allow the myo-D-[³H]inositol to be incorporated into phosphoinositide 4,5-bisphosphate (PIP₂) within the cell.
3. Remove the media containing the inositol, and incubate the cells in 1 ml Krebs-bicarbonate solution (120 mM NaCl, 3.1mM KCl, 1.2 mM $MgSO_4$, 2.6 mM $CaCl_2$, 10mM glucose, and 25 mM $NaHCO_3$, pH 7.4) containing 10 mM LiCl, that has been preincubated for 1 h at 37°C, 95% O_2, 5% CO_2. A Krebs 10 × stock solution is made up without Ca^{2+} and Mg^{2+}. On the day of the assay, to 435 ml prewarmed H_2O (CO_2 incubator, loose cap overnight), add 50 ml 10× Krebs, 5 ml 100 × Ca^{2+}, 5 ml 100 × Mg^{2+}, 5 ml 100 × Li^+.
4. Incubate the cells at 37°C in 95% O_2, 5% CO_2 for 30 min.
5. Remove the Krebs-LiCl solution, and replace by a series of ACh concentrations, usually in the range of 10^{-10} to 10^{-2} M, again made up in pre-equilibrated Krebs-LiCl. Continue the incubation for a further 30 min. This is within the linear range of the assay.[42]
6. Terminate the stimulation of PI hydrolysis by removal of the agonist-containing medium. Add 0.5 ml of ice-cold 5% perchloric acid to the wells, and incubate for 20 min at 4°C.
7. Add 0.4 ml of the lysate to 0.1 ml of 10 mM EDTA, pH 8.0, in Eppendorf tubes, and neutralize by adding 0.5 ml 1:1 (v/v) tri-n-octylamine:1,1,2 trichlorotrifluoroethane (freon). Vortex-mix the samples, and centrifuge them at 4000 rpm for 10 min in a swing-out rotor.
8. Add 0.3 ml of the aqueous upper phase to columns containing 1 ml Dowex AG 1 X 8 resin (formate form, dry mesh size 100 to 200). Wash the columns

with 10 ml dH$_2$O, then 10 ml 25 mM NH$_4$COOH, and finally elute the [^{3}H]-inositol phosphates with 10 ml 1 M NH$_4$COOH, 0.1 M HCOOH.

9. Count 1 ml of the eluted sample.
10. Regenerate the columns for further use by washing with 10 ml 2 M NH$_4^+$COO$^-$, 0.1 M HCOOH, followed by two washes with 10 ml H$_2$0.

6.4.9 IMMUNOCYTOCHEMISTRY AND IMMUNOFLUORESCENCE

We used the following procedure to visualize wild-type and mutant M$_1$ mAChRs in COS-7 cells using an anti-C-terminal antibody. It can be used with antibodies against an N-terminal tag without the inclusion of permeabilizing agents in the fixative.

PROTOCOL 6-6:   IMMUNOSTAINING OF M$_1$ MACHRS

1. Transfect near confluent COS-7 cells with receptor DNA, and plate onto sterile coverslips contained in the wells of 24-well plates. The coverslips should have been pretreated with 25 μg/ml fibronectin for at least 2 h.
2. After 3 days, aspirate the medium. Wash the cells three times with ice-cold PBS.
3. Fix the cells with ice-cold 4% paraformaldehyde in PBS containing 0.05% Triton X-100/Nonidet P-40 (1:1 v/v) for 5 min. Wash three times with PBS. This and subsequent wash steps must be conducted very carefully to avoid the loss of sheets of fixed cells.
4. Block the cells with 5% nonfat dry milk in PBS for 60 min at room temperature. Wash three times with PBS.
5. Incubate the coverslips for 60 min at 37°C with 0.3 ml/well of a 1:200 dilution of an immuno-purified rabbit antibody raised against the carboxyl-terminal 13 amino acids of the M$_1$ mAChR (which are identical in rat and human sequences[43]). Cover and humidify during the incubation. The antibody is diluted in PBS containing 3% bovine serum albumin. Wash three times with PBS.
6. Incubate the cells with a 1:5000 dilution of a goat anti-rabbit IgG/alkaline phosphatase conjugate (Promega, Madison, WI) for 60 min at 37°C. Wash three times in Tris-buffered saline (100 mM NaCl, 50 mM Tris-Cl pH 7.5) to remove phosphate.
7. Develop the color by incubating with a solution of nitro blue tetrazolium (NBT)/5-bromo-4-chloro-3-indolyl phosphate (BCIP) at room temperature. For 10 ml, needed is alkaline Tris-buffered saline (100 mM NaCl, 100 mM Tris-Cl pH 9.5) 9.5 ml; 1 M MgCl$_2$, 0.5 ml; NBT (75 mg/ml in 70% dimethyl formamide, stored in the dark at 4°C) 45 μl; BCIP (50 mg/ml in 100% dimethyformamide, stored in dark at 4°C), 35 μl. Exclude light during color development.
8. Wash three times in PBS.
9. Mount coverslips on a drop of 50% glycerol in PBS with 0.02% azide.
10. Note that this protocol can readily be adapted for immunofluorescence, mounting the coverslips in CitiFluor.

6.4.10 DATA ANALYSIS

Data from binding experiments were analyzed using SigmaPlot® for windows (SPSS Scientific, Chicago, IL). Data points were fitted to the models by nonlinear least-squares fitting, the points being weighted according to the inverse of their standard error squared.

Direct saturation binding curves were fitted to a one-site model of binding that takes into account depletion of the free ligand concentration as a result of receptor binding, as described previously.[44] This allows total and nonspecific binding curves to be fitted simultaneously, and provides an estimate of the affinity constant of radioligand binding, and the total concentration of receptor binding sites per milli-leter assay volume that can be used to estimate expression level and specific activity.

Competition binding curves were fitted either to the Hill equation, providing an estimate of the apparent affinity constant of the competing ligand, and the Hill coefficient (slope factor), or, where appropriate, to a one- or two-site model of binding. Apparent affinity constants were corrected for radioligand occupancy using the Cheng–Prusoff correction factor $(1 + K_{bin}([^3H]ligand)$.

PI dose–response curves were fitted to a four-parameter logistic function, providing estimates of basal and maximal signaling activity, an EC_{50} value, and a slope factor.

In all fitting equations, apparent affinity constants and ligand concentrations were expressed in a logarithmic form, in conformity with the log-normal distribution of binding constants. The equations used were described in detail previously[45] and are available from the author on request, in SigmaPlot format.

6.5 KEY CONSIDERATIONS IN PERFORMING MUTAGENESIS EXPERIMENTS

This section illustrates some of the factors that must be considered, and problems that may be encountered, during the functional characterization of mutant receptors.

6.5.1 MEASURING THE EXPRESSION OF FUNCTIONAL RECEPTORS

In overexpressing cells lines, such as COS-7 cells, a significant fraction of the receptor protein produced in the cells is not trafficked to the cell-surface membranes. Much of the protein is retained within the cell and is visualized in a perinuclear compartment, presumably representing receptors in transit through the endoplasmic reticulum and the Golgi apparatus.[43] Because the relative concentration of functional receptors is an important parameter for understanding function, it is necessary to ensure that binding sites measured in radioligand binding experiments *in vitro* represent functional receptors.

When a non-membrane-permeant radioligand, such as $[^3H]NMS$ is available, this can be ascertained by comparing the whole cell with membrane-binding measurements. An example is shown in Figure 6.2, which compares the relative concentration of binding sites for $[^3H]NMS$ measured in nonpermeabilized whole cells with that measured in membrane preparations made from cells expressing different

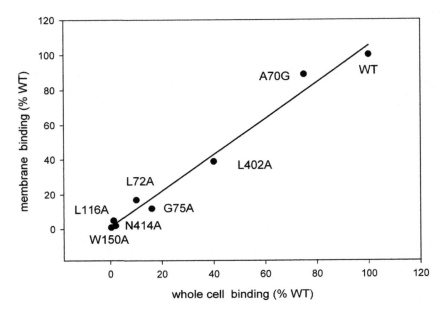

FIGURE 6.2 The relationship between receptor expression levels measured in membrane preparations and on the surface of whole cells for a number of mutants of the M_1 mAChR expressed in COS-7 cells. The results are normalized to the level of expression given by the wild-type receptor.

mutant receptors, with expression levels ranging from 2 to 100% of the wild-type value. A 1:1 correlation is evident, giving us confidence that the levels of binding measured on membrane preparations represent the expression of functional receptors. In principle, this measurement is more quantitative than an immunocytochemical measurement using an N-terminal epitope tag that may still include nonfunctional, though surface-expressed, proteins. The measurement of binding curves for agonists to intact cells may be complicated by the occurrence of agonist-induced internalization.

6.5.2 How Should Functional Data Be Expressed? The Contribution of Basal Signaling Activity

The problem is how to estimate the receptor-dependent basal signaling activity of mutant receptors, and how to correct for this in measurements of the maximum agonist-induced signaling activity.

The measured basal and maximal PI signals are estimated by fitting a four-parameter logistic function to the experimental PI dose–response curve. It is usual to express the results as a ratio of maximal to basal activity, $r = PI_{max}/PI_{basal}$, to facilitate the comparison of experiments and the pooling of data.

We may write

$$PI_{basal} = \alpha\gamma B_{asal} + \alpha\beta$$

$$PI_{max} = \alpha\gamma E_{max} + \alpha\beta$$

In these expressions, Basel and E_{max} are the receptor-dependent maximal and basal signaling activity per cell, β is a constant depending on cell type and represents the intrinsic background PI signal per cell irrespective of transfection, α is a constant proportional to cell density and [^{3}H]inositol uptake, and γ is proportional to the fraction of cells transfected. Thus,

$$r = PI_{max}/PI_{basal} = \left(\gamma E_{max} + \beta\right)/\left(\beta + \gamma B_{asal}\right)$$

The virtue of expressing the result as a ratio is that it becomes independent of experiment-to-experiment variability related to cell density, or radiolabeling of the phosphoinositide pool. Thus, the values of r obtained from a set of experiments have a lower variance than the absolute values of the maximal and basal signals. However, the value of r is influenced by the basal activity of the receptor.

To obtain an undistorted estimate of the receptor-dependent signal, it must be calculated relative to the untransfected cellular background, PI_{bgd}. In evaluating the effect of mutations on E_{max}, we are interested in the value of this parameter relative to the wild-type value.

$$\frac{E_{max.mut}}{E_{max.wt}} = \frac{r_{mut} \cdot \left(PI_{basal.mut}/PI_{bgd}\right) - 1}{r_{wt} \cdot \left(PI_{basal.wt}/PI_{bgd}\right) - 1}$$

If we write $PI_{basal} = (1 + \delta).PI_{bgd}$ so that δ represents the fractional increase in the background signal attributable to the transfected receptor, then

$$\frac{E_{max.mut}}{E_{max.wt}} = \frac{r_{mut} \cdot \left(1 + \delta_{mut}\right) - 1}{r_{wt} \cdot \left(1 + \delta_{wt}\right) - 1} \tag{6.17}$$

The basal activity of the mutants is also expressed relative to the wild-type receptor. Thus,

$$PI_{basal.mut}/PI_{basal.wt} = \left(1 + \delta_{mut}\right)/\left(1 + \delta_{wt}\right) \tag{6.18}$$

and the value of δ_{mut} can be calculated provided that the value for δ_{wt} is known. The ratio δ_{mut}/δ_{wt} is a direct estimate of the basal signaling of the mutant relative to the wild-type receptor. The values of δ_{mut} and δ_{wt} can be inserted into Equation (6.17)

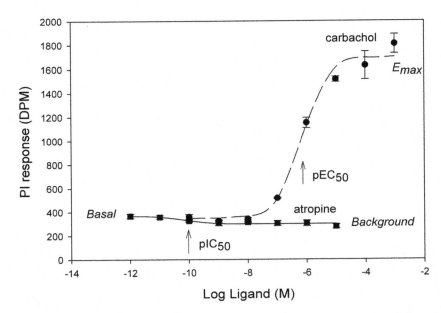

FIGURE 6.3 The phosphoinositide signaling response elicited by the wild-type M_1 mAChR expressed in COS-7 cells. The definitions of basal, maximal, and background signaling and the EC_{50} of the agonist and antagonist are illustrated.

to yield a corrected estimate of the ratio of the E_{max} of the mutant relative to that of the wild-type receptor. Likewise, the corrected value of the E_{max} of the wild-type receptor, relative to the untransfected cellular background, is $r_{wt}(1 + \delta_{wt})$.

δ_{wt} can be estimated (a) by the systematic comparison of the PI signal in cells transfected with the wild-type receptor and mock-transfected cells, or cells transfected with signaling-dead mutants or (b) by inhibition of the basal PI signal by an inverse agonist.

In the case of the wild-type M_1 mAChR in COS-7 cells, both approaches provided similar estimates. An example is shown in Figure 6.3, in which inhibition of the basal PI signal by atropine is compared to the signal elicited by the full agonist carbachol. Atropine inhibited the basal PI signal by 19%, giving a value of δ_{wt} of 0.23. Comparison of the wild-type basal signal with that of a series of inactive mutants suggested a mean δ_{wt} of 0.3. The wild-type basal signaling activity is in the range of 4 to 6% of the E_{max} value. Using this value in association with Equation (6.13), and an $\underline{R}_T$ value of 20 (see next section), we obtain an estimate of the basal signaling efficacy of the wild-type receptor, e_0, of 2.6×10^{-3}. This substantiates the earlier statement that the value of $\underline{K}_G$ for the wild-type receptor is much less than one.

It is evident that in order to compare the signaling properties of mutant receptors with the wild type, a wild-type control must be included in every experiment. Individual values of r_{mut} and r_{wt} are then calculated, and the ratio, r_{mut}/r_{wt} is computed, as is the ratio $PI_{basal,mut}/PI_{basal,wt}$. The means of these values are then calculated from repetitions of the experiment. The value of δ_{wt} is estimated from a separate set of experiments. The values of δ_{mut} and $E_{max,mut}/E_{max,wt}$ are then calculated using Equations (6.17) and (6.18).

6.5.3 THE EFFECT OF VARIABLE RECEPTOR EXPRESSION ON THE FUNCTIONAL DOSE–RESPONSE CURVE: THE EFFECTIVE RECEPTOR:G PROTEIN RATIO

The ternary complex model of receptor–G protein interaction shows that changes in the level of expression of functional receptor relative to G protein are expected to affect the measured potency of the agonist in stimulating a functional response, irrespective of the induction of the underlying conformational change in the receptor. It is necessary to compensate for this effect if values for the signaling efficacies of different mutants are to be computed.

In transient transfection experiments with mutant receptors, variations may occur in both the fraction of cells transfected and the expression of functional receptor protein within the transfected cells. Binding studies only measure the overall expression of receptor binding sites and do not distinguish between these possibilities. In the case of COS-7 cells transfected with mutant M_1 mAChRs, immunocytochemical staining of the transfected cells with a receptor-specific antibody established that the fraction of cells expressing the receptor was approximately the same under optimum conditions, irrespective of the mutant being expressed.[43,46] Transfection with suboptimal amounts of plasmid DNA merely reduced the fraction of cells transfected. When a PI dose–response curve to ACh was measured, the maximum response, but not the potency, was reduced.[43]

Reducing the concentration of functional receptors expressed in the cell membrane has a different outcome. In the case of the M_1 mAChR, this was accomplished by blocking receptors in the membranes of intact cells by treatment with an irreversible antagonist, propylbenzilylcholine mustard aziridinium ion (PrBCM). An alternative approach would be to use an expression vector with an inducible promoter.

An example is shown in Figure 6.4. The data shown in Figure 6.4a show that the ACh dose–response curve elicited from the wild-type M_1 mAChR was shifted at least 10-fold to lower potency (from log $K_{act} = 7.3$ to log $K_{act} = 6.3$), but that the E_{max} was reduced by only 10%. The corresponding log K_{bin} value was 5.1, measured by radioligand binding. This set of curves could be fitted by the ternary complex model [Equation (6.15)] with a common e_A (efficacy) value of 8.0, and different $\underline{R}_T$ values. To fit the data, it was necessary to assume that $\underline{R}_T$, the functional ratio of receptor:G protein, is of the order of 20. A similar set of data for a mutant receptor, D122H (log $K_{bin} = 4.8$) that initially gave 95% of the wild-type E_{max} could be fitted with lower values of e_A (4.0) and of $\underline{R}_T$ (Figure 6.4b). Measurement of [³H]NMS saturation curves on membranes prepared from these PrBCM-pretreated cells (Figure 6.4c) gave B_{max} values in agreement with the functional estimates (Figure 6.4d).

These considerations show that the simple ternary complex model can describe functional dose–response curves for M_1 mAChRs transiently expressed in COS-7 cells, and that there is a significant "spare receptor" ratio for the wild-type receptor in these cells. Variations in this ratio must be taken into account when interpreting

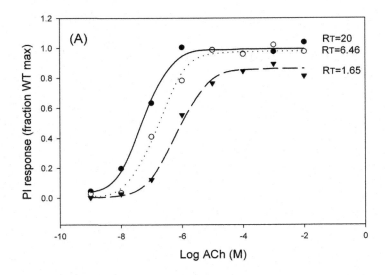

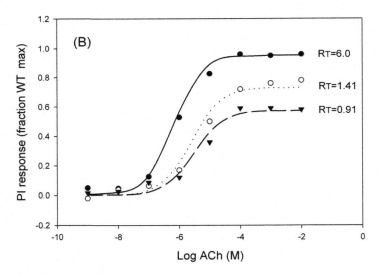

FIGURE 6.4 Estimation of the functional receptor:G protein ratio by irreversible receptor blockade. The effect of pretreatment of COS-7 cells expressing M_1 mAChRs with the irreversible blocking agent propylbenzilylcholine mustard aziridinium on (a) the PI response of the wild-type receptor; and (b) the PI response of the D122H mutant. For each data set, the curves show the best fit of the simple ternary complex model to the data using the experimentally measured estimates of the affinity constant for ACh, a common value of e_{A_1}, and different values of $\underline{R}_T$, the ratio of functional receptor to functional G protein. (c) [^{3}H]NMS saturation binding to the D122H mutant measured subsequently in membrane preparations. (d) The agreement between $\underline{R}_T$ estimates from functional and binding measurements for the wild type, and the D122H.

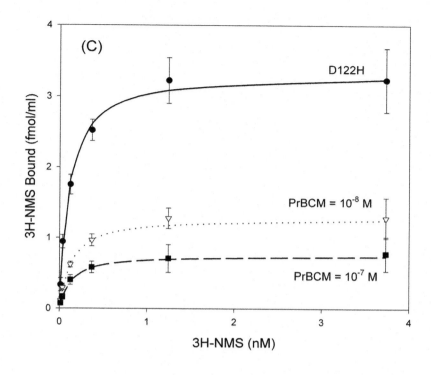

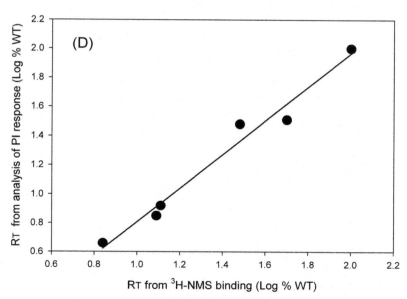

FIGURE 6.4 (CONTINUED)

the effects of mutations on dose–response relationships. Furthermore, the ratio can be estimated by measuring the total concentration of [^{3}H]NMS binding sites in membranes prepared from the transfected cells, using radioligand saturation binding assays.

6.5.4 MUTANTS THAT GIVE NO MEASURABLE BINDING OF [^{3}H]NMS

Approximately 5% of the Ala-substitution mutants that we studied showed strongly reduced binding of [^{3}H]NMS, to such an extent that the quantitative measurement of a radioligand saturation curve became difficult or impossible. Two eventualities were examined: the expression level of the receptor had been reduced, or the amino acid residue targeted was essential for the binding of the particular radioligand. Different tactics were needed to tackle these two situations.

6.5.4.1 Reduced Expression

Reduced expression levels usually follow the deletion of a side chain that faces into the core of the receptor structure and makes contacts that are important for the stability of the protein fold. In this case, the incubation of the cells expressing the mutant receptor with the high-affinity cell-permeant antagonist atropine (log K_{bin} = 9.0) for 48 h before harvesting often rescues the expression of the active receptor protein enough to allow binding measurements to be made. It is as though the binding energy of the antagonist replaces the free energy lost as a result of the deletion of the amino acid side chain, allowing the small molecule to act as a pharmacological chaperone.

The effects can be dramatic. Figure 6.5a shows the effect of atropine (10^{-6} M) pretreatment of COS-7 cells expressing the mutant L116A on the binding of [^{3}H]NMS; this residue is in the endofacial (inner leaflet of the membrane bilayer) section of transmembrane helix 3. As a result of atropine treatment, the B_{max} increased from 2% to about 90% of wild type, making it possible to characterize the ligand binding properties of the receptor. Atropine rescue before conducting PI assays revealed that this mutant had strongly raised basal signaling activity equivalent to 60% of the wild-type E_{max} value (Figure 6.5b).[47] This was not detectable without rescue because of the low expression level of the mutant. Using Equation (6.13), it follows that the basal signaling efficacy, e_0, of the unoccupied L116A mutant is 0.08, an increase of 30-fold over the value for the wild-type receptor. Without atropine rescue, the E_{max} of the L116A mutant was only 40 % of the wild-type level, which is consistent with the $\underline{R}_T$ value of 0.4 calculated from the 2% expression level. This suggests that there is only enough of the L116A mutant to activate 40% of the fraction of the G protein pool activated by the wild type, even when it is fully occupied with ACh.

Atropine rescue seems to be generally applicable to low-expressing mutants. For instance, it has also been applied to Cys-substitution mutants used in cross-linking experiments.[48] In order to be effective, the antagonist must be present at a concentration that gives high occupancy of the binding site (in this case, the con-

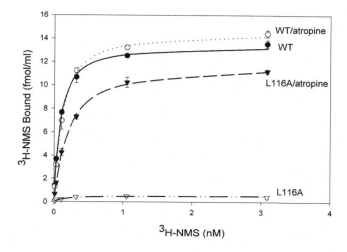

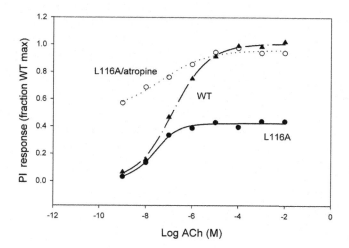

FIGURE 6.5 Atropine rescue of a poorly expressed mutant M_1 mAChR, L116A. COS-7 cells expressing wild-type and mutant M_1 mAChRs were pretreated for 48 h in culture with 10^{-6} M atropine and washed extensively before assay. (a) Effect on the [³H]NMS saturation binding curve and (b) effect on the PI response — note the uncovering of a high level of basal signaling activity by the L116A mutant.

centration is $1000 \times K_d$), and to have rapid enough dissociation kinetics to be washed out before the assays of interest are performed; in the case of atropine, the half-time of dissociation is approximately 2 min at 30°C. Four washes over a period of 20 min are needed. It is still advisable to include an atropine-pretreated wild-type control in the experiment, especially if a PI dose–response curve is to be measured, because there is sometimes a residual right-shift of up to threefold, which may reflect the ability of atropine to partition into phospholipids.

6.5.4.2 Reduced Ligand Affinity

Several low-binding Ala-substitution mutants failed to yield measurable specific binding of [³H]NMS, even after atropine rescue. This suggested that the targeted side chains might be part of the NMS binding site. In these cases, the use of a second high-affinity radioligand, [³H]QNB, proved to be valuable.

NMS and QNB have distinct chemical structures, the first having a quaternary amine methyl-scopine headgroup, and a tropic acid side chain, which has a single aromatic ring, and a polar hydroxymethyl substituent, while the second has a tertiary amine 3-hydroxy-quinuclidine headgroup and a benzilic acid side chain that bears two aromatic rings. These ligands have significantly different modes of binding. Figure 6.6a illustrates that [³H]QNB bound to the Y381A mutant in the exofacial (outer leaflet of the membrane bilayer) section of transmembrane helix 6 with undiminished affinity relative to the wild type, although with reduced expression level. A competition binding assay using [³H]QNB as the tracer ligand (Figure 6.6b) shows that NMS bound with 1000-fold lower affinity to the Y381A mutant than to the wild-type receptor, accounting for the absence of binding of [³H]NMS.

Figure 6.6b shows an unexpected feature of the NMS inhibition curve at the wild-type receptor — the presence of a minor population of low-affinity binding sites. These may arise from the ability of the tertiary amine ligand [³H]QNB, which, like atropine, can exist in an unprotonated state, to penetrate sealed membrane vesicles containing sequestered receptors poorly accessible to the charged [³H]NMS. Consistent with this, a low-affinity population of sites was not seen when the tertiary analogue scopolamine was used in the inhibition experiment. Furthermore, the apparent B_{max} for [³H]QNB was up to 40% higher than for [³H]NMS in some experiments. When this phenomenon was seen, we used a two-site model of binding to analyze the competition curves. The high-affinity component corresponds to that measured in direct saturation assays with [³H]NMS.

6.5.5 SHOULD WE BE USING A DIMERIC RECEPTOR TERNARY COMPLEX MODEL TO ANALYZE BINDING AND DOSE–RESPONSE DATA?

A subject of lively debate is whether rhodopsin-like GPCRs signal as monomers or dimers. The poorly expressed but high signaling efficacy mutant L116A may help to provide some insight into this question.

ACh bound to the L116A mutant with 40-fold higher affinity than to the wild-type receptor. At the same time, the Hill coefficient of the binding curve was significantly reduced, from 0.85 to 0.66 (Figure 6.7a). This might be consistent with the mutation inducing increased negative cooperativity within a dimeric receptor unit. An alternative possibility is that the free energy difference separating the inactive from the active state of the receptor is diminished, which is consistent with an increased affinity for the agonist and the raised basal signaling activity of the mutant, but that the energetics of the conformational change are influenced by variations in the membrane microenvironment of the receptor population.

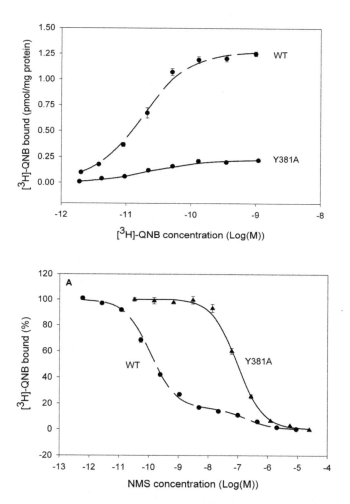

FIGURE 6.6 Use of an alternative radioligand to determine the properties of Y381A, a mutant M_1 mAChR that failed to bind [³H]NMS. (a) saturation curves for [³H]QNB; the affinity constant, K_{bin}, was 6×10^{10} M^{-1} for both wild-type and Y381A; B_{max} for Y381A was 16% of the wild-type value; (b) NMS inhibition of the binding of [³H]QNB to wild-type and Y381A showing a 1000-fold reduction in affinity. Note the small population of apparent low-affinity binding sites for NMS in the wild-type receptor preparation. These would not be detected in binding studies with [³H]NMS.

Because of its low level of expression (without atropine rescue), even full occupancy of the L116A mutant only elicits a fractional PI response relative to the wild-type receptor. Does the slope of the PI dose–response curve follow that of the binding curve, as the dimeric receptor ternary complex model suggests that it should? The analysis shown in Figure 6.7b suggests that the simpler monomeric receptor ternary complex model using a unitary K_{bin} value derived from Hill analysis of the ACh binding curve provides a better description of the activation curve than the

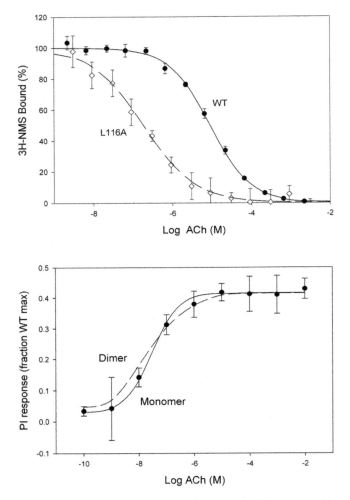

FIGURE 6.7 Comparison of the simple ternary complex model and the dimeric receptor ternary complex model for analyzing the PI response of the L116A mutant expressed at 2% of the wild-type level. (a) ACh binding curves measured by inhibition of [³H]NMS binding; fitting the binding curve to a two-site model yielded high- and low-affinity binding constants of $2.8 \times 10^7 \, M^{-1}$ and $1.2 \times 10^6 \, M^{-1}$ for L116A, increases of 93- and 28-fold over the wild type; (b) analysis of the PI dose–response curve using a dimeric receptor model compared to a monomeric receptor model.

dimeric receptor ternary complex model using separate high- and low-affinity ACh binding constants. While this analysis is by no means definitive, there seems to be no pressing reason to embark on the use of more complex models for the analysis of the M_1 mAChR mutants.

Based on the analysis of Figure 6.7, the signaling efficacy of ACh at the L116A mutant, e_A, is about 1.5-fold the value for the wild-type control in the same series of experiments. The mutation, therefore, seems to increase the basal signaling efficacy of the receptor more than the ACh signaling efficacy. While the mutation may decrease the free energy difference between the inactive and active states of the receptor, it may, at the same time, decrease the cooperativity of ACh binding to the activated state. In effect, the L116A mutation may break the link between the agonist binding site and the G protein binding site of the receptor.

6.6 THE INTERPRETATION OF ALA-SCANNING MUTAGENESIS STUDIES

6.6.1 DETERMINING THE FUNCTIONAL PHENOTYPE OF MUTANT RECEPTORS

The product of an Ala scan of a muscarinic receptor domain is a set of primary measurements of receptor expression level, affinity constants K_{bin} for NMS (and preferably other antagonists such as [^{3}H]QNB) and for the agonist ACh, measurements of basal and maximal PI signaling activity relative to the wild-type receptor, and of ACh potency, K_{act}. The expression levels are used to calculate $\underline{R}_T$, the estimated ratio of functional receptor to functional G protein; for the wild-type M$_1$ mAChR transiently expressed in COS-7 cells, this spare receptor ratio is about 20. Provided that the resultant value is not less than two (10% of wild-type expression level), the observations are combined using Equations (6.11) through (6.13) to yield estimates of the basal signaling efficacy of the unoccupied receptor and of the signaling efficacy of the ACh-occupied receptor. In the context of the ternary complex model of agonist–receptor–G protein interaction, these values are proportional to $\underline{K}_G$ and $\alpha\underline{K}_G$, respectively, where $\underline{K}_G$ is the affinity of the G protein for the ACh–receptor complex weighted by multiplication by the effective concentration of receptor-accessible G protein, and α is the cooperativity between agonist and receptor for binding to the G protein. It should be noted that comparisons of signaling efficacy assume that the expression of the different mutants does not have differential effects on G protein levels in the cells.

If the expression level drops below 10%, analysis of the dose–response curves is performed by fitting Equation (6.15) to the PI dose–response data using the experimentally determined values of $\underline{R}_T$, K_{bin}, and $\underline{K}_G$ as fixed parameters, and the value of α as unknown (see Figure 6.7b). In addition, atropine rescue experiments are performed. If no well-determined value of $\underline{K}_G$ is available, the default is to set it at the value for the wild-type receptor; in the case of the wild-type M$_1$ mAChR expressed in COS-7 cells, this value is about 2.6×10^{-3}. Note that even if $\underline{K}_G$ is poorly determined because it is defined by the basal signaling activity which is usually low, the product $\alpha\,\underline{K}_G$, the agonist signaling efficacy, is well-defined by the agonist dose–response curve.

The resulting set of parameters, $\underline{R}_T$, K_{bin}, e_0, and e_A provide a phenotypic description of each mutated side chain that can help to define its function in the context of the entire domain, and of the receptor structure.

6.6.2 Phenotypes Predicted by the Extended Ternary Complex Model of Agonist–Receptor–G Protein Interaction

The extended ternary model of receptor activation suggests that we may anticipate five limiting classes of function for amino acid side chains within the receptor molecule[49,50]:

> *Null*: Filling gaps in the receptor structure without making specific intra- or intermolecular interactions. The mutation of these nondescript side chains should have little effect.
>
> *Stabilizing*: Making interactions that contribute to both the ground state and the activated state of the receptor. Ala-substitution of a purely stabilizing side chain would be expected to reduce the stability of the protein fold, if its interaction is within the receptor structure, or to reduce ligand affinity (through a reduction in K_A), if the interaction is with the ligand molecule. Because there is no change in the free energy of the active state relative to the ground state, there should be no net effect of the mutation on the signaling efficacy.
>
> *Constraining*: Making interactions that selectively stabilize the inactive state of the receptor but that are weakened or broken in the activated state. Deletion of these side chains would be expected to elicit two correlated effects: first, a reduction in the stability of the ground state structure, and second, an increase in the population of the activated state relative to the ground state (because the decreased free energy difference leads to an increase in the isomerization constant K), reflected both in increased agonist affinity $K_A(1 + \alpha^*K)$, and increased signaling efficacy $\underline{K}^*_{G \cdot \alpha}{}^*K/(1 + \alpha^*K)$.
>
> *Activating*: Making interactions that selectively stabilize the activated state. The primary effect of mutating a pure activator side chain should be a reduction in signaling efficacy. Three subcategories can be anticipated: first, side chains with interactions that are strictly intramolecular — Ala-substitution of these should reduce the isomerization constant K; second, those that make selective interactions with the agonist in the activated state, but not the ground state, so acting to transduce the binding of the agonist into receptor activation — mutation of these should reduce the cooperativity factor, α^*; third, those that directly or indirectly mediate selective binding of the G protein in the activated state, so helping to transmit the conformational change in the receptor into G protein binding — mutation of these side chains should decrease $\underline{K}^*_G$. Ala-substitution of an activator side chain would not be expected to exert a large effect on receptor stability, because it should not affect the integrity of the ground state. There might be variable reductions in the overall affinity of the agonist, and the basal signaling activity, depending upon the precise role of the side chain, the energetics of the activation process, and the receptor:G protein ratio.
>
> *Catalytic:* Making interactions that are important for the conformational changes needed for catalysis of GDP release or GTP binding after the formation of the receptor–G protein complex. Because the primary effect

is on K_{cat}, mutating a catalytic side chain should alter maximal and basal signaling activity without affecting agonist binding or potency.

Clearly, it is possible to anticipate more complex superpositions of these primary roles for the side chains of particular residues. An example might be a side chain that progresses from a constraining to a transducing role as the activation cycle proceeds.

6.6.3 A Case Study: Transmembrane Helix 7

We investigated residues in TM 2, 3, 4, 5, 6, and 7 of the M_1 mAChR.[43,46,50–53] In a recent review, I attempted to interpret these results in a structural context, defining the primary binding sites for NMS and ACh, and identifying some of the interhelical contacts that may help to stabilize the inactive and active conformations of the receptor.[50] Drawing upon these studies, Figure 6.8 summarizes the outcome of a scanning mutagenesis study on TM 7, illustrating a typical range of mutant phenotypes.

Mutation of three residues decreased receptor expression levels to less than 10% of wild type, necessitating atropine rescue for their full characterization. Two of these, N414 and P415, are part of the canonical NPxxY motif characteristic of TM 7 of the rhodopsin-like GPCRs. The third, Y408, is homologous to the retinal-attachment lysine in rhodopsin and is conserved in other cationic amine receptors. Based on the effects of their mutation on receptor stability, we proposed that these residues make important intramolecular contacts within the ground state structure of the receptor.[53]

Three residues were identified whose mutation substantially (>10-fold) reduced NMS affinity, namely, Y404, C407, and Y408. These are all in the exofacial segment of TM 7, a pattern repeated in several of the other TM helices. The magnitudes of the reductions suggested that the side chains of these amino acids may be part of the binding pocket for NMS. Effects smaller than this, though statistically significant, might point to a second-shell role. The mutation of N110 evoked a threefold increase in NMS affinity. This probably reflects an indirect perturbation of the binding site. Interestingly, as in the case of Y381A, these mutations did not affect the binding of [³H]QNB.

Mutation of these residues in the outer segment of TM 7 had a parallel effect on the affinity of ACh, suggesting that, in this domain of the receptor, the NMS and ACh binding sites may overlap. This is also true in TM 3, but there are perceptible differences in TM 5, TM 6, and the second extracellular loop (A. Goodwin, unpublished data). As the mutagenic scan extended into the endofacial segment of TM 7, another characteristic phenomenon emerged, namely, increases in ACh affinity. These were 20-fold for N414, P415, and Y418 of the NPxxY sequence. Coupled with the decreases in receptor expression levels, these observations suggested that intramolecular contacts made by the side chains of these residues help to stabilize the signaling inactive ground state of the receptor, so that their deletion promotes the high-affinity agonist binding state.

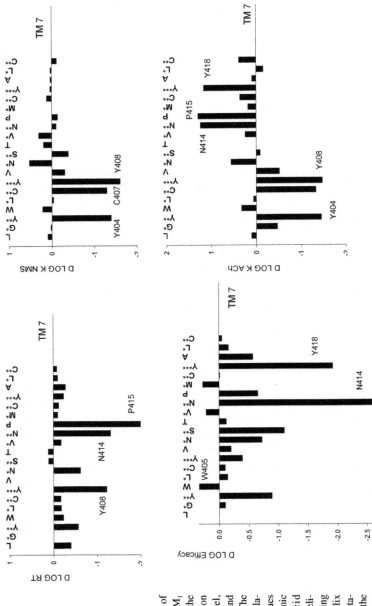

FIGURE 6.8 Summary of Ala-scan of TM 7 of the M₁ mAChR. The plots show the effects of the mutations on receptor expression level, NMS and ACh affinity, and ACh signaling efficacy. The effects were expressed relative to the wild-type values and plotted on a logarithmic scale. The amino acid sequence is plotted as a helical net, with residues facing into the core of the helix bundle lowest. The orientation of the plot is from the exofacial to the endofacial end of the helix.

Calculation of the changes in the ACh signaling efficacy suggested that only one of the presumptive binding site residues, Y404, makes a strong contribution to the transduction of binding energy into receptor activation. A follow-up substitution of Y404 with Phe suggested that the aromatic ring of the Tyr residue was the important functional moiety in this respect, while removal of the hydroxyl group was responsible for the reduction in binding affinity,[53] most likely by perturbation of a hydrogen-bonding network, a pattern parallel to that seen with Y381.[46]

Mutation of the more deeply buried residues, particularly N414 and Y418, caused the most dramatic reductions in ACh signaling efficacy. In the case of N414A, the E_{max} of the ACh-induced PI response only achieved 16% of the wild-type value, even after atropine rescue to the wild-type level. A calculation based on Equation (6.12) indicated an efficacy value e_A of only 0.01 for this mutant, a reduction of about 500-fold with respect to the wild-type control in these experiments. These observations indicate a dual role for N414, P415, and Y418: Not only is their integrity important for the stability of the inactive state of the receptor, it is also vital for the formation of the G protein binding conformation. In this respect, these amino acids differ from L116 in TM 3.

Interestingly, mutation of several residues, W405, V413, and M416, caused an increase in the calculated signaling efficacy of the receptor. Consistent with this, these mutants showed somewhat raised basal PI signaling activity. In the case of W405, this amounted to 15% of the E_{max}, value, giving an e_0 value of 0.015, approximately sixfold higher than that of the wild type.

Nearly all of the amino acid residues that we studied by Ala-substitution mutagenesis have given phenotypes that can be interpreted within the context of the equilibrium extended ternary complex model of agonist–receptor–G protein interaction.

Interestingly, a handful of amino acids failed to conform to this pattern. One way to detect them is to look for deviations from the relationship between E_{max}, K_{bin}, and K_{act} predicted by Equation (6.14). The plot shown in Figure 6.9 is drawn from data from TM 7. In general, the linear relationship is remarkably well obeyed, with one exception, L402, with a mutation that caused a larger reduction in E_{max} (to 50% of the wild-type level) than the changes in agonist potency and affinity predict. Similar deficits occur for N115 (TM 3) and P157 (TM 4), while one mutant, of the disulfide-bonded C178 (e2 loop), may signal better than would be predicted. Reference to Equation (6.5) suggests that mutation of these residues may affect the ability of the ACh–receptor–G protein complex to undergo GDP release. They may provide an indication of the parts of the receptor structure that undergo conformational changes during the process of catalysis, rather than agonist or G protein binding.

6.7   CONCLUSIONS

A weakness of all mutagenesis studies is that they cannot examine the contribution of the protein backbone to the function under investigation. Ala-scanning may also underestimate the free energy contributed by a particular side chain if its role can be replaced by a water molecule in the mutant receptor. However, scanning mutagenesis

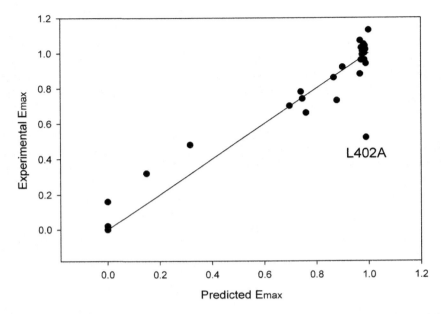

FIGURE 6.9 Correlation between experimentally measured maximum PI response, and the maximum PI response predicted from changes in receptor expression level, ACh affinity, and ACh signaling potency for TM 7 mutants of the M_1 mAChR. The full line is the line of equivalence. The L402A mutant shows a large reduction in E_{max} that is not predicted by the changes in these parameters.

studies are, by definition, comparative, allowing each side chain to be assessed relative to its neighbors. Thus, they provide information that can be used to interrogate or refine a homology model of the protein, in the absence of direct structural information.

Ala-scanning mutagenesis pinpoints important residues whose functions can be evaluated further by a series of point mutations designed to dissect the roles of the constituent moieties of the side chain; an example used above is comparison of the effects of a Tyr-to-Phe with a Tyr-to-Ala mutation. Another variation is steric hindrance scanning mutagenesis, in which side chains are replaced by bulkier side chains, occluding particular volumes within a binding site.[54] A recent development is the substitution of novel synthetic amino acids using amber codon suppression in combination with a tRNA charged *in vitro*.[55]

If intramolecular interactions are implied, they can be investigated by combinatorial mutagenesis techniques such as histidine or cysteine-substitution mutagenesis for probing by metal binding, or oxidative cross-linking.[56,57] A creative use of oxidative cross-linking techniques is to probe for the conformational changes that follow ligand binding.[49,59] If a particular side chain is thought to interact with a ligand, it can be subjected to cysteine substitution, rendering it susceptible to reaction with generic or site-directed sulfhydryl reagents that may differ in their polarity properties, and thus give additional information about the environment of the residue.[51,59,60] A subject that has been largely neglected in studies of receptor mutants, with honorable exceptions,[61,62] is how the performance of kinetic studies of ligand association and dissociation to mutant receptors may shed additional light on the function of particular residues.

To get the most out of mutagenesis studies, a good three-dimensional model of the target protein is needed. For the mAChRs, the crystal structure of rhodopsin has provided an excellent starting point for a homology model,[53,63] but this may not be true for other GPCRs. Ideally, the modeling process and mutagenesis studies should proceed in cycles of refinement, one feeding back upon the other.

ACKNOWLEDGMENTS

The author is grateful to past members of the laboratory, particularly Dr. Zhiliang Lu, and Dr. Stuart Ward for data used in some of the illustrations, and to Dr. Nigel Birdsall for comments on the manuscript.
This work was supported by the Medical Research Council, U.K.

REFERENCES

1. Vassilatis, D.K. et al., The G protein-coupled receptor repertoires of human and mouse, *Proc. Natl. Acad. Sci. USA*, 100, 4903–4908, 2003.
2. Becker, O.M. et al., G protein-coupled receptors: *in silico* drug discovery in 3D, *Proc. Natl. Acad. Sci. USA*, 101, 11,304–11,309, 2004.
3. Mirzadegan, T., Benko, G., Filipek, S., and Palczewski, K., Sequence analyses of G-protein-coupled receptors: similarities to rhodopsin, *Biochemistry*, 42, 2759–2767, 2003.
4. Waelbroeck, M., Activation of guanosine 5'-[-³⁵S]thio-triphosphate binding through M₁ muscarinic receptors in transfected Chinese hamster ovary cell membranes: 1. Mathematical analysis of catalytic G protein activation, *Mol. Pharmacol.*, 59, 875–885, 2001.
5. Waelbroeck, M., Activation of guanosine 5'-[-³⁵S]thio-triphosphate binding through M₁ muscarinic receptors in transfected Chinese hamster ovary cell membranes: 2. Testing the "two-states" model of receptor activation, *Mol. Pharmacol.*, 59, 886–893, 2001.
6. Baldwin, J.M., Schertler, G.F.X., and Unger, V.M., An alpha-carbon template for the transmembrane helices in the rhodopsin family of G-protein-coupled receptors, *J. Mol. Biol.*, 272, 144–164, 1997.
7. Unger, V.M., Hargrave, P.A., Baldwin, J.M., and Schertler, G.F., Arrangement of rhodopsin transmembrane alpha-helices, *Nature*, 389, 203–206, 1997.
8. Palczewski, K. et al., Crystal structure of rhodopsin: a G protein-coupled receptor, *Science*, 289, 739–745, 2000.
9. Teller, D.C., Okada, T., Behnke, C.A., Palczewski, K., and Stenkamp, R.E., Advances in determination of a high-resolution three-dimensional structure of rhodopsin, a model of G-protein-coupled receptors (GPCRs), *Biochemistry*, 40, 7761–7772, 2001.
10. Okada, T. et al., The retinal conformation and its environment in rhodopsin in light of a new 2.2 A crystal structure, *J. Mol. Biol.*, 342, 571–583, 2004.

11. Lapinsh, M. et al., Classification of G-protein coupled receptors by alignment-independent extraction of principal chemical properties of primary amino acid sequences, *Protein Sci.*, 11, 795–805, 2002.

12. Gether, U., Uncovering molecular mechanisms involved in activation of G protein-coupled receptors, *Endocrine Rev.*, 21, 90–113, 2000.

13. Hulme, E.C., Birdsall, N.J., and Buckley, N.J., Muscarinic receptor subtypes, *Annu. Rev. Pharmacol. Toxicol.*, 30, 633–673, 1990.

14. Ovchinnikov, Y.A., Rhodopsin and bacteriorhodopsin: structure–function relationships, *FEBS Lett.*, 148, 179–190, 1982.

15. Nathanson, N.M., A multiplicity of muscarinic mechanisms: enough signaling pathways to take your breath away, *Proc. Natl. Acad. Sci. USA*, 97, 6245–6247, 2000.

16. Wess, J., Muscarinic acetylcholine receptor knockout mice: novel phenotypes and clinical implications, *Annu. Rev. Pharmacol. Toxicol.*, 44, 423–450, 2004.

17. Rosenblum, K., Futter, M., Jones, M., Hulme, E.C., and Bliss, T.V., ERKI/II regulation by the muscarinic acetylcholine receptors in neurons, *J. Neurosci.*, 20, 977–985, 2000.

18. Caulfield, M.P. and Birdsall, N.J., International Union of Pharmacology. XVII. Classification of muscarinic acetylcholine receptors, *Pharmacol. Rev.*, 50, 279–290, 1998.

19. Birdsall, N.J., Lazareno, S., Popham, A., and Saldanha, J., Multiple allosteric sites on muscarinic receptors, *Life Sci.*, 68, 2517–2524, 2001.

20. Ford, D.J., Essex, A., Spalding, T.A., Burstein, E.S., and Ellis, J., Homologous mutations near the junction of the sixth transmembrane domain and the third extracellular loop lead to constitutive activity and enhanced agonist affinity at all muscarinic receptor subtypes, *J. Pharmacol. Exp. Ther.*, 300, 810–817, 2002.

21. Hill-Eubanks, D., Burstein, E.S., Spalding, T.A., Bräuner-Osborne, H., and Brann, M.R., Structure of a G-protein-coupling domain of a muscarinic receptor predicted by random saturation mutagenesis, *J. Biol. Chem.*, 271, 3058–3065, 1996.

22. Spalding, T.A., Burstein, E.S., Henderson, S.C., Ducote, K.R., and Brann, M.R., Identification of a ligand-dependent switch within a muscarinic receptor, *J. Biol. Chem.*, 273, 21,563–21,568, 1998.

23. Schmidt, C. et al., Random mutagenesis of the M3 muscarinic acetylcholine receptor expressed in yeast. Identification of point mutations that "silence" a constitutively active mutant M3 receptor and greatly impair receptor/G protein coupling, *J. Biol. Chem.*, 278, 30,248–30,260, 2003.

24. Wells, J.A., Systematic mutational analysis of protein–protein interfaces, *Methods in Enzymol.*, 202, 390–411, 1991.

25. Wells, J.A., Binding in the growth hormone receptor complex, *Proc. Natl. Acad. Sci. USA*, 93, 1–6, 1996.

26. Vaughan, C.K., Buckle, A.M., and Fersht, A.R., Structural response to mutation at a protein–protein interface, *J. Mol. Biol.*, 286, 1487–1506, 1999.

27. DeLano, W.L., Unraveling hot spots in binding interfaces: progress and challenges, *Curr. Opin. Struct. Biol.*, 12, 14–20, 2002.

28. Kortemme, T., Kim, E., and Baker, D., Computational alanine scanning of protein–protein interfaces, *Sci. STKE*, 2004, 12, 2004.

29. De Lean, A., Stadel, J.M., and Lefkowitz, R.J., A ternary complex model explains the agonist-specific binding properties of the adenylate cyclase-coupled beta-adrenergic receptor, *J. Biol. Chem.*, 255, 7108, 1980.

30. Biddlecome, G.H., Berstein, G., and Ross, E.M., Regulation of phospholipase C-$\beta 1$ by G_q and M_1 muscarinic cholinergic receptor, *J. Biol. Chem.*, 271, 7999–8007, 1996.

31. Mukhopadhyay, S. and Ross, E.M., Rapid GTP binding and hydrolysis by G_q promoted by receptor and GTPase-activating proteins, *Proc. Natl. Acad. Sci. USA*, 96, 9539–9544, 1999.

32. Black, J.W. and Leff, P., Operational models of pharmacological agonism, *Proc. R. Soc. Lond. B.*, 220, 141–162, 1983.

33. Whaley, B.S., Yuan, N., Birnbaumer, L., Clark, B., and Barber, R., Differential expression of the β-adrenergic receptor modifies agonist stimulation of adenylyl cyclase: a quantitative evaluation, *Mol. Pharmacol.*, 45, 481–489, 1994.

34. Hulme, E.C. and Lu, Z.-L., Scanning mutagenesis of transmembrane domain 3 of the M₁ muscarinic acetylcholine receptor, *J. Physiol. (Paris)*, 92, 269–274, 1998.

35. Samama, P., Cotecchia, Costa, T., and Lefkowitz, R.J., A mutation-induced activated state of the β2-adrenergic receptor, *J. Biol. Chem.*, 268, 4625–4636, 1993.

36. Ghanouni, P. et al., Functionally different agonists induce distinct conformations in the G protein coupling domain of the beta 2 adrenergic receptor, *J. Biol. Chem.*, 276, 24,433–24,436, 2001.

37. Kniazeff, J. et al., Closed state of both binding domains of homodimeric mGlu receptors is required for full activity, *Nat. Struct. Mol. Biol.*, 11, 706–713, 2004.

38. Angers, S., Salahpour, A., and Bouvier, M., Dimerization: an emerging concept for G protein-coupled receptor ontogeny and function, *Annu. Rev. Pharmacol. Toxicol.*, 42, 409–435, 1903.

39. Salahpour, A. et al., Homodimerization of the β2-adrenergic receptor as a prerequisite for cell surface targeting, *J. Biol. Chem.*, 279, 33,390–33,397, 2004.

40. Okayama, H. and Berg, P., A cDNA cloning vector that permits expression of cDNA inserts in mammalian cells, *Mol. Cell. Biol.*, 3, 280–289, 1983.

41. Ho, S.N., Hunt, H.D., Horton, R.M., Pullen, K., and Pease, L.R., Site-directed mutagenesis by overlap extension using the polymerase chain reaction, *Gene*, 77, 51–59, 1989.

42. Jones, P.G., Curtis, C.A.M., and Hulme, E.C., The function of a highly-conserved arginine residue in the activation of the muscarinic M₁ receptor, *Eur. J. Pharmacol.*, 288, 251–257, 1995.

43. Lu, Z.-L., Curtis, C.A.M., Jones, P.G., Pavia, J., and Hulme, E.C., The role of the aspartate-arginine-tyrosine triad in the M₁ muscarinic receptor: mutations of aspartate 122 and tyrosine 124 decrease receptor expression but do not abolish signaling, *Mol. Pharmacol.*, 51, 234–241, 1997.

44. Hulme, E.C. and Birdsall, N.J.M., Receptor–ligand interactions — a practical approach, in Hulme, E.C., Ed., IRL Press, Oxford, 1992, pp. 63–176.

45. Hulme, E.C., Methods in molecular biology: receptor binding techniques, in Keen, M., Ed., Humana Press, Totowa, NJ, 1999, pp. 139–184.

46. Ward, S.D.C., Curtis, C.A.M., and Hulme, E.C., Alanine-scanning mutagenesis of transmembrane domain 6 of the M₁ muscarinic acetylcholine receptor suggests that Tyr381 plays key roles in receptor function, *Mol. Pharmacol.*, 56, 1031–1041, 1999.

47. Lu, Z.-L. and Hulme, E.C., The functional topography of transmembrane domain 3 of the M₁ muscarinic acetylcholine receptor, revealed by scanning mutagenesis, *J. Biol. Chem.*, 274, 7309–7315, 1999.

48. Ward, S.D., Hamdan, F.F., Bloodworth, L.M., and Wess, J., Conformational changes that occur during M₃ muscarinic acetylcholine receptor activation probed by the use of an *in situ* disulfide cross-linking strategy, *J. Biol. Chem.*, 277, 2247–2257, 2002.

49. Hulme, E.C., Lu, Z.-L., Ward, S.D.C., Allman, K., and Curtis, C.A.M., The conformational switch in 7-transmembrane receptors: the muscarinic receptor paradigm, *Eur. J. Pharmacol.*, 375, 247–260, 1999.

50. Hulme, E.C., Lu, Z.L., and Bee, M.S., Scanning mutagenesis studies of the M_1 muscarinic acetylcholine receptor, *Receptors and Channels*, 9, 215–228, 2003.

51. Allman, K., Page, K.M., Curtis, C.A.M., and Hulme, E.C., Scanning mutagenesis identifies amino acid side chains in transmembrane domain 5 of the M_1 muscarinic receptor that participate in binding the acetyl methyl group of acetylcholine, *Mol. Pharmacol.*, 58, 175–184, 2000.

52. Hulme, E.C., Lu, Z.-L., Bee, M., Curtis, C.A.M., and Saldanha, J., The conformational switch in muscarinic acetylcholine receptors, *Life Sci.*, 68, 2495–2500, 2001.

53. Lu, Z.-L., Saldanha, J., and Hulme, E.C., Transmembrane domains 4 and 7 of the M_1 muscarinic actylcholine receptor are critical for ligand binding and the receptor activation switch, *J. Biol. Chem.*, 276, 34,098–34,104, 2001.

54. Holst, B., Zoffmann, S., Elling, C.E., Hjorth, S.A., and Schwartz, T.W., Steric hindrance mutagenesis versus alanine scan mapping of ligand binding sites in the tachykinin NK_1 receptor, *Mol. Pharmacol.*, 53, 166–175, 1998.

55. Monahan, S.L., Lester, H.A., and Dougherty, D.A., Site-specific incorporation of unnatural amino acids into receptors expressed in mammalian cells, *Chem. Biol.*, 10, 573–580, 2003.

56. Elling, C.E., Raffetseder, U., Nielsen, S.M., and Schwartz, T.W., Disulfide bridge engineering in the tachykinin NK1 receptor, *Biochemistry*, 39, 667–675, 2000.

57. Lu, Z.-L. and Hulme, E.C., A network of conserved intramolecular contacts defines the off-state of the transmembrane switch mechanism in a seven-transmembrane receptor, *J. Biol. Chem.*, 275, 5682–5686, 2000.

58. Yu, H., Kono, M., and Oprian, D.D., State-dependent disulfide cross-linking in rhodopsin, *Biochemistry*, 38, 12,028–12,032, 1999.

59. Javitch, J.A., Fu, D., Chen, J., and Karlin, A., Mapping the binding-site crevice of the dopamine D2 receptor by the substituted-cysteine accessibility method, *Neuron*, 14, 825–831, 1995.

60. Marjamäki, A. et al., Chloroethylclonidine and 2-aminoethylmethanethiosulfonate recognize two different conformations of the human α_2A-adrenergic receptor, *J. Biol. Chem.*, 274, 21,867–21,872, 1999.

61. Matsui, H., Lazareno, S., and Birdsall, N.J.M., Probing of the location of the allosteric site on M_1 muscarinic receptors by site-directed mutagenesis, *Mol. Pharmacol.*, 47, 88–98, 1995.

62. Jakubik, J., El-Fakahany, E.E., and Tucek, S., Evidence for a tandem two-site model of ligand binding to muscarinic acetylcholine receptors, *J. Biol. Chem.*, 275, 18,836–18,844, 2000.

63. Filipek, S., Teller, D.C., Palczewski, K., and Stenkamp, R., The crystallographic model of rhodopsin and its use in studies of other G protein-coupled receptors, *Annu. Rev. Biophys. Biomol. Struct.*, 32, 375–397, 2003.

64. Archer, E., Maigret, B., Escrieut, C., Pradayrol, L., and Fourmy, D., Rhodopsin crystal: new template yielding realistic models of G-protein-coupled receptors? *Trends Pharmacol. Sci.*, 24, 36–40, 2003.

7 Analysis of the Regulation of Muscarinic Acetylcholine Receptors

Chris J. van Koppen

CONTENTS

7.1 INTRODUCTION

The signaling pathways of the five members of the muscarinic acetylcholine receptor (mAChR) family are regulated by a number of pathways. These regulatory pathways protect cells against too strong or persistent mAChR stimulation and allow mAChRs to return to normal responsiveness after receptor desensitization.[1]

Strong or persistent activation of mAChRs for minutes to hours can induce receptor internalization and downregulation, which proceed in a highly regulated manner depending on receptor subtype and cell type.[2,3] Following agonist binding and receptor stimulation, mAChRs are phosphorylated by receptor kinases (e.g., G protein-coupled receptor kinases and casein kinase 1α (see for review, Van Koppen and Kaiser[3]). Phosphorylation of the receptors by G protein-receptor kinases allows binding of β-arrestin to the phosphorylated receptor. The binding of ß-arrestin to the receptor not only sterically hinders further coupling to the heterotrimeric G proteins but also initiates the process of receptor internalization. The β-arrestin–receptor complex interacts with clathrin and the clathrin adapter complex AP-2 in the clathrin-coated pit.[1] The pinching of the clathrin-coated pit from the plasma membrane is mediated through the GTPase dynamin. Activated dynamin molecules polymerize as a collar around the neck of the endocytic pit containing the mAChRs and catalyze the fission of the clathrin-coated vesicle from the plasma membrane. Depending on the receptor involved, ß-arrestin dissociates with different rates from the receptor. The dissociation of ß-arrestin allows receptors to be dephosphorylated by putative receptor phosphatases and to return to the plasma membrane as resensitized mAChRs for another round of receptor activation, desensitization, and resensitization. In many cells, persistent activation of mAChRs over hours induces a loss of the total number of cellular mAChRs. This loss of receptor number, also termed receptor downregulation, can only be overcome by *de novo* receptor synthesis, indicating receptor degradation. In this chapter, we describe various methods to analyze the function and regulation of mAChRs in intact cells and cell homogenates.

7.2 RADIOLIGAND BINDING ASSAYS TO MEASURE mAChR INTERNALIZATION AND DOWNREGULATION

In this chapter, two different binding assays with the muscarinic radioligands and antagonists [³H]*N*-methylscopolamine ([³H]NMS) and [³H]Quinuclidinyl benzilate ([³H]QNB) are described. These binding assays are utilized to determine mAChR internalization and downregulation, respectively.

7.2.1 ANALYSIS OF mAChR INTERNALIZATION IN INTACT CELLS USING [³H]NMS BINDING ASSAY

Analysis of the number of cell-surface mAChRs is performed by measuring the specific binding of the positively charged mAChR radioligand [³H]NMS to the cell surface of intact cells. Because [³H]NMS is a quaternary, positively charged radioligand, it is not able to cross the plasma membrane of intact cells to bind to intracellular (e.g., internalized) mAChRs. The loss of specific [³H]NMS binding to intact cells is a measure of agonist-induced receptor internalization.

Materials Required

1. [^{3}H]N-methylscopolamine (specific radioactivity of ~80 Ci/mmol)
2. 24 well plates [While HEK-293 cells easily dislodge during washing steps, HEK-293 cells should only be plated out on poly-L-lysine-coated plates. For this, the wells are coated with 25 μg/ml poly-L-lysine (molecular weight >300,000; Sigma-Aldrich, St. Louis, MO) in dH$_2$O for 30 to 60 min at room temperature under sterile conditions before plating out the cells. Air dry the plates in the laminar flow. Poly-L-lysine coating of the plates is not required for CHO cells. A stock solution of poly-L-lysine solution (1 mg/ml in dH$_2$O) can be stored for a couple of weeks at 4°C or even longer at −20°C.]
3. Hank's balanced salt solution (HBSS): 118 mM NaCl, 5 mM KCl, 1 mM CaCl$_2$, 1 mM MgCl$_2$, 15 mM HEPES, 5 mM D-glucose, pH 7.4
4. Carbachol
5. Atropine
6. 1% Triton X-100 in dH$_2$O

PROTOCOL 7-1

1. HEK-293 or CHO cells heterogenously expressing mAChRs are subcultured on (poly-L-lysine-precoated) 24-well plates.
2. On the day of the binding experiment, the culture medium is removed, and the cells are washed once with HBSS (37°C).
3. After incubation of the cells in 500 μl HBSS (37°C) with or without carbachol (1 μM to 10 mM) for 0 to 60 min at 37°C in a humidified incubator, the medium is rapidly removed.
4. Cells are rinsed three times with 500 μl ice-cold HBSS.
5. Thereafter, cells are incubated in 500 μl HBSS containing a receptor-saturating concentration of 2 nM [^{3}H]NMS. Atropine (final concentration 10 to 50 μM) is included in parallel incubations to measure the nonspecific binding of [^{3}H]NMS. It is important that incubation with [^{3}H]NMS and subsequent washing be carried out at 4°C, because internalized receptors may return to the cell surface at incubation temperatures higher than 10 to 15°C.
6. After 4 h, cells are washed two times with 500 μl of ice-cold HBSS to wash away nonspecifically bound [^{3}H]NMS.
7. Cells are then solubilized in 500 μl 1% Triton X-100 and scraped off the plates. The cell lysates are transferred into scintillations vials, which receive 3.5 ml scintillation fluid (Emulsifier-Scintillator Plus, Packard Instruments, Boston, MA). After vigorous vortexing for 10 s, radioactivity is measured in a liquid scintillation counter.
8. Total [^{3}H]NMS binding is determined in quadruplicate, whereas nonspecific binding of [^{3}H]NMS is measured in duplicate. Care should be taken that depletion of radioligand by binding to specific and nonspecific

binding sites is less than 15% of total radioligand added. Internalization of mAChRs is expressed as (1 − quotient of cell surface receptors of carbachol-treated and untreated cells) × 100%.

7.2.2 ANALYSIS OF mAChR DOWNREGULATION USING [³H]QNB BINDING ASSAY TO CELL MEMBRANES

Usually, incubation of cells with mAChR agonists for longer than 60 min reduces the total mAChR number. This receptor loss can vary between 20 and 80% after 2 to 8 h of incubation with a receptor-saturating concentration of mAChR agonist. Downregulation of mAChRs may result from increased degradation of receptor protein in lysosomes and proteasomes, loss of mAChR mRNA synthesis, or increased mRNA degradation.[3]

It is important to note that in heterologous expression systems like in HEK-293 and CHO cells, which essentially do not endogenously express mAChRs, mAChR mRNA synthesis is not under the control of the endogenous mAChR promoter but is under an artificial one (usually viral). Therefore, if agonist-induced downregulation of mAChRs is to be studied in a heterologous mAChR cell system, cells should be pretreated with cycloheximide (final concentration of 350 μM in the case of CHO or HEK-293 cells) for 15 min, and incubation with the agonist should be done in the presence of cycloheximide. If other heterologous expression systems are used, it is wise to test whether the cycloheximide concentration is sufficiently high. This can be done by determining the inhibition of [³H]leucine incorporation into protein by cycloheximide. For this, cultures on 35- or 60-mm dishes are incubated with 1 μCi/ml [³H]leucine (specific activity ~60 Ci/mmol) in cell culture medium in the presence or absence of cycloheximide for 12 h in a humidified incubator at 37°C. The incorporation is terminated by removal of medium, followed by two washes with ice-cold HBSS, and addition of 4 ml of ice-cold 5% trichloroacetic acid. Cells are scraped into polypropylene tubes, incubated for 5 min on ice, and filtered through presoaked GF/C filters. Following two 4 ml washes with ice-cold trichloroacetic acid and one 4 ml wash with 95% ethanol, radioactivity on the filters is determined by liquid scintillation counting. The inclusion of cycloheximide should result in more than a 95% decrease in the incorporation of [³H]leucine.

PROTOCOL 7-2:   [³H]QNB BINDING ASSAY

Materials Required

1. [³H]Quinuclidinyl benzilate (specific radioactivity of ~45 Ci/mmol)
2. (poly-L-lysine-coated) six-well plates
3. HBSS
4. Ground glass homogenizer
5. Carbachol
6. Atropine
7. GF/C glass-fiber filters

Protocol

1. Following culturing, cells on (poly-L-lysine-coated) six-well plates are treated with or without carbachol (1 μM to 10 mM) for 0 to 12 h at 37°C in HBSS followed by three washes with ice-cold HBSS to remove agonist.
2. Cells from each treatment are scraped from the plates with 1 ml/well ice-cold HBSS and are homogenized together in a ground glass homogenizer by hand with 10 to 20 strokes.
3. The crude cell homogenates are placed on ice for immediate [³H]QNB binding assay. Alternatively, the cell membranes may be stored as pellet at −80°C after centrifugation at 10,000 g at 4°C.
4. To perform a [³H]QNB binding assay, 800 μl of crude cell homogenate is added to a 3.5 ml polypropylene tube and mixed with 100 μl [³H]QNB (receptor-saturating, final concentration of ~600 pM) in HBSS, 100 μl HBSS with or without atropine (final concentration 10 to 50 μM), and are incubated while shaking in a water bath at 37°C.
5. After 60 to 90 min, the binding reaction is terminated by vacuum filtration of the incubation mixture through presoaked Whatman GF/C glass-fiber filters and washing of the filters twice with 6 ml ice-cold HBSS.
6. Filters are transferred into scintillation vials containing 3.5 ml scintillation fluid (Packard Emulsifier-Scintillator Plus). Radioactivity in the filters is determined after vigorous vortexing of the vials for 10 s. Total and non-specific binding of [³H]QNB is determined in quadruplicate and duplicate, respectively. Care should be taken that radioligand depletion by binding to specific and nonspecific binding sites is less than 15% of total radio-ligand added.
7. Protein content of the membrane homogenates can be determined according to method of Bradford[4] or Peterson.[5] The total mAChR number is to be expressed as fmol receptor/mg protein.

7.3 [³⁵S]GTPγS ASSAY TO MEASURE COUPLING OF mAChR TO HETEROTRIMERIC G PROTEINS

Most mAChR signaling pathways are initiated by stimulation of specific heterotrimeric G proteins. The odd-numbered M_1, M_3, and M_5 mAChRs couple preferentially to the G_q family of G proteins, whereas the M_2 and M_4 mAChRs preferentially couple to the G_i family of G proteins.[3] Activation of the heterotrimeric G proteins is initiated by receptor-induced dissociation of guanosine 5-diphosphate (GDP) from the α subunit and association of guanosine 5′-triphosphate (GTP) to the α subunit.[6] Following binding of GTP to the α subunit, the α subunit can stimulate or inhibit specific enzymes or ion channels. The active state of the α subunit is terminated by GTP hydrolysis through the GTPase activity of the Gα subunit. This GTPase activity can be enhanced by interaction of the subunit with RGS proteins (regulators of G protein signaling).

In this chapter, a quantitive method to study heterotrimeric G protein activation by mAChRs is described. It measures the binding of the radiolabeled GTP analogue guanosine 5'-O-(γ-[^{35}S]thio)triphosphate ([^{35}S]GTPγS, specific radioactivity of ~1250 Ci/mmol) to the receptor-activated Gα subunit in crude cell membranes. In contrast to GTP, GTPγS is highly resistant to hydrolysis. GTPγS has a high affinity for all types of G protein α subunits. In this section, the isolation of partially purified membranes of HEK-293 or CHO cells heterologously expressing mAChRs is described, followed by presentation of the experimental procedure to measure mAChR-stimulated binding of [^{35}S]GTPγS to heterotrimeric G proteins. It is evident that this assay is also suitable for determination of desensitization of mAChR-mediated activation of the heterotrimeric G proteins. For this, intact cells on the plates are first pretreated with or without 1 μM to 1 mM carbachol for 1 to 15 min in Hank's buffered salt solution (HBSS) or any physiological buffer. Thereafter, the cells are rinsed four times with 10 ml ice-cold HBSS to remove any agonist, and the membranes are collected as described below. For more background details on the [^{35}S]GTPγS binding assay, the reader is refered to Wieland and Jakobs.[7]

PROTOCOL 7-3:   ISOLATION OF PARTIALLY PURIFIED CELL MEMBRANES OF
                 HEK-293 OR CHO CELLS

Materials Required

1. HBSS
2. Buffer A: 140 mM NaCl, 10 mM triethanolamine hydrochloride (TEA), pH 7.4
3. Buffer B: 250 mM sucrose, 10 mM Tris-HCl 1.5 mM MgCl$_2$, 1 mM ATP, 3 mM benzamidine, 100 μM phenylmethylsulfonyl fluoride, 1 μM leupeptin, pH 7.5
4. Buffer C: 20 mM Tris-HCl, 1 mM EDTA, 1 mM ATP, 3 mM benzamidine, 100 μM phenylmethylsulfonyl fluoride, 1 μM leupeptin, 1 mM dithiothreitol, pH 7.5
5. 50 mM EGTA in dH$_2$O
6. Cheesecloth
7. Nitrogen cavitation chamber

Protocol

1. Three 150 mm tissue culture dishes with HEK-293 or CHO cells (80 to 100% confluency) are necessary to isolate sufficient cell membranes. The plates are rinsed three to four times with 10 ml ice-cold HBSS, and the cells are collected in a total volume of 18 ml ice-cold HBSS and centrifuged for 10 min at 2000 g at 4°C.
2. The pellets are resuspended in 5 ml buffer A. After centrifugation for 10 min at 2000 g, the pellets are resuspended in 20 ml buffer B.

3. The cells are cracked in the nitrogen cavitation chamber under high pressure (20 bar) for 30 min at 4°C.

4. Next, ice-cold 500 μl 50 mM EGTA (final concentration 1.25 mM) are added to the cell homogenate, and the tubes are centrifuged for 15 min at 1000 g at 4°C.

5. Then, the supernatant is filtered over two layers of cheesecloth and centrifuged for 15 min at 10,000 g at 4°C. The pellets are resuspended in 2 ml buffer C.

6. The membrane suspension is centrifuged again for 15 min at 10,000 g at 4°C, taken up in 1 ml buffer C, recentrifuged for 15 min at 10,000 g at 4°C, and then resuspended in 350 μl buffer C. Aliquots of 50 μl are quick-frozen at –70°C.

PROTOCOL 7-4:   [^{35}S]GTPγS ASSAY

Materials Required

1. Buffer D: 50 mM TEA, 150 mM NaCl, 5 mM $MgCl_2$, 1 mM EDTA, 1 mM dithiothreitol, pH 7.4
2. Carbachol
3. Unlabeled GTPγS and GDP (Roche)
4. [^{35}S]GTPγS (specific radioactivity ~1250 Ci/mmol)

Protocol

1. Cell membranes are resuspended in freshly made buffer D to a final volume of 200 μl (final membrane protein concentration 0.5 to 1.0 μg/μl) and are held on ice.

2. The reaction mixture for measuring [^{35}S]GTPγS binding contains buffer D, 1 μM GDP, 0.4 nM GTP containing about 1×10^6 cpm [^{35}S]GTPγS, with or without 1 mM carbachol and 10 μl of the diluted membranes in a total volume of 100 μl.

3. The incubation in 3 ml polypropylene reaction tubes is started by adding a membrane suspension to the prewarmed reaction mixture. This is carried out in quadruplicate for 15 to 20 min at 30°C. Nonspecific binding is measured in the presence of 10 μM unlabeled GTPγS in the reaction mixture.

4. The reaction is terminated by rapid filtration of the incubation mixture through presoaked Whatman GF/C glass-fiber filters. The filters are rapidly washed with 2×12 ml ice-cold 50 mM Tris, 5 mM $MgCl_2$, pH 7.4.

5. The filters are put into 8 ml scintillation vials, after which 3.5 ml scintillation fluid is added. After vigorous shaking for 30 min at room temperature, radioactivity is measured in a liquid scintillation counter. The results are expressed as fmol specific [^{35}S]GTPγS binding per mg protein or, optionally, per fmol mAChRs.

7.4 ANALYSIS OF mAChR SIGNAL TRANSDUCTION PATHWAYS

7.4.1 PHOSPHOLIPASE C STIMULATION

Stimulation of phospholipase C (PLC) leads to the breakdown of phosphatidylinositol 4,5-bisphosphate in the plasma membrane into the two second messengers, diacylglycerol and inositol 1,4,5-trisphosphate (IP_3). Diacylglycerol is able to stimulate protein kinase C. IP_3 opens IP_3-regulated calcium channels in the endoplasmic reticulum, leading to elevation of the intracellular calcium concentration. IP_3 is rapidly dephosphorylated by inositol phosphatases into inositol. For this reason, PLC assays are routinely performed with cells pretreated with 10 mM LiCl for 5 to 15 min. LiCl inhibits inositol phosphatases and allows accumulation of inositol 1-phosphate (IP_1), inositol 1,4-bisphosphate (IP_2), and IP_3 over the period of receptor stimulation. In the present experimental procedure, we describe an anion-exchange chromatography method to either isolate total inositol phosphates (= IP_3 + IP_2 + IP_1) or IP_3 alone. The PLC assay is particularly useful for the $G_{q/11}$-coupled M_1, M_3, and M_5 mAChRs, which robustly activate PLC at low agonist concentrations. Activation of PLC by M_2 and M_4 mAChRs is much weaker and requires ~100-fold higher concentrations of mAChR agonist. If desensitization of mAChR-stimulated PLC is to be analyzed, [^{3}H]*myo*-inositol-labeled cells are first stimulated with 1 µM to 1 mM carbachol for 0 to 60 min in HBSS without LiCl. The plates are then washed three times with warm (37°C) HBSS without LiCl to remove muscarinic receptor agonist. Then, cells are preincubated for 5 to 10 min in HBSS + 10 mM LiCl to inhibit inositol phosphatases. Then, carbachol is added to the plates.

Materials Required

1. Dulbecco's modified Eagle's/F-12 medium
2. [^{3}H]*myo*-inositol (specific activity 10 to 25 Ci/mmol)
3. HBSS
4. Carbachol
5. Methanol, chloroform, and dH_2O (all ice-cold)
6. Polypropylene tubes, cell scrapers
7. 50 mM ammonium formate solution
8. 1 M ammonium formate + 100 mM formic acid solution
9. 2 M ammonium formate + 100 mM formic acid solution
10. Bio-Rad AGX1-8 anion exchange resin, ionic form: formate 200 to 400 mesh
11. Columns, column holders, and cell scrapers (rubber policemen)
12. Packard UltimaGold scintillation fluid, glass scintillation vials
13. 1 M lithium chloride in H_2O
14. (Poly-L-lysine precoated) six-well plates or 35 mm cell culture dishes

Preparing the Bio-Rad AGX1-8 Anion Exchange Columns

1. Wash a spoon of resin with 300 ml dH_2O in a 1 l beaker, let resin sink down, and decant the water.
2. Repeat the washing of the resin. Resuspend the resin in 300 ml H_2O, and stir the solution while filling the columns with 1 ml resin bed volume (corresponding to 1 to 2 ml resin suspension).

Regeneration of the Columns

1. The columns containing 1 ml of resin can be reused after washing with 5 ml 2 M ammonium formate + 0.1 M formic acid solution, followed by washing with 10 ml dH_2O.
2. Regenerated columns can be stored, dried, at 4°C. The columns can be regenerated six times before disposal.

PROTOCOL 7-5

1. Harvest a 150 mm plate with 80% confluent cells.
2. After centrifugation, cells are taken up in 45 ml cell culture medium with 5% fetal or bovine calf serum. Two wells on a six-well plate receive 2 ml of cell suspension. These wells are later used for protein determination (see below). Add to the remaining 40 ml cell suspension the following: 40 to 80 µl [^{3}H]*myo*-inositol (final concentration of 1 to 2 µCi/ml). Cells are incubated in a humidified 5% CO_2 incubator. If mAChR-mediated PLC activaton is expected to be weak, incubation of the cells with [^{3}H]*myo*-inositol can be extended from 16 h to 48 h, or custom-made inositol-free DME medium (available from Invitrogen/BRL, Carlsbad, CA) with [^{3}H] *myo*-inositol can be used.
3. To each 35 mm dish or well on a six-well plate, 2 ml of this suspension is added and incubated at 37°C under 5% CO_2 in a humidified incubator. Immediately following plating, a 100 µl aliquot from the cell suspension is taken for measurement of total radioactivity added. Remember that HEK-293 cells should be grown on poly-L-lysine-coated plates.
4. After 16 to 48 h, before washing the plates, another aliquot of 100 µl for measurement of free radioactivity is taken. Cellular uptake of radioactivity should be 40 to 80%. The cells are washed twice with 2 ml of HBSS (37°C) to remove free [^{3}H]*myo*-inositol.
5. Cells are then incubated for 5 to 15 min with 2 ml 10 mM LiCl in HBSS in a humidified incubator at 37°C. Stimulation is started with the addition of 20 µl carbachol (final concentration between 1 nM and 1 mM) or vehicle at 37°C in a humidified incubator.
6. After 0 to 60 min, medium is rapidly removed, and the dishes or six-well plates are put on ice. Then, 0.5 ml of ice-cold methanol is added immediately to the dishes/wells.

7. After scratching the cells from the dishes/plates, the homogenate is transferred to the polypropylene tubes on ice. The plates are rinsed once again with 0.5 ml ice-cold methanol, which is transferred to the same tube.

8. To the tubes, 1 ml of ice-cold chloroform and 0.5 ml ice-cold dH_2O are added. The tubes are capped and then vigorously vortexed for 10 s at maximum speed. The tubes are centrifuged for 10 to 15 min at 2400 rpm at 4°C (Heraeus Megafuge 1.0 R). During centrifugation, regenerated columns are washed with dH_2O.

9. The complete (1 ml) upper-phase water fractions are transferred to the columns.

10. Following elution of 1 ml water fractions from the columns, the columns are washed with 6 ml H_2O followed by 5 ml of 50 mM ammonium formate. From the total eluant (12 ml), an aliquot of 1 ml may be taken for measurement radioactivity to determine cellular [³H]*myo*-inositol and [³H]-glycerophosphoinositols.

11. Then, new collection vials are placed under the columns, and 6 ml of 1 M ammonium formate + 0.1 M formic acid solution are added to the columns. From this eluant, 1 ml of eluant is measured for radioactivity. This eluant contains total [³H]inositol phosphates.

12. For protein determination, 1 ml of HBSS is added to the wells reserved for protein determination (see above). Cells are scratched from the plates and transferred to an Eppendorf microfuge tube for protein determination.

Alternative Protocol for Separate Collection of [³H]IP₃ and [³H]IP₂ + [³H]IP₁

1. After elution with 6 ml H_2O and 5 ml of 50 mM ammonium formate, change the collection vials.

2. Add 10 ml of 0.4 M ammonium formate + 0.1 M formic acid solution to the column to elute IP₁ and IP₂ from the column.

3. Then, after changing collection vials, 6 ml 1 M ammonium formate + 0.1 M formic acid solution are added, and IP₃ is eluted from the column. Take 1 ml aliquot from the eluant for radioactivity counting.

7.4.2 PHOSPHOLIPASE D ACTIVATION

As is the case with PLC stimulation, M_1, M_3, and M_5 mAChRs couple efficiently to phospholipase D (PLD), whereas M_2 and M_4 mAChRs are much less able to stimulate PLD. However, in contrast to PLC stimulation, the precise mechanisms by which mAChRs activate PLD is far from understood. PLD preferentially hydrolyzes the main plasma-membrane-embedded phospholipid phosphatidylcholine into phosphatidic acid (PA) and choline. PLD can also catalyze a transphosphatidylation reaction in which H_2O is substituted for by a primary alcohol, leading to the generation of metabolically stable phosphatidylalcohols. This transphosphatidylation reaction is generally utilized to determine PLD activation in the presence of 400 mM ethanol, resulting in the formation of the relatively stable phosphatidylethanol (PtdEtOH).

Materials Required

1. [³H]oleic acid (specific radioactivity of ~24 Ci/mmol)
2. (Poly-L-lysine-coated) 35 mm dishes or six-well plates
3. HBSS
4. Methanol and chloroform (both ice cold)
5. Chloroform/methanol (1/1 by volume)
6. Polypropylene tubes (3.5 and 6 ml), cell scrapers
7. Precoated silica gel 60C plates (Merck, Whitehouse Station, NJ)
8. Phosphatidic acid (Sigma)
9. Phosphatidylethanol (Avanti, Alabaster, AL)
10. Iodine crystals (Merck)
11. Ethylacetate/isooctane/acetic acid/water (13:2:3:10 by volume)

PROTOCOL 7-6

1. Harvest a 150 mm plate with 80% confluent cells.
2. After centrifugation, cells are taken up in 45 ml cell culture medium with 5% fetal or bovine calf serum. Two dishes or two wells on a six-well plate receive 2 ml of cell suspension. These wells are later used for protein determination (see below). Add to the remaining 40 ml cell suspension, [³H]oleic acid (final concentration of 2 μCi/ml). Immediately following plating on six-well plates or 35 mm dishes, 100 μl aliquot from the cell suspension is taken for measurement of total radioactivity added. Remember that HEK-293 cells should be grown on poly-L-lysine-coated plates.
3. To each 35 mm dish or well of a six-well plate, 2 ml of this suspension is added and incubated at 37°C in a humidified 5% CO_2 incubator. The assay should be performed with triplicate or quadruplicate samples.
4. After 16 to 24 h, before washing the dishes/plates, another aliquot of 100 μl for measurement of free radioactivity is taken. The incorporation of radioactivity should be 40 to 80%. Cells are washed twice with 2 ml of HBSS (37°C) to remove free [³H]oleic acid. Then, the cells are equilibrated in HBSS for 10 min in a humidified incubator at 37°C.
5. Fresh HBSS containing 400 mM ethanol at 37°C is added with or without carbachol to measure [³H]PtdEtOH formation. Also incubate two dishes/plates in the absence of ethanol and carbachol. The dishes/plates are placed in a humidified 37°C incubator.
6. After 2 to 10 min, the reaction is terminated by rapid removal of the incubation medium and addition of 1 ml ice-cold methanol to the dishes/plates.
7. Cells are scraped off from the culture dishes and transferred into 6.5 ml polypropylene tubes on ice.

8. Cell lysates are mixed thoroughly with 0.5 ml of ice-cold water and 1 ml of ice-cold chloroform, followed by centrifugation in an Heraeus table centrifuge for 10 min at 3400 rpm (4°C). The lower organic phase containing [^{3}H]PtdEtOH is pipetted into a 3.5 ml reaction tube and dried by vacuum centrifugation at room temperature. Store samples at –20°C.

9. Pencil-mark the precoated Silica Gel 60C TLC plates (Merck) approximately 2 cm from the bottom of the plate to indicate where the samples are to be applied. Preload each lane on the TLC plate with the lipid standards PA (2 µl of a 5 mg/ml stock solution in chloroform–methanol) and PtdEtOH (2 µl of a 5 mg/ml stock solution in chloroform–methanol).

10. Then, the samples are dissolved in 25 µl chloroform/methanol (1:1) and spotted in 10 µl aliquots on the plate.

11. Freshly prepare an organic mixture of ethylacetate/isooctane/acetic acid/water (13:2:3:10 by volume) in a separatory funnel. After mixing, the upper phase is collected as TLC solvent and added to the solvent tank, which contains a piece of Whatman paper for saturation of the tank atmosphere. Take care that the level of the solvent is about 0.5 cm below the sample origin on the plate.

12. After 60 min, place the TLC plate in the equilibrated solvent tank, and cover the tank.

13. After 1.5 to 2 h (i.e., until the solvent has migrated within 2 to 3 cm from the top of the plate), the liquid front on each lane should be marked by a pencil, and the plates are dried in the hood.

14. Then, the plates are placed in an iodine tank. After 15 min, PA (lower band) and PtdEtOH (upper band) standards can be detected by yellow-brown staining and should be pencil-marked. As iodine staining disappears fast, pencil-marking should be done quickly.

15. The silica gel plates are sprayed with dH$_2$O. PtdEtOH samples are scraped from the plates onto folded paper (use filter mask) and mixed with 3.5 ml liquid scintillation fluid for radioactivity counting. The remaining part of each lane (i.e., until the pencil-marked sample origin) is also collected for radioactivity counting.

16. Measurement of [^{3}H]PtdEtOH formation in the absence of ethanol is used to determine background radioactivity. Radioactivity in these samples should be lower than or equal to the radioactivity determined in samples of plates incubated in the presence of ethanol without carbachol. The protein amount is measured by Bradford's method[4] in separate culture dishes that contain no radioactivity. The formation of [^{3}H]PtdEtOH is expressed as a percentage of total labeled phospholipids collected from each lane. Agonist-induced [^{3}H]PtdEtOH formation in stable HEK-293 cells expressing M$_1$ or M$_3$ mAChRs usually amounts to approximately 0.3% of total phospholipids.

7.4.3 Elevation of $[Ca^{2+}]_i$

The $[Ca^{2+}]_i$ assay is used for the measurement of the intracellular free calcium levels in various cell types. For this, cells are loaded with Fura-2 AM, the acetoxymethylester of the calcium indicator Fura-2. Following cellular uptake, Fura-2 AM is hydrolyzed into Fura-2 by intracellular esterases. The charged Fura-2 is much less able to cross the plasma membrane. The binding of Ca^{2+} ions to Fura-2 changes the excitation/fluorescence spectrum of Fura-2. As stimulation of mAChRs in a variety of cells leads to rapid changes in $[Ca^{2+}]_i$, these changes will be reflected in the fluorescence signal of Fura-2. At the end of each measurement, the maximum fluorescence signal is determined by digitonin-induced plasma membrane permeabilization and influx of extracellular Ca^{2+} ions (concentration of 1 to 2 mM). The minimum signal is determined following the addition of EGTA (final concentration of at least 5 mM) to the digitonin-permeabilized cells. EGTA binds all free calcium ions. The intracellular calcium concentration is subsequently calculated using the computer program provided by the spectrofluorimeter vendor. It is important to note that M_1, M_3, and M_5 mAChRs potently increase $[Ca^{2+}]_i$ through coupling to PLC. However, M_2 and M_4 mAChRs are also able to increase $[Ca^{2+}]_i$, through coupling with PLC (albeit at higher agonist concentrations) and other enzymes, like, for example, sphingosine kinase.[8]

Materials Required

1. Phosphate-buffered saline without Ca^{2+} and Mg^{2+} (Invitrogen/BRL)
2. HBSS
3. 1 mM Fura-2 AM (Molecular Probes, Eugene, OR) [Dissolve 50 μg in 50 μl of dimethylsulfoxide under dimmed light, and store 10 μl aliquots wrapped in aluminum foil at −20°C.]
4. Carbachol
5. 50 ml blue cap Falcon tubes, aluminum foil, polystyrene cuvettes (Sarstedt, Nürnberg, Germany)
6. Spectrofluorimeter
7. Digitonin stock solution (1.0 to 1.5 g/100 ml H_2O, warm to 100°C to dissolve, store at 4°C)
8. 500 mM EGTA, pH 8.0

Protocol 7-7: Preparation and Fura-2 Loading of Cells

1. Harvest a 150 mm plate with 80% confluent cells.
2. Centrifuge cell suspension at 1000 rpm for 5 min at room temperature in a Heraeus table centrifuge.
3. Remove supernatant, and resuspend the cell pellet in 10 ml HBSS (room temperature).

4. Add 10 µl of 1 mM Fura-2 AM (final concentration of 1 µM) under dimmed light. Wrap the centrifuge tubes in aluminum foil. Place the centrifuge tubes in a secured vertical position in the dark at room temperature.
5. After 60 min, centrifugate the cells at 1000 rpm for 5 min at room temperature in a Heraeus centrifuge. Avoid direct sunlight.
6. Resuspend the cell pellet in 1 ml of HBSS. Add 9 ml of HBSS (room temperature), and repeat centrifugation at 1000 rpm for 5 min.
7. Resuspend the cell pellet in 1 ml of HBSS (room temperature). Add 9 ml of HBSS (room temperature) to the cell suspension (cell concentration approximately 1E6 cells/ml). Keep the cell suspension stored in the dark in a cabinet at room temperature.
8. Add 1 ml of HBSS (room temperature) to the cuvette. Then, add 1 ml of cell suspension to the cuvette, and place the cuvette in the spectrofluorimeter. Stimulate cells with carbachol. To measure 100% value (maximal fluorescence due to permeabilization of the plasma membrane and influx of extracellular calcium ions), add 100 µl of digitonin solution to the cuvette. After a constant level of fluorescence is obtained (usually within 10 to 20 s), add 100 µl 500 mM EGTA, pH 8.0, to the cuvette to determine 0% value.

7.4.4 INHIBITION OF cAMP ACCUMULATION

The M_2 and M_4 mAChRs couple efficiently to the inhibitory G_i proteins to decrease the activity of adenylyl cyclase at low concentrations. A relatively simple method to determine the functional responsiveness of these mAChR subtypes is to assess the inhibition of forskolin-stimulated accumulation of intracellular cAMP by agonists in the presence of theophylline. Forskolin activates adenylyl cyclase directly, independent of the stimulatory G_s protein. The cAMP phosphodiesterase inhibitor theophylline blocks hydrolysis of cAMP produced upon forskolin treatment. It should be kept in mind that G_q-coupled mAChR subtypes (M_1 and M_3 mAChRs) may increase intracellular cAMP levels through Ca^{2+}-calmodulin stimulation of adenylyl cyclase in some cell types, but may also decrease cAMP levels in the absence of a cAMP phosphodiesterase inhibitor, due to stimulation of a calmodulin-activated phosphodiesterase in other cell types. Thus, experiments on alterations in cAMP accumulation should be interpreted in the context of the experimental conditions used. The first part of this section describes preparation of the cation-exchange columns and stimulation of the cells in the presence of forskolin and muscarinic agonist. The second part describes the isolation of cAMP from the cellular extract by cation-exchange chromatography and quantification of cAMP by a competitive cAMP protein-binding assay, as originally published by Gilman.[9]

Materials Required

1. [^{3}H]cAMP (specific radioactivity of ~30 Ci/mmol)
2. Forskolin

3. Theophylline
4. HBSS
5. Bio-Rad AG 50W-X4, 200 to 400 mesh
6. 35 or 60 mm (poly-L-lysine-coated) tissue culture dishes

Preparing the Bio-Rad AG 50W-X4 Columns

1. Wash a spoon of resin with 300 ml H_2O in a 1 l beaker, let resin sink down, and decant the water.
2. Repeat the washing of the resin. Resuspend the resin in 300 ml H_2O, and stir the solution while filling the columns with 1 ml resin bed volume (corresponding with 1 to 2 ml resin suspension).

Regeneration of the Bio-Rad AG 50W-X4 Column

The 1 ml Bio-Rad AG 50W-X4 columns are regenerated shortly before use by 2 × 1 ml 1 M HCl, 2 × 1 ml 0.5 M NaOH, and 6 × 1 ml dH_2O. The columns can be regenerated at least three times.

PROTOCOL 7-8

1. Cells are cultured to near confluency on 35 or 60 mm tissue culture dishes (poly-L-lysine-coated in case of HEK-293 cells, see above).
2. Cells are washed three times with 2 ml HBSS at 37°C. Then, 2 ml of HBSS containing 5 mM theophylline is added, and the plates are placed in a 37°C humidified incubator for 20 min. Then, forskolin (dissolved in 50% ethanol) is added to the plates (final concentration of 50 to 100 μM). Receptor agonists (e.g., carbachol) are added to the cells simultaneously with forskolin in the same solution.
3. After incubation for 5 to 20 min at 37°C, cells are rinsed twice with 2 ml of ice-cold HBSS. This is followed by the addition of ice-cold 2 ml of 5% (w/v) trichloroacetic acid to the plates.
4. Cell lysates are scraped off the plates and transferred into test tubes on ice.
5. A recovery standard consisting of 0.7 nCi [3H]cAMP is added to each tube, and the cell lysates are centrifuged at 2700 g for 15 to 30 min at room temperature.
6. The supernatant of the cell lysates is then poured over the Bio-Rad AG 50W-X4 cation-exchange columns to partially purify cellular cAMP from the extract. The pellet is saved for protein determination.
7. The columns are subsequently washed with 3 ml dH_2O, and cAMP is eluted with further addition of 3 ml dH_2O.
8. Then, 1 ml of column eluant is mixed with scintillation fluid to determine the efficiency of [3H]cAMP recovery from the column by comparing

radioactivity in the eluant with radioactivity from an aliquot of the [³H]cAMP recovery standard.

9. The protein pellet is hydrolyzed with 1 to 2 ml of 1 M NaOH at 60 to 70°C for 30 to 60 min, and the protein amount is measured by a modification of the Lowry protein assay method.[5]

Protocol 7-9:   cAMP Competitive Protein-Binding Assay

Materials Required

1. 0.3 M sodium acetate, pH 4.0
2. Regulatory subunit of protein kinase A (Sigma)
3. [³H]cAMP (specific radioactivity of ~30 Ci/mmol)
4. cAMP (Roche)
5. 20 mM KH_2PO_4, pH 6.0
6. 0.45 μm nitrocellulose filters

Protocol

1. In a final assay volume of 200 μl, 10 to 160 μl of eluant should be added. The correct volume of the eluant is dependent on the cell type, length of time that the cells were incubated with forskolin, and the agonist used. As the eluant should remove 75% of [³H]cAMP added to the assay tube, a pilot study should be performed to determine the required volume.

2. In addition to column eluant, 30 μl of an assay mix containing 3.3 mg/ml bovine serum albumin (BSA), 0.3 M sodium acetate, pH 4.0, and 1 pmol [³H]cAMP is added. The reaction is initiated by the addition of 10 μl containing approximately 3 μg of the regulatory subunit of protein kinase A (Sigma). The regulatory subunit of protein kinase A should be able to bind approximately 30% of the [³H]cAMP in the assay mix in the absence of competing cAMP.

3. In parallel, a standard curve is made by substituting column eluant with varying amounts of unlabeled cAMP of known concentration (0.5 to 30 pmol) in the assay tubes. The amount of accumulated cAMP can be determined by using this standard curve.

4. The assay tubes are incubated on ice for 2 to 4 h, and the reaction is terminated by the addition of 1 ml of ice-cold 20 mM KH_2PO_4, pH 6.0.

5. The assay volume is then filtered over 0.45 μm nitrocellulose filters followed by three washes of 3 ml ice-cold 20 mM KH_2PO_4, pH 6.0.

6. Filters are subsequently dissolved in Packard Emulsifier-Scintillator Plus scintillation fluid, and radioactivity is counted in a liquid scintillation counter. Triplicate plates are used for each treatment, and each plate is assayed in triplicate in the binding assay.

The procedure described here can also be used to determine agonist-induced mAChR desensitization. For this, cells are first pretreated with muscarinic agonist in the absence of theophylline or forskolin for 0 to 60 min at 37°C. Then, cells are washed three times to remove the agonist, and cells are pretreated with 5 mM theophylline for 20 min and further incubated in the presence of carbachol and forskolin, as described above.

ACKNOWLEDGMENTS

This work was supported by the Deutsche Forschungsgemeinschaft and an intramural grant of the Universitatsklinikum Essen.

REFERENCES

1. Shenoy, S.K. and Lefkowitz, R.J., Multifaceted roles of ß-arrestins in the regulation of seven-membrane-spanning receptor trafficking and signaling, *Biochem. J.*, 375, 503, 2003.
2. Ferguson, S.S.G., Evolving concepts in G protein-coupled receptor endocytosis: the role in receptor desensitization and signaling, *Pharmacol. Rev.*, 53, 1, 2001.
3. Van Koppen, C.J. and Kaiser, B., Regulation of muscarinic acetylcholine receptor signaling, *Pharmacol. Ther.*, 98, 197, 2003.
4. Bradford, M.,M., A rapid and sensitive method for the quantitation of microgram quantities of protein utilizing the principle of protein-dye binding, *Anal. Biochem.*, 72, 248, 1976.
5. Peterson, G.L., A simplification of the protein assay method of Lowry *et al.* which is more generally applicable, *Anal. Biochem.*, 83, 346, 1977.
6. Neer, E.J., Heterotrimeric G proteins: organizers of transmembrane signals, *Cell*, 80, 249, 1995.
7. Wieland, T. and Jakobs, K.H., Measurement of receptor-stimulated guanosine 5'-O-(γ-thio)triphosphate binding by G proteins, *Meth. Enzymol.*, 237, 3, 1994.
8. Meyer zu Heringdorf, D., Lass, H., Alemany, R., Laser, K.T., Neumann, E., Zhang, C., Schmidt, M., Rauen, U., Jakobs, K.H., and Van Koppen, C.J., Sphingosine kinase-mediated Ca^{2+} signaling by G-protein-coupled receptors, *EMBO J.*, 17, 2830, 1998.
9. Gilman, A.G., A protein binding assay for adenosine 3':5'-cyclic monophosphate, *Proc. Natl. Acad. Sci. USA*, 67, 305, 1970.

8 Single-Molecule Analysis of Chemotactic Signaling Mediated by cAMP Receptor on Living Cells

*Masahiro Ueda, Yukihiro Miyanaga,
and Toshio Yanagida*

CONTENTS

8.1 INTRODUCTION

8.1.1 SINGLE-MOLECULE DETECTION TECHNIQUES

Single-molecule detection techniques (SMDs) have made it possible to visualize individual biomolecules in real time, both *in vitro* and in living cells.[1–3] SMDs have been successfully applied to a variety of biomolecules, and the results derived from this approach have yielded new insights into the molecular mechanisms of biomolecules.[4–8] In this chapter, we will review SMDs as applied to G protein-coupled receptors (GPCRs).[9–10]

The ligand-binding characteristics of GPCRs have been elucidated using radiolabeled ligand-binding techniques. The techniques use radioisotopes, which can prepare the probes so they are identical to the native ligands in their chemical properties. However, radioligand-binding techniques require large populations of cells to detect the radiation signals. This sometimes imposes limitations on its application to small populations of cells, such as those prepared primarily from tissues. Using fluorescence-labeled ligand analogs is one approach to overcoming this obstacle. By combining this technique with fluorescence microscopy or fluorescence correlation spectroscopy, ligand–receptor interactions can be visualized in single cells.[11] Furthermore, using novel optical microscopes and novel fluorescent probes, we succeeded in visualizing single ligand molecules bound to GPCRs in living cells.[9] The locations, movements, and association/dissociation events of the ligand–GPCR complexes can be detected quantitatively on individual cells at the single-molecule level. This technical progress would be critical to elucidate the molecular mechanisms governing GPCR signaling because spatial and temporal changes in ligand-binding characteristics, receptor localization, cytoskeletal organization, and membrane organization, such as microdomains, may be important in determining the exact functions of GPCRs in living cells.

SMDs have been successfully used to visualize unitary reactions in signal transduction, including ligand binding, dimerization, complex formation, phosphorylation, diffusion, and conformational changes of signaling molecules in living cells.[4–10] One can follow such reactions in the context of the cellular environments. When cells respond to environmental stimuli, intracellular signaling networks undergo dynamic changes to process signals, giving rise to spatial and temporal modulation in intracellular environments, such as ion concentrations, lipid compositions, and cytoskeletal organization. Even without stimulation, living cells have intrinsic heterogeneity in their intracellular environments. Such heterogeneity could be the molecular basis for the diversity in the properties of biomolecules in living cells. In ensemble measurements, which include a large number of cells and molecules, the variations in molecular properties are averaged, and then individual variations are obscured. SMDs reveal the distributions of molecular properties in relation to temporal and spatial changes of cells' physiological states.

SMDs also have the potential to detect transient intermediates in biochemical reactions, because of their ability to follow the time course of reactions of individual molecules. SMDs do not require synchronization of biochemical reactions to capture transient intermediates, which is in contrast to ensemble measurements, in which the intermediates can be detected only when a large number of molecules are synchronized. In fact, SMDs have successfully detected new intermediate states in some biomolecules, such as molecular chaperon, ribozyme, flavin enzyme, and F1-ATPase *in vitro*, offering new insights on molecular mechanisms.[12–18]

This chapter begins with a brief introduction of chemotactic responses by taking *Dictyostelium* cells as an example. This is followed by technical considerations, including theory and optical configuration for single molecule imaging, especially total internal reflection fluorescence microscopy (TIRFM). In the next sections, single-molecule ligand-binding analysis using this technique is reviewed.

8.1.2 CHEMOTACTIC SIGNALING AND GPCRS

Chemotaxis is a fascinating phenomenon in which cells move toward the source of attractant molecules (Figure 8.1a). This directional response of cells has been found in a range of biological processes, including immunity, neuronal patterning, morphogenesis, and nutrient finding. The cellular slime mold, *Dictyostelium discoideum*, is a model organism for elucidating molecular mechanisms of chemotaxis.[19–23] *Dictyostelium* cells exhibit chemotaxis to cyclic adenosine 3,5-monophosphate (cAMP), which binds to cAMP receptors (cARs), a family of GPCRs (Figure 8.1b). The binding of cAMP molecules to cARs leads to the activation of their coupled trimeric G proteins, which is followed by the activation and signaling through a number of pathways, including Guanylyl cyclase, Ras, PI3-kinase, PTEN, and Akt/PKB (Figure 8.1b). Receptor stimulation ultimately gives rise to actin polymerization at the leading edge of the cells for pseudopod formation, and myosin II assembly at the rear for tail retraction (Figure 8.1c). Therefore, the chemical gradient of extracellular signals is converted into intracellular signals to form anterior–posterior polarity in the motile apparatus of cells.

From a viewpoint of single-molecule nanobiology, we will describe the chemotactic signaling in *Dictyostelium* cells. *Dictyostelium* cells are extremely sensitive to cAMP gradients. At the threshold stimulation, the mean cAMP concentration around the cells is estimated to be about 1 nM, with a spatial gradient of 4 pM/μm.[24–25] *Dictyostelium* cells are 10 to 20 μm in size and contain about 40,000 receptors on a basal cell surface with a Kd of about 100 nM.[19,25] This suggests that, on average, 396 receptors are occupied with cAMP, and the differences in receptor occupancy between the front and rear half of the cell are about 4 to 8 molecules. Because ligand binding to the receptors is a stochastic process, receptor occupancy should fluctuate with both time and space. If ligand–receptor binding is a Poisson process, fluctuations in receptor occupancy are estimated to be a root of the averaged occupancy, therefore, about 20. That is, the fluctuations in receptor occupancy are greater than the spatial differences in receptor occupancy across the cell body. This implies that receptor occupancy could be transiently reversed, with respect to the direction of chemical gradients. Such reversal in receptor occupancy would be noise for directional sensing. However, *Dictyostelium* cells can exhibit chemotaxis in these noisy environments. Thus, *Dictyostelium* cells can detect a small signal with high accuracy under the strong influence of stochastic noise. Moreover, G protein-linked signal transduction processes would also be noisy, because the signaling molecules are a few tenths of a nanometer in size, and relatively small copies of each signaling

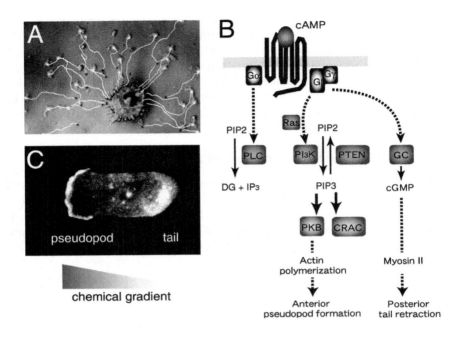

FIGURE 8.1 Chemotaxis of *Dictyostelium discoideum* cells. (a) Cells migrate toward an aggregation center. The attractant molecules, cAMP, are secreted from the center. White lines are trajectories of individual cells. Yellow and red dots represent start and end points, respectively. (b) GPCRs and G protein-linked signaling pathways for chemotaxis of *Dictyostelium* cells. PLC, phospholipase C; PI3K, PtdIns-3-kinase; PTEN, PtdIns-3-phosphatase; GC, guanylyl cyclase; CRAC, cytosolic regulator of adenylyl cyclase; PKB, protein kinase B. (c) Localization of actin-GFP. Receptor stimulation activates their coupled G protein, and ultimately gives rise to the assembly of F-actin at the pseudopod and myosin II at the side and tail regions.

molecules are involved in signal processing. Therefore, signal transduction is carried out with stochastic transducers that process noisy signals. How cells obtain information about gradient direction from such noisy inputs is a critical question for chemotaxis. The signal reception system must have the abilities to either utilize or exclude stochastic fluctuations. In general, it is important to ask how signal transduction is carried out at the verge of stochastic and thermal noise, because many types of cells show extreme sensitivity to environmental stimuli.

To reveal the mechanisms of stochastic signaling processes, it is important to monitor directly how signaling molecules behave in living cells. Recently, we succeeded in monitoring single fluorescent-labeled cAMP molecules bound to the receptors in living cells.[9] Using this technique, we can examine how chemotactic signals are input into cells. As described below, the signaling activities of the cAMP receptors were monitored and localized on the living cells undergoing chemotaxis.

8.2 TOTAL INTERNAL REFLECTION FLUORESCENCE MICROSCOPY (TIRFM)

In this section, we deal briefly with the theoretical and technical aspects of TIRFM. Currently, TIRFM is one of the most popular microscopy techniques used for single molecule imaging.

There are two types of TIRFM, an objective-type and a prism-type, that would be selected according to the experiments to be performed. For single-molecule imaging experiments in living cells, we usually used an objective-type TIRFM. Figure 8.2 illustrates a typical configuration of an objective-type TIRFM on an inverted microscope. This type of microscopes has free space above the specimen. This configuration can be applied to thicker samples, such as living cells. Also, it is relatively easy to combine with other techniques utilized in cell biology, such as microinjection, micromanipulation, and electric recording.[26–28] On the other hand, for experiments that require extremely low background and complex optical systems,

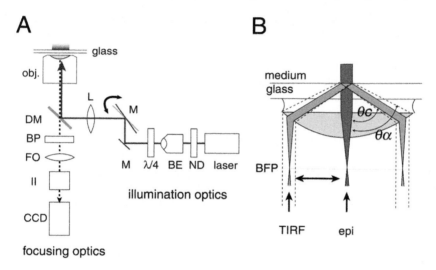

FIGURE 8.2 Objective-type total internal reflection fluorescence microscopy (TIRFM). (a) Configuration of objective-type TIRFM. ND, neutral density filter; BE, beam expander; $\lambda/4$, quarter-wave plate; M and M, mirrors; L, focusing lens; DM, dichroic mirror; obj, objective lens; BP, bandpass filter; FO, focusing optics; II, image intensifier; CCD, charge-coupled device camera. Switching between epifluorescence microscopy and TIRFM can be performed by tilting a single mirror (M), which is located at the focus of the lens (L). (b) Light path of the incident laser. The laser beam is focused on the back focal plane (BFP) of the objective lens. θ c is the critical angle of the glass–water interface. θ α is defined as NA = n $\sin\theta$ α, where n is the reflective index of glass and NA is the numerical aperture of the objective. When the incident beam is positioned at the objective edge between θ α and θ c, the beam is totally internally reflected, generating an evanescent field at the glass surface. The illumination mode can be switched from TIR to standard epi by shifting the position of the beam focus at the BFP from the edge to the center.

such as single-molecule fluorescence spectroscopy, fluorescence resonance energy transfer (FRET), and simultaneous measurements of fluorometry and nanometry,[29–31] we used a prism-type TIRFM because this type of microscope has fewer restrictions on the design of optical systems. Because this review is intended to present applications of single-molecule imaging techniques to cell biology, we focus here on the objective-type TIRFM. The configurations of a prism-type TIRFM and the applications are described elsewhere.[10]

8.2.1 THEORY: EVANESCENT FIELD

TIRFM can be used to observe fluorophores near a glass surface. To illuminate fluorophores selectively near a glass surface by the excitation lights, TIRFM utilizes an "evanescent field" generated at the interface between the aqueous solution and the glass surface. When the excitation light for fluorophores is incident above some "critical angle" upon the glass/water interface, the light is totally internally reflected and generates a thin electromagnetic field in the water (Figure 8.2b). This field is called the evanescent field. The intensity of the evanescent field decays exponentially with the distance from the glass surface; therefore, fluorophores further from the surface are not excited. That is, TIR provides a means to excite fluorophores near the glass surface. The intensity and penetration depth of the evanescent field are critical on single-molecule imaging of fluorophores. For the intensity of the evanescent wave I_{eva}, we have,

$$I_{eva} = I_o \exp\left[-z / d\right] \qquad (8.1)$$

where z and d are the perpendicular distance from the glass surface and the penetration depth [1/e value in Equation (8.1)], respectively; I_o is the intensity of the evanescent wave at $z = 0$. The penetration depth d can be written as

$$d = \frac{\lambda}{4\pi\sqrt{n_1^2 \sin^2 \theta - n_2^2}} \qquad (8.2)$$

where l is the wavelength of the incident light in vacuum, and n_1 and n_2 are refractive indices for glass ($n_1 = 1.52$) and water ($n_2 = 1.33$), respectively. For cells, the refractive index n_2 is about 1.37. And, θ is the incidence angle measured from the norm (z axis). If we take $n_1 = 1.52$ (glass), $n_2 = 1.33$ (water), l = 532 nm, and $\theta = 64°$, then the penetration depth $d = 132$ nm. Thus, the fluorophores can be selectively excited near the glass/water interface. Note that a higher-power laser would be required as a source of the incident light to observe fluorophores further from the glass surface. According to Equations (8.1) and (8.2), the intensity of the evanescent field near the glass/water interface is maximal at $\theta = \sin 1 (n_2/n_1)$. This angle is known as the critical angle. In practice, the excitation light is incident at a near-critical angle to obtain a brighter field of evanescent wave. The theoretical aspects of evanescent field are discussed in detail by Axelrod[32–34] and Fornel.[35]

8.2.2 CONFIGURATION OF OBJECTIVE-TYPE TIRFM

Figure 8.2 illustrates the optical configuration of objective-type TIRFM on an inverted microscope. An objective lens with high numerical aperture is mounted on the inverted microscope (e.g., IX-71, Olympus, Japan). A laser beam is passed through a beam expander (e.g., LBED, Sigma Koki, Japan) to adjust its diameter. When the laser polarized linearly is used, the polarization of the beam is converted from linear to circular by a quarter-wave plate (e.g., WPQ 5900-4M, Sigma Koki, Japan). The incident laser beam is focused on the back focal plane (BFP) of the objective, so that specimens are illuminated uniformly. When the incident beam is focused at the center of the objective, the microscope can be used as a standard epi-fluorescence microscope. When the path is shifted from the center to the edge between $\theta\alpha$ and θc, the laser beam is incident above a critical angle at the glass/water interface, where the beam is totally internally reflected, generating an evanescent field in the water. Thus, the illumination of the excitation light can be switched from standard epi to TIR simply by shifting the position of the beam focus at BFP from center to edge. In our microscope, this shift is carried out by adjusting the position of the single mirror as shown in Figure 8.2a. The laser beam is decayed by each pass through optic materials. To perform single-molecule imaging of commonly used dyes, such as Cy3, Cy5, and green fluorescent protein (GFP), a laser beam is incident to specimen at a power of 1 mW on a circular area of 40 mm in diameter.

Objective-type TIRFM requires an objective lens with a very high NA, practically larger than 1.4. Objective lenses with a NA > 1.4 are commercially available, e.g., NA1.65 Olympus Apo 100∞ oil High Reso, NA1.45 Olympus PlanApo 100× oil, and NA1.45 Olympus PlanApo 60× oil. These lenses work well for observing living cells. Because the NA1.45 objectives use standard glass and oil with a refractive index of 1.52, practically, we use the NA1.45 objectives for single-molecule observation in living cells. The NA1.65 objective requires special glass and oil with a refractive index of 1.78.

Fluorescence signals from fluorophores are collected in the same manner. The scattered light from the incident laser beam is excluded with suitable bandpass filters. The dichroic mirrors and filters should be carefully selected to minimize the loss of fluorescence intensity and to maximize the specificity of fluorescence wavelength. In our apparatus, to observe Cy3-labeled molecules in living *Dictyostelium* cells, we use HQ565/30M (Chroma) for excitation filter, DM580 (Olympus) for dichroic mirror, HQ620/60M (Chroma) for emission filter, and Ar/Kr laser (568 nm, air-cooled laser 643-RYB, Melles Griot) as a light source. For GFP observation, we used BP470-490 (Olympus) for excitation filter, DM500 (Olympus) for dichroic mirror, BA510-550 (Olympus) for emission filter, and Ar/Kr laser (488 nm, 643-RYB, Melles Griot) as a light source.

The fluorescent images can be intensified by an image intensifier (e.g., GaAsP C8600-05, Hamamatsu Photonics, Shizuoka Pref., Japan) and acquired by a highly sensitive charge-coupled device (CCD) camera (e.g., EB-CCD, C7190-23, Hamamatsu Photonics). The images can be recorded with a digital video recorder at the video rates (33 msec intervals).

8.2.3 MEASUREMENT OF ANGLE OF INCIDENT LASER BEAM

The angle of incident laser beam can be measured using a simple method. When a thicker glass slide is placed on the objective of TIRFM, the leaser beam repeats total internal reflection several times, as illustrated in Figure 8.3a and Figure 8.3b. By measuring the thickness of the glass slide and the intervals between the laser spots on the surface of glass slide, the angle can be determined (Figure 8.3c). Precise control of the incident beam angle is important to generate an evanescent field in a reproducible manner, because the intensity and depth of the evanescent field depend on the angle [Equations (8.1) and (8.2)]. The intensity of the excitation light influences both the fluorescent intensity and lifetime of fluorophores.

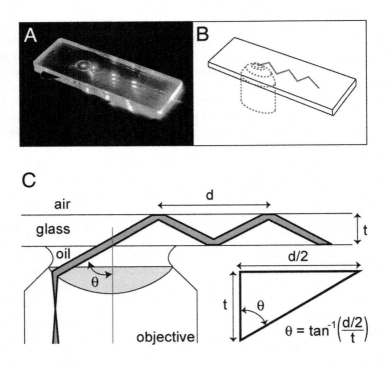

FIGURE 8.3 Angle measurements of incident laser. (a) Laser angle measure. The glass surface is highlighted by fluorescent marker to detect reflective points easily. (b) Schematic drawing of laser angle measure. (c) Principle of measurements of laser angle. The incident laser beam reaches the upper surface of the glass and reflects the total internally at the interface between air and glass. The beam repeats total internal reflection. The incident angle can be calculated from the thickness of glass (t) and the distance between the reflection spots (d) by taking the inverse tangent.

8.3 LIGAND-BINDING ANALYSIS

8.3.1 PREPARATION OF FLUORESCENT-LABELED cAMP ANALOGUE

To observe cAMP binding to the receptors on living *Dictyostelium* cells, we prepared a fluorescent-labeled cAMP (Figure 8.4). An orange fluorescent cyanine dye, Cy3, was conjugated to the 2'-OH of the ribose moiety of cAMP (Cy3-cAMP). Cy3-cAMP functions as a chemoattractant for *Dictyostelium* cells.[9,36] Cy3B and Cy5 can also be used for fluorophores. The synthesis consists mainly of two steps: the first step is preparation of Cy3-ethylendiamine, and the second is the conjugation between Cy3-ethylendiamine and 2'-O-monosuccinyl cAMP with carbodiimide.

PROTOCOL 8-1: SYNTHESIS OF CY3-cAMP

1. Materials and buffers required:
   a. Ethylendiamine (EDA) (Wako, 059-00933)
   b. N,N-dimethylformamide, dehydrated (dry DMF) (Wako, 041-25473)
   c. N-hydroxysuccinimide ester of Cy3 (Amersham, PA23001)
   d. Trifluoroacetic acid (TFA) (Wako, 206-1073)
   e. Acetonitrile (Wako, 019-08631)
   f. 1-Ethyl-3-(3-dimethylaminopropyl)-carbodiimide (WSC) (Dojindo, 348-03631)
   g. 2-O-monosuccinyladenosine 3'5'-cyclic monophosphate (2-O-succinyl cAMP) (Sigma, M9131)
   h. Buffer A: 0.1% TFA in distilled water (DW)
   i. Buffer B: 0.1% TFA, 60% acetonitrile in DW
   j. Buffer C: 10 mM KPO4 (pH 5)
   k. Buffer D: 0.6M KCl, 10 mM KPO4 (pH 5)
   l. RPC resource 3 ml column (Pharmacia LKB)
   m. Mono Q 1 ml column (Pharmacia LKB)

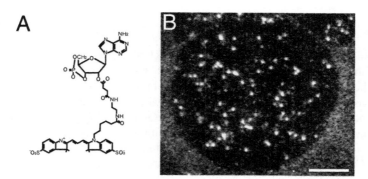

FIGURE 8.4 Fluorescent analogue of cAMP. (a) Structure of Cy3-labeled cAMP. (b) Typical example of single-molecule imaging of Cy3-cAMP bound to the receptors in living *Dictyostelium* cells. White spots are single molecules of Cy3-cAMP. Bar, 5 mm.

2. Dissolve N-hydroxysuccinimide ester of Cy3 (four vials) in 80 ml dry DMF.
3. Add 20 ml of EDA, and incubate for 1 h at room temperature.
4. Dilute the reaction mixture with 1 ml of buffer A, and load the mixture onto the RPC resource column, which is pre-equilibrated with buffer A.
5. Elute with a linear gradient produced by buffer A and buffer B, and collect the fractions containing Cy3-EDA. At about 18% of acetonitrile, Cy3-EDA would be eluted with a major peak.
6. Dry the fractions with a vacuum evaporator, and dissolve in 300 ml of DW. Store the Cy3-EDA solution at 80°C for further use.
7. Dissolve 10 mg of 2-*O*-succinyl cAMP in 310 l of the Cy3-EDA solution.
8. Add 10 mg of WSC, and incubate for 3 h at 30°C.
9. Load the reaction mixture onto an RPC resource column pre-equilibrated with buffer A, and elute with a linear gradient produced by buffer A and buffer B. Collect the fractions containing Cy3-cAMP that would be eluted at 20% of acetonitrile.
10. Load the RPC column elutes onto Mono Q column pre-equilibrated with buffer C.
11. Wash the column with 30 ml of buffer C, and elute with a linear gradient produced by buffer C and buffer D.
12. Collect the fractions around 0.3 M KCl, and dry them with an evaporator.
13. Resuspend the powder of Cy3-cAMP in 1 ml of DW, and repeat step #9.
14. Dry the fractions with an evaporator, and store the Cy3-cAMP powder at −80°C.
15. Note that approximately 400 nmoles of Cy3-cAMP can be obtained in this protocol.

8.3.2 SINGLE-MOLECULE IMAGING ON LIVING CELLS AND CRUDE MEMBRANES

8.3.2.1 Coverslip Preparation

Single molecules cannot be observed under high background noise. Thus, the cleanness of the glass surface critically affects the background level. Also, contaminations should be minimized because single-molecule imaging is extremely sensitive to low-level fluorescence light. Therefore, coverslips are ultrasonicated in laboratory-grade detergents for 30 min, in 0.1 M NaOH (or KOH) for 30 min, and in pure ethanol for 30 min, followed by an exhaustive rinse in ultrapure water. The washed coverslips can be stored in pure ethanol or ultrapure water at room temperature. In our experience, Cy3 molecules do not bind to the coverslips washed using this method. One of the preferred materials for the glass is a fused silica with low autofluorescence. However, in some commercially available borosilicate glass, we observed significant fluorescence around 650 to 850 nm using an excitation light of 457.9 nm.[10]

8.3.2.2 Cell Preparation and Observation

Dictyostelium discoideum cells (strain AX2) are cultured axenically using standard techniques.[9] Cells are resuspended in Sörensen's phosphate buffer (14.6 mM KH_2PO_4, 2mM Na_2HPO_4, pH 6.0) and incubated for 5 to 8 h at 21°C and then treated with 2 to 5 mM caffeine for 30 min to maximize the cell's response to cAMP. After washing cells in DB supplemented with caffeine, an aliquot is placed on a coverslip and covered with a thin agarose sheet.[37] By using the agar-overlay method, the surface of the cells comes in close proximity to the surface of the coverslip. The agar-overlay is critical in *Dictyostelium* cells for single-molecule imaging by TIRFM. For cells that adhere well onto a glass surface, the agar-overlay would not be required. For single-molecule observation of ligand binding, a small droplet of Cy3-cAMP solution (10nM to 1 μM) was placed on the agar sheet. Before the addition of Cy3-cAMP solution, cells should be observed by TIRFM for autofluorescence. In our experience, fluorescent materials are sometimes observed in cell vesicles or vacuoles. The culture medium, including yeast extract, bactopeptone, proteose peptone, and serum, sometimes contains fluorescent materials.

Figure 8.4b shows a typical example of single-molecule imaging of Cy3-cAMP bound to the surface of living *Dictyostelium* cells. When Cy3-cAMP solution is added uniformly to cells, Cy3-cAMP molecules can be seen as spots on the surface of the cells. The free single Cy3-cAMP molecule in aqueous solutions cannot be clearly imaged as a fluorescent spot at the video rate because the fluorescent cAMP molecules undergo rapid Brownian motion. However, when the fluorescent cAMP molecule binds to the receptors, the fluorescence arising from the Cy3 molecule becomes visible as a single spot. After the fluorescent cAMP molecule is released from the receptor, the fluorescent spot suddenly disappears because of rapid Brownian motion. Thus, the association/dissociation events of single-ligand molecules with the receptors can be visualized on living cells. Lateral diffusions of the Cy3-cAMP–receptor complexes can also be observed. By tracking individual Cy3-cAMP spots, we can determine the lifetimes (dissociation rates) and the diffusion constants of cAMP–receptor complexes, as described below.

8.3.2.3 Crude Membrane Preparation

A crude membrane preparation was developed by Van Haastert and colleagues.[38] These membranes can be used to observe Cy3-cAMP binding to the receptors at a single-molecule level. By measuring the binding time of Cy3-cAMP to membranes, we can determine the dissociation rates of Cy3-cAMP–receptor complexes. Furthermore, the ability of cAMP receptors to couple to G proteins can be determined by examining the sensitivity of Cy3-cAMP binding to GTP.

PROTOCOL 8-2:   DICTY MEMBRANE PREPARATION

1. Suspend cells in 10^8 per ml in Sörensen's phosphate buffer (14.6 mM KH_2PO_4, 2mM Na_2HPO_4, pH 6.0) on ice.
2. Break cells by passing through a Millipore filter (pore size, 5 mm).

3. Spin down crude membranes at $15,000 \times$ g for 5 min.
4. Remove supernatants, and resuspend pellets in 10^8 cell equivalents per ml in Sörensen's phosphate buffer.
5. Place an aliquot on a coverslip, and incubate for 10 min on ice.
6. Wash the coverslip with Sörensen's phosphate buffer briefly.
7. Add Cy3-cAMP solution (1 to 10 nM) on the coverslip, and observe Cy3-cAMP molecules bound to the surface of the coverslip by using TIRFM.

8.3.2.4 Verification of Single-Molecule Imaging

We demonstrated that the fluorescent spots represent single molecules in two different ways. First, we measured the fluorescence intensity and plotted the data as a histogram of fluorescence intensities to obtain the distributions of the intensities (Figure 8.5). When single molecules are visualized, the distributions of intensities should have several peaks that occur in regular intervals. For control experiments, the samples containing a mixture of single molecules and two molecules in known ratios may be observed (Figure 8.5). The first peak and the second peak in the histogram would correspond to single and two molecules, respectively, and the ratios of the peaks would be consistent with those of the mixture. Furthermore, the fluorescent intensities are brighter, but the number of the visible spots would be unchanged when the incident laser power is increased. Additionally, we used measurements of photobleaching (Figure 8.6a). Photobleaching would result in the spots corresponding to the first peak in the histogram disappearing in a single stepwise manner, and the spots corresponding to the second peak should be bleached in two steps. Such stepwise photobleaching is a good indicator for successful imaging of single molecule. Also, single fluorescent molecules sometimes exhibit an on/off stepwise pattern of flickering, which is known as blinking.[39] In the paper by Funatsu, which is the first report of single-molecule imaging of fluorophores in aqueous solution, the demonstration was further achieved by electron microscopy.[1] They found that the spots observed by fluorescent microscopy correspond exactly to individual molecules observed by electron microscopy.

8.3.3 Dissociation Rate Analysis

In a simple kinetic scheme, ligand binding to receptors can be written as follows:

$$R + L \underset{k-}{\overset{k+}{\longleftrightarrow}} RL$$

where R, L, k_+, and k_- represent receptors, ligands, association rates, and dissociation rates, respectively. The dissociation constant (affinity) is the ratio of the dissociation rates to the association rates (k_-/k_+). The dissociation rates k reflect on the interactions between ligands and receptors. When both molecules interact tightly with each other, the ligand-binding durations become longer, resulting in slower dissociation rates. When the interactions are relatively weak, the ligand-binding durations are shorter,

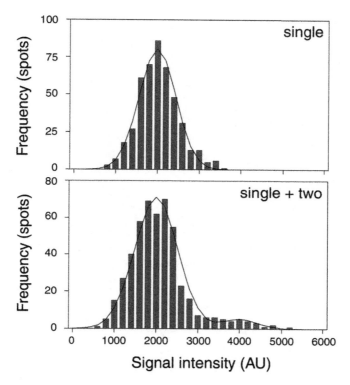

FIGURE 8.5 Distribution of fluorescent intensities of Cy3-cAMP absorbed randomly on the glass surface. The histogram can be fitted by the sum of Gaussian distributions. When the samples containing dimer of Cy3 were observed, a secondary peak was detected at the position with double intensity (bottom).

resulting in faster dissociation rates. In GPCRs, the dissociation rate depends on the states of the receptors interacting with their coupled G proteins. GPCRs have slower dissociation rates when the receptors are bound to the G proteins. When the G proteins dissociate from the receptors, the receptors have faster dissociation rates. These rates and the dependency on G protein coupling can be measured using membrane fractions containing receptors and G proteins (see below).

Using SMDs, we can determine the dissociation rates of the receptors on living cells and membranes. The binding durations of individual Cy3-cAMP molecules are the time difference between the appearance and the disappearance of the Cy3-cAMP spots on the cells (Figure 8.6a). From the histogram of the binding duration, we can obtain the distribution of the binding duration (Figure 8.6b and Figure 8.6c). To determine the dissociation rates, these histograms are fitted with a sum of exponential functions, $\exp[-k_1 t]$. The k_1 values are the dissociation rates of the ligands bound to its receptors.

There are two ways to produce the histograms shown in Figure 8.6b and Figure 8.6c. Figure 8.6a shows the distribution in the number of Cy3-cAMP spots with a

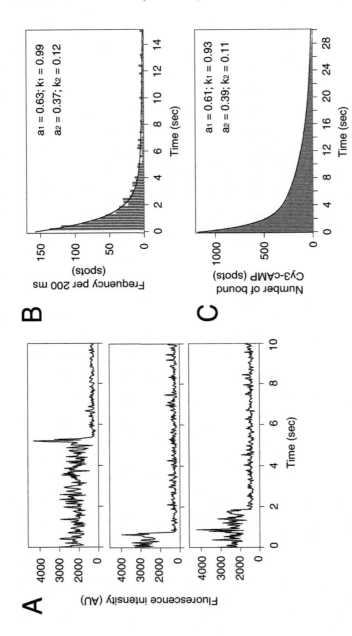

FIGURE 8.6 Single-molecule analysis of the dissociation rates of Cy3-cAMP bound to the receptors. (a) Typical time course of fluorescence intensity of Cy3-cAMP spots bound to the receptors on living cells, showing the disappearance of the fluorescence signal as a single step. (b) The histogram of binding duration of Cy3-cAMP bound to cAMP receptors. The line represents the fitting of data to a sum of two exponential functions: a_1, a_2, k_1, and k_2 are fitting parameters. (c) Cumulative frequency histogram of binding duration of Cy3-cAMP. In both methods, the same values of the dissociation rates a_i and k_i can be obtained.

binding duration per bin of 200 msec width. In this case, the general form used for the fitting analysis is

$$f(t) = a_1 k_1 \exp\left[-k_1 t\right] + a_2 k_2 \exp\left[-k_2 t\right] + \ldots\ldots \tag{8.3}$$

where

$$\sum a_i = 1 \tag{8.4}$$

and the a_i represents the relative amounts of the ith components, k_i. On the other hand, Figure 8.6c shows the time course of decline of receptor occupancy. In this second case, the number of receptor occupancy at $t = 0$ is the total number of Cy3-cAMP spots measured. Because receptors that are occupied at $t = 0$ release ligands after a variable length of time, the number in the histogram decays with time. The general form used for fitting analysis is

$$f(t) = a_1 \exp\left[-k_1 t\right] + a_2 \exp\left[-k_2 t\right] + \ldots\ldots \tag{8.5}$$

With both methods, we obtain the same values of the dissociation rates a_i and k_i. Equation (8.3) [or Equation (8.4)] can be fitted to the histograms by the least-squares method using the Levenberg–Marquardt algorithm. Practically, commercially available software (e.g., KaleidaGraph, Synergy Software) can be used for the fitting analysis. In Figure 8.6b and Figure 8.6c, the estimated exponential functions have been multiplied by the number of observations.

For a reversible kinetic scheme, the histograms of ligand binding should be an exponential function with rate constants k_i as described above. Considering that receptors have an intermediate state before release of ligands; they then have at least two rate-limiting steps as follows:

$$R + L \rightarrow RL \xrightarrow{k_1} R^*L \xrightarrow{k_2} R + L$$

where k_1 and k_2 are constants of rate-limiting steps. R^*L represents an intermediate state before release of a ligand. In this case, the distribution of ligand binding is the following:

$$f(t) = \frac{k_1 k_2}{(k_2 - k_1)}\left(\exp\left[-k_1 t\right] - \exp\left[-k_2 t\right]\right) \tag{8.6}$$

which is the convolution function of two exponentials (k_1 and k_2). The histogram of this distribution will adopt a convex shape with a peak. When k^1 is not a rate-limiting step, and the rate is very fast, Equation (8.6) becomes Equation 8.3 ($i = 1$). Thus, transient intermediates can be detected by making a histogram of the time duration when the rate constants for the transition between kinetic states are slow enough to be measured. Again, note that SMDs do not require synchronization of biochemical reactions to detect the transient intermediates.

Histograms and statistic analysis have been extensively used in single-molecule experiments, such as single-channel recording.[16,40]

8.3.3.1   GTP Sensitivity

For most GPCRs, ligand binding in membranes is sensitive to the presence of GTP. The addition of GTP causes a shift in the dissociation rates of the receptors from slower states (high-affinity states) to faster states (low-affinity states). This shift reflects on the dissociation of G protein from the receptor in the presence of GTP. In membranes prepared from *Dictyostelium* cells, the addition of GTP induced an increase in dissociation rates of Cy3-cAMP. Such a GTP-dependent increase in the dissociation rates was not found in the membranes prepared from the mutant cells lacking functional G protein, either α- or β-subunits, indicating that the differences in the dissociation rates result from altered interactions between the receptors and their coupled G proteins. The rate constants were about 1.3 and 0.4 1/s in the presence and absence of GTP, respectively. This suggests that the cAMP–receptor–G protein complex adopts the slower state with the dissociation rate of 0.4 1/s, and the dissociation of G protein shifts the receptor to the faster state with 1.3 1/s. In addition to these components, we found another component with a very slow rate constant (about 0.06 to 0.16 1/s), although the kinetic state was not identified. These values agree well with the rate constants measured by using radiolabeled cAMP in previous studies.[19,38,41] It is important to confirm the consistency between ensemble measurements and single-molecule experiments.

8.3.3.2   Receptor States on Living Cells

When the *Dictyostelium* cells undergo chemotaxis in a gradient of cAMP, the cells adopt a polarized shape with an anterior pseudopod at the leading edge and a posterior tail at the rear end. The signaling events downstream of the activated G proteins are initiated locally at the leading edge of the cells facing the higher concentration of attractant, although the cAMP receptors and their coupled G proteins are uniformly distributed in the polarized cells.[22–23] Therefore, it is important to elucidate the regulatory mechanisms that spatially localized the signaling events to produce a directional response, or chemotaxis. It is unknown whether the receptors–G proteins coupling is localized spatially or not under the gradient of cAMP.

To probe receptor states in the cells during chemotaxis, we applied SMDs to real-time imaging of cAMP binding to the receptors on living cells[9] (Figure 8.7a). The dissociation rate analysis was performed for Cy3-cAMP bound to the anterior pseudopod and the posterior tails of the *Dictyostelium* cells. As shown in Figure 8.7a, the dissociation rates of Cy3-cAMP in the anterior region were about three times faster than those in the posterior region. Fits of the dissociation suggested that the majority of the receptors in the anterior and the posterior portions of the cell have dissociation rates of 1.1 and 0.4 1/s, respectively. That is, the faster dissociation form of the receptors is localized preferentially at the anterior region, while the slower dissociation form of the receptors is localized at the posterior region. Such polarity in the receptor states was not found in the mutant cells lacking the a- and b-subunits of the G protein, suggesting that the differences in the receptor states reflect on the difference of the coupling with G protein between the anterior and the posterior regions. The differences in the dissociation rates between the anterior and posterior regions resemble the differences in cAMP binding to membranes in the presence and absence of GTP (Figure 8.7b). This implies that the receptor–G protein complex at the anterior region spends less time in the intermediate coupling states before activation and dissociation of G proteins by GTP. This suggests anterior–posterior polarity in the efficiency of G protein activation along with the chemotaxing cells (Figure 8.7d). Thus, SMDs in living cells allow us to directly monitor the behavior of signaling molecules in relation to intracellular environments, cell polarity, and cell response.

Anterior–posterior polarity in receptor states can explain the observations by Swanson and Taylor.[42] They found that the *Dictyostelium* cells have a polarity in responsiveness along the length of the cells. When the cells are stimulated locally by a micropipette containing cAMP solutions, the cells form a pseudopod toward the tip of the micropipette. The anterior region of the cells responds immediately to positioning of the pipette within a few seconds, but the tail (posterior) region requires about 40 s or more to respond. Our findings suggest that polarity in receptor–G protein coupling is involved in the polarized responsiveness of the *Dictyostelium* cells.[9]

Such polarity in receptor states may be a molecular basis of robustness against stochastic fluctuations in receptor occupancy in the polarized cells. When the polarized cells undergo chemotaxis toward the source of chemoattractant, receptor occupancy would be reversed spontaneously and transiently with respect to the direction of the gradient. If the cells have an anterior–posterior polarity in the efficiency of G protein activation along the length of the cells, the ligand binding at the anterior pseudopod (or posterior tail) would preferentially affect cell behavior.

8.4 GREEN FLUORESCENT PROTEIN (GFP) IMAGING AT SINGLE-MOLECULE LEVEL

Fluorescence from single GFP (or YFP) molecules is strong enough to be observed at the single-molecule level in living cells under TIRFM.[43–46] GFP-tagging techniques have been widely used to investigate the dynamics of signaling molecules in living cells. When combined with SMDs, information about behaviors of individual signaling molecules can be obtained. We observed YFP-tagged cAMP-receptors by TIRFM and analyzed the lateral diffusion on the plasma membrane (Figure 8.8a and Figure 8.8b). The receptor had diffusion constants in the range of 0.005–0.03 $\mu m^2/s$, which is similar to the diffusion constants of Cy3-cAMP–receptor complexes (Figure 8.8c). Most of the receptors showed a simple diffusion. However, some of GFP-tagged receptors stopped transiently, whereas others moved in a relatively linear

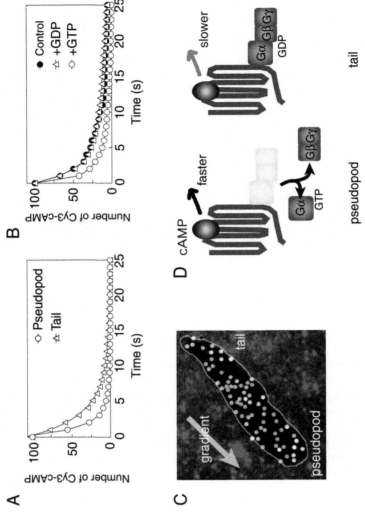

FIGURE 8.7 Receptor states in chemotaxing cells. (a) The release curves of spots bound to the anterior pseudopods or the posterior tails. The lines represent the fitting of data to a sum of exponential functions. At the pseudopod, $k_1 = 1.1$ 1/s (71%), and $k_2 = 0.4$ 1/s (29%). At the tail, $k_1 = 0.4$ 1/s (76%), and $k_2 = 0.16$ 1/s (24%). (b) The release curves of Cy3-cAMP spots bound to membranes. ●, no addition of guanine nucleotides; △, in the presence of 100 μM GTP; ○, in the presence of 100μM GDP. For the comparison, the release curves were fitted to three exponentials, with $k_1 = 1.3$, $k_2 = 0.4$, and $k_3 = 0.08$ 1/s. Control: 21, 46, and 33%; + GTP: 64, 17, and 20%; + GDP: 60, 19, and 21%, respectively. G proteins dissociate from the receptors in the presence of GTP, resulting in a fast-dissociation form of the receptors. (c) Receptor occupancy of a *Dictyostelium* cell under a gradient of Cy3-cAMP. (d) A receptor forms at the anterior pseudopods or at the posterior tails. The majority of the binding sites at the anterior pseudopods facing the higher concentration adopted the fast-dissociation form of cAMP receptors. The posterior tails facing the lower concentration adopted the slow-dissociation form.

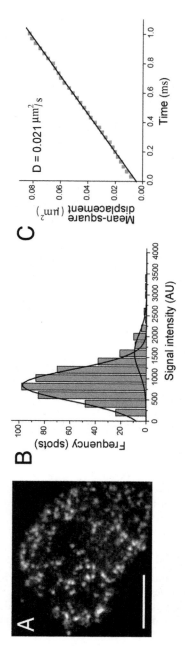

FIGURE 8.8 Single-molecule imaging of YFP-tagged cAMP-receptor proteins in living cells. (a) Cells were observed by objective-type TIRFM. White spots are single molecules of YFP fused to the cAMP receptor. Scale bar, 5 μm. (b) The distribution of fluorescence intensity of a YFP-cAMP receptor. This histogram suggests that most fluorescence spots are single molecules. (c) Plot of the mean-square displacement (MSD) versus time, indicating that the receptor diffused by simple Brownian motion. D, diffusion coefficient.

path. This suggests possible interactions of the receptors with some microdomains on plasma membranes and with underlying structures, such as cortical cytoskeletons, although the biological significance remains to be resolved.

Compared with cyanine dyes, such as Cy3 and Cy5, which are suitable for single-molecule imaging, fluorescence emitted from GFP is relatively weak and quickly photobleached under the continuous illumination of excitation light. In our microscopes, we can observe the Cy3 molecule for about 20 to 30 s, while GFP molecules disappear within about 5 s (1/e value of photobleaching decay curve) on living cells. Furthermore, because GFP molecules have higher molecular weights than the fluorescent dyes, steric inhibitions can lead to a loss of function of GFP-tagged proteins. An assay system is required to obtain functional GFP-tagged proteins. We used GFP-tagged proteins only when the fusion genes can rescue the mutants lacking the intrinsic genes. Nevertheless, GFP is useful, because it is easy to prepare cells expressing GFP-tagging proteins using genetic engineering, and the labeling ratio of GFP to proteins can be 100%.

ACKNOWLEDGMENTS

The authors would like to thank Dr. F. Brozovich for critically reading the manuscript, and our colleagues at Osaka University, ERATO, ICORP, and CREST for helpful discussion. This study is supported by MEXT's Leading Project.

ABBREVIATIONS

cAMP	cyclic adenosine 3 5-monophosphate
cARs	cAMP receptors
FRET	fluorescence resonance energy transfer
GFP	green fluorescent protein
GPCR	G protein-coupled receptors
SMDs	single-molecule detection techniques
TIR	total internal reflection
TIRFM	total internal reflection fluorescence microscopy

REFERENCES

1. Funatsu, T. et al., Imaging of single fluorescent molecules and individual ATP turn-overs by single myosin molecules in aqueous solution, *Nature*, 374, 555, 1995.
2. Tokunaga, M. et al., Single-molecule imaging of fluorophores and enzymatic reactions achieved by objective-type total internal reflection fluorescence microscopy, *Biochem. Biophys. Res. Comm.*, 235, 47, 1997.
3. Sako, Y., Minoguchi, S., and Yanagida, T., Single-molecule imaging of EGFR signaling on the surface of living cells, *Nature Cell Biol.*, 2, 168, 2000.
4. Weiss, S., Fluorescence spectroscopy of single biomolecules, *Science*, 283, 1676, 1999.

5. Forkey, J.N., Quinlan, M.E., and Goldman, Y.E., Protein structural dynamics by single-molecule fluorescence polarization, *Prog. Biophys. Mol. Biol.*, 74, 1, 2000.
6. Ishii, Y., Ishijima, A., and Yanagida, T., Single molecule nanomanipulation of bio-molecules, *Trends in Biotech.*, 19, 211, 2001.
7. Ishijima, A. and Yanagida, T., Single molecule nanobioscience, *Trends Biochem. Sci.*, 26, 438, 2001.
8. Sako, Y. and Yanagida, T., Single-molecule visualization in cell biology, *Nat. Rev. Mol. Cell Biol.*, SS1, 2003.
9. Ueda, M. et al., Single-molecule analysis of chemotactic signaling in *Dictyostelium* cells, *Science*, 294, 864, 2001.
10. Wazawa, T. and Ueda, M., Total internal reflection fluorescence microscopy in single molecule nanobioscience, in *Advances in Biochemical Engineering/Biotechnology: Microscopic Techniques*, Rietdorf, J., Ed., Springer-Verlag, Heidelberg, 2004 .
11. Briddon, S.J. et al., Quantitative analysis of the formation and diffusion of A1-adenosine receptor-antagonist complexes in single living cells, *Proc. Natl. Acad. Sci. USA*, 101, 4673, 2004.
12. Taguchi, H. et al., Single-molecule observation of protein–protein interactions in the chaperonin system, *Nat. Biotechnol.*, 19, 861, 2001.
13. Ueno, T. et al., GroEL mediates protein folding with a two successive timer mecha-nism, *Mol. Cell*, 14, 423, 2004.
14. Zhuang, X. et al., Correlating structural dynamics and function in single ribozyme molecules, *Science*, 296, 1473, 2002.
15. Lu, H.P., Xun, L., and Xie, S., Single-molecule enzymatic dynamics, *Science*, 282, 1877, 1998.
16. Xie, S., Single-molecule approach to enzymology, *Single Mol.*, 4, 229, 2001.
17. Yasuda, R. et al., Resolution of distinct rotational substeps by submillisecond kinetic analysis of F1-ATPase, *Nature*, 410, 898, 2001.
18. Kinosita, K. Jr., Yasuda, R., and Noji, H., F1-ATPase: a highly efficient rotary ATP machine, *Essays Biochem.*, 35, 3, 2000.
19. Janssens, P.M.W. and Van Haastert, P.J.M., Molecular basis of transmembrane signal transduction in *Dictyostelium discoideum*, *Microbiol. Rev.*, 51, 396, 1987.
20. Devreotes, P.N. and Zigmond, S.H., Chemotaxis in eukaryotic cells: a focus on leukocytes and *Dictyostelium*, *Annu. Rev. Cell Biol.*, 4, 649, 1988.
21. Parent, C.A. and Devreotes, P.N., Molecular genetics of signal transduction in *Dic-tyostelium*, *Annu. Rev. Biochem.*, 65, 411, 1996.
22. Parent, C.A. and Devreotes, P.N., A cell's sense of direction, *Science*, 284, 765, 1999.
23. Kimmel, A.R. and Parent, C.A., The signal to move: *D. discoideum* go orienteering, *Science*, 300, 1525, 2003.
24. Mato, J.M. et al., Signal input for a chemotactic response in the cellular slime mold *Dictyostelium discoideum*, *Proc. Natl. Acad. Sci. USA*, 72, 4991, 1975.
25. Van Haastert, P.J.M., Transduction of the chemotactic cAMP signal across the plasma membrane, in *Dictyostelium*, Maeda, Y. et al., Ed., Universal Academy Press, Tokyo, 1997, p. 173 .
26. Ide, T. and Yanagida, T., An artificial lipid bilayer formed on an agarose-coated glass for simultaneous electrical and optical measurement of single ion-channels, *Biochem. Biophys. Res. Commun.*, 265, 595, 1999.
27. Ide, T. et al., Simultaneous optical and electrical recording of a single ion-channel, *Jpn. J. Physiol.*, 52, 429, 2002.
28. Kitamura, K. et al., A single myosin head moves along an actin filament with regular steps of 5.3 nm, *Nature*, 397, 129, 1999.

29. Ishii, Y. et al., Fluorescence resonance energy transfer between single fluorophores attached to a coiled-coil protein in aqueous solution, *Chemical Phys,*, 247, 163, 1999.

30. Wazawa, T. et al., Spectral fluctuation of a single fluorophore conjugated to a protein molecule, *Biophys. J.*, 78, 1561, 2000.

31. Ishijima, A. et al., Simultaneous observation of individual ATPase and mechanical events by a single myosin molecule during interation with actin, *Cell*, 92, 161, 1998.

32. Axelrod, D., Total internal reflection fluorescence microscopy, *Methods Cell Biol.*, 30, 245, 1989.

33. Axelrod, D., Hellen, E.H., and Fulbright, R.M., Total internal reflection fluorescence, in Lakowicz, J.R., Ed., *Topics in Fluorescence Spectroscopy, Vol. 3: Biochemical Applications*, Plenum Press, New York, 1992, p. 289 .

34. Axelrod, D., Total internal reflection fluorescence microscopy in cell biology, *Traffic*, 2, 764, 2001.

35. Fornel, F., *Evanescent Waves*, Springer-Verlag, Heidelberg, 2001.

36. Janetopoulos, C. et al., Chemoattractant-induced phosphatidylinositol 3,4,5-trisphosphate accumulation is spatially amplified and adapts, independent of the actin cytoskeleton, *Proc. Natl. Acad. Sci. USA*, 101, 8951, 2004.

37. Fukui, Y., Yumura, S., and Yumura, T.K., Agar-overlay immunofluorescence: high-resolution studies of cytoskeletal components and their changes during chemotaxis, *Methods Cell Biol.*, 28, 347, 1987.

38. Van Haastert, P.J.M. et al., G protein-mediated interconversions of cell-surface cAMP receptors and their involvement in excitation and desensitization of guanylate cyclase in *Dictyostelium discoideum*, *J. Biol. Chem.*, 261, 6904, 1986.

39. Dickson, R.M. et al., On/off blinking and switching behaviour of single molecules of green fluorescent protein, *Nature*, 388, 355, 1997.

40. Sakmann, B. and Neher, E., Eds., *Single Channel Recording*, Plenum, New York, 1995.

41. Van Haastert, P.J.M. and de Wit, R.J.W., Demonstration of receptor heterogeneity and affinity modulation by nonequilibrium binding experiments, *J. Biol. Chem.*, 259, 13,321, 1984.

42. Swanson, J.A. and Taylor, D.L., Local and spatially coordinated movements in *Dictyostelium discoideum* amoebae during chemotaxis, *Cell*, 28, 225, 1982.

43. Sako, Y. et al., Single-molecule imaging of signaling molecules in living cells, *Single Mol.*, 1, 159, 2000.

44. Iino, R., Koyama, I., and Kusumi, A., Single molecule imaging of green fluorescent proteins in living cells: E-cadherin forms oligomers on the free cell surface, *Biophys. J.*, 80, 2667, 2001.

45. Watanabe, N. and Mitchison, T.J., Single-molecule speckle analysis of actin filament turnover in lamellipodia, *Science*, 295. 1083, 2001.

46. Hibino, K. et al., Single- and multiple-molecule dynamics of the signaling from H-Ras to cRaf-1 visualized on the plasma membrane of living cells, *Chemphyschem.*, 4, 748, 2003.

9 Oligomerization of G Protein-Coupled Purinergic Receptors

Hiroyasu Nakata, Kazuaki Yoshioka, and Toshio Kamiya

CONTENTS

9.1 INTRODUCTION

9.1.1 OLIGOMERIZATION OF G PROTEIN-COUPLED RECEPTORS

It is well recognized that protein–protein interactions are fundamental processes for many biological systems between not only cytosolic proteins but also membrane-bound signaling proteins in the cell. In particular, the existence of homo- and hetero-oligomers between G protein-coupled receptors (GPCRs) that show distinct pharmacological and functional properties has been demonstrated.[1–5] Such GPCR oligomerization adds another level of complexity or diversity to how GPCRs are activated, as well as to signaling and trafficking in the cells; although the traditional view of GPCR signaling incorporates a monomeric GPCR interacting through specific intracellular domains with a single heterotrimeric G protein. There is no doubt that the functional implications of GPCR oligomerization, such as trafficking to the plasma membrane, agonist binding activity, signal transduction, and receptor downregulation, which will have an enormous impact on GPCR biology, and the existing evidence, support the notion that dimerization is a general feature of this important class of receptors.[6] Therefore, studies on how GPCRs complex as dimeric (or oligomeric) will be of considerable importance, not only for our understanding of GPCR cellular function, but also for novel drug design in the future.

Most studies of GPCR oligomerization employ a combination of biochemical immunoprecipitation, immunohistochemistry, functional studies, and resonance energy transfer techniques, performed in heterologous expression systems, i.e., co-expression of cloned and appropriately tagged (immunological epitopes or fluorescent reporters) GPCRs in cultured cells. In this chapter, these techniques are introduced to establish the prevalence and physiological significance of oligomer formation of GPCRs, employing purinergic receptors as a model receptor system.

9.1.2 PURINERGIC RECEPTORS

Adenosine and adenine nucleotides, including ATP and ADP, mediate a wide variety of physiological processes, including the regulation of neural transmission via purinergic receptors, which are divided into adenosine (P1) and P2 receptors.[7] Molecular cloning and pharmacological studies identified four types of adenosine receptors: adenosine A_1, A_{2A}, A_{2B}, and A_3 receptors, which are all G protein-coupled receptors. P2 receptors (or ATP receptors) are subclassified into P2X and P2Y types: seven mammalian P2X receptors, which are ligand-gated ion channels, and eight mammalian P2Y receptors ($P2Y_{1, 2, 4, 6, 11, 12, 13, 14}$), all of which are GPCRs, have been cloned.

Evidence for the existence of homo- and hetero-oligomerization between purinergic receptors and between purinergic and other GPCRs of distinct families has been accumulated (Table 9.1). Adenosine A_1 receptor (A_1R) functionally couples to pertussis toxin (PTX)-sensitive $G_{i/o}$ proteins, and its activation modulates several effectors, such as adenylyl cyclase and K^+ channels. A_1R is reported to form hetero-oligomers with $P2Y_1$ receptor ($P2Y_1R$), altering its ligand-binding pharmacology.[8] $P2Y_1R$ is a subtype of G_q protein-coupled P2 receptor that activates phospholipase

TABLE 9.1

Oligomerization of G Protein-Coupled Purinergic Receptors

Oligomerization Pair	Methods	Ref.
Homo-oligomerization		
Adenosine A_1	SDS-PAGE, BRET	23, 30
Adenosine A_{2A}	Immunoprecipitation, BRET	12, 31
Hetero-oligomerization		
Adenosine A_1 and dopamine D_1	Immunoprecipitation	9
Adenosine A_{2A} and dopamine D_2	Immunoprecipitation, BRET	12, 13
Adenosine A_1 and P2Y$_1$	Immunoprecipitation, BRET	8, 23
Adenosine A_1 and P2Y$_2$	Immunoprecipitation	8
Adenosine A_1 and mGluR1	Immunoprecipitation	10
Adenosine A_{2A} and mGluR5	Immunoprecipitation	11

Note: mGluR, metabotropic glutamate receptor.

C to form inositol trisphosphate and causes calcium to be released from intracellular stores. A_1R is also known to form a heterocomplex with other GPCRs, such as the dopamine D_1 receptor altering D_1 receptor signaling,[9] and metabotropic glutamate receptor 1 altering cellular signaling.[10] A_{2A} adenosine receptor ($A_{2A}R$), another subtype of the adenosine receptor, which couples with G_s protein to activate adenylate cyclase, is reported to form hetero-oligomers with metabotropic glutamate receptor 5 to enhance mitogen-activated protein kinase activity.[11] It was recently demonstrated that $A_{2A}R$ exists as a homomeric or heteromeric oligomer in living cells with the dopamine D_2 receptor (D_2R) by co-immunoprecipitation and resonance energy transfer methods.[12] The latter complex is likely to be involved in the co-desensitization and co-internalization of these receptors upon treatment with one or both agonists.[13] These findings may be important for the treatment of Parkinson's disease, because the A_{2A} receptor-mediated antagonism of D_2R via hetero-oligomerization may provide a possible explanation for the reduced effect of L-DOPA, a D_2-acting drug used to treat Parkinson's disease.[14]

9.2 ANALYSIS OF OLIGOMERIZATION OF PURINERGIC RECEPTORS

9.2.1 Receptor Pharmacology

Hetero-oligomerization of GPCRs can affect various aspects of receptors, including the alteration of ligand-binding specificity and signal transduction and cellular trafficking. Hetero-oligomerization between A_1R and P2Y$_1$R in co-transfected cells was, therefore, analyzed by ligand-binding and adenylate cyclase assays as described below.

9.2.1.1 Transfection and Expression in HEK293T Cells

Transient transfection of HEK293T cells with epitope tagged-receptors (HA- or Myc-) in expression plasmids is routinely performed. Epitopes enable the detection of their expression levels without interference from the immunoreactivity of endogenous receptors. The incorporation of sequences encoding the HA epitope tag (YPY-DVPDYA) and the Myc epitope tag (EQKLISEEDL) into rat A_1R and rat $P2Y_1R$, respectively, was performed by polymerase chain reaction (PCR). Each epitope was positioned immediately before the first methionine of the appropriate gene. Because GPCR interactions with G proteins involve the carboxyl termini of GPCRs, it is preferable to place the epitope in the N-terminal site of GPCR. Purified full-length complementary DNA (cDNA) of HA-A_1R was subcloned into pcDNA3.1 (Stratagene, La Jolla, CA), and purified full-length cDNA of Myc-$P2Y_1R$ was subcloned into pcDNA3 (Stratagene). The generation of each construct was confirmed by sequencing analysis.

There are many methods for the transient transfection of cultured mammalian cell lines, including the DEAE-dextran method, calcium phosphate co-precipitation, electropolation, and commercially available polycationic transfection methods. We routinely transfect HEK293T cells using the Effectene transfection system (Qiagen, Hilden, Germany). We express GPCR cDNAs under the control of the cytomegalovirus promoter in the pcDNA3 series (Stratagene) and prepare plasmids using the Promega Wizard MagneSil Tfx™ System (Promega, Madison, WI). The expression of transfected receptors can be examined by ligand-binding, functional assays such as the adenylate cyclase assay, or immunoblotting, as described in Protocol 9-1.

PROTOCOL 9-1:   CO-TRANSFECTION OF HEK293T CELLS WITH PURINERGIC RECEPTOR cDNA

1. On day 1, detach HEK293T cells grown in Dulbecco's Modified Eagle Medium (DMEM) supplemented with 10% fetal bovine serum (FBS) and kanamycin (0.1 mg/ml) by trypsin-EDTA solution, and seed them into 100 mm dishes at a density of 2×10^6 cells/dish.

2. On day 2, transfect the cells with purinergic receptor cDNAs, HA-A_1R subcloned into pcDNA3.1 and Myc-$P2Y_2R$ subcloned into pcDNA3, using the Effectene transfection reagent kit (Qiagen). Add 300 µl of Buffer EC to a 5 ml plastic tube, and mix aseptically with one or two DNA(s) of interest for a total quantity of 2 µg with or without carrier DNA (usually, empty vector). Add Enhancer reagent (16 µl) to this tube, and vortex for 1 s. Allow mixture to incubate at room temperature for 5 min.

3. Add 10 µl of Effectene transfection reagent per µg DNA to this mixture, and vortex for 10 s (volumes of Effectene reagent per unit mass of DNA may vary according to the cell type to be transfected). Allow mixture to incubate at room temperature for 6 to 10 min while the 30 to 50% confluent HEK293T cell monolayers plated on 100 mm dishes are rinsed once with

DMEM (10 ml/dish) and replaced with 7 ml of DMEM supplemented with 10% FBS.

4. Dilute the DNA mixture with 3 ml of DMEM supplemented with 10% FBS and mix twice by pipetting. Add the DNA mixture to the HEK293T cells dropwise, and swirl the dish gently. Return the cells to the incubator.

5. Culture the transfected HEK293T cells in a 37°C, humidified, 5% CO_2 atmosphere. Prepare cell membranes for immunoprecipitation and Western blotting from the cells 48 h after transfection. For adenylyl cyclase assay, the cells are passaged to 24-well plates, 48 h after transfection, and are cultured for another 24 h at 37°C.

9.2.1.2 Ligand Binding

The ligand-binding assay of A_1R expressed in HEK293T cells can be performed using 10 µg of cell membranes prepared as described in Protocol 9-2 with 2 nM [³H]8-cyclopentyl-1,3-dipropylxanthine (DPCPX, an A_1R specific antagonist, PerkinElmer, Wellesley, MA) containing 2 U/ml adenosine deaminase (Sigma Chemicals, Perth, Western Australia), 5 mM $MgCl_2$, and 50 mM Tris-acetate buffer (pH 7.4) for 60 min at 25°C in the absence or presence of various concentrations of unlabeled ligands. For agonist binding, 30 to 50 µg of membrane proteins are incubated with 40 nM [³H]5-*N*-ethylcarboxamidoadenosine (NECA, a nonselective adenosine receptor agonist, Amersham Pharmacia Biotech, Piscataway, NJ) under the same conditions as those described above. Dissociation constants (K_i) and B_{max} are determined from saturation or displacement curves using GraphPad Prism 2.0 (GraphPad Software, San Diego, CA). As A_1R is a $G_{i/o}$-coupled receptor, the functional activity can be assessed by attenuation of the adenylate cyclase assay.

Cell membranes expressing HA-A_1R show [³H]DPCPX and [³H]NECA binding activities similar to those of cell membranes expressing intact A_1R, suggesting that N-terminal modification of A_1R with HA-tag does not alter ligand-binding activities. No apparent significant differences in [³H]NECA binding activity are observed between HA-A_1R-transfected and HA-A_1R/Myc-$P2Y_1R$-transfected cell membranes. However, [³H]NECA-binding pharmacology of the co-transfected cell membranes by competition experiments with other purinergic ligands shows a unique change in the ligand specificity. The apparent binding potency and efficacy of A_1R-selective agonist cyclopentyladenosine (CPA) (Figure 9.1a, left) to the [³H]NECA binding site were reduced in the co-transfected cells. Selective $P2Y_1R$ antagonist N^6-methyl-2-deoxyadenosine-3,5-bisphosphate (MRS2179) failed to displace [³H]NECA bound to HA-A_1R-transfected and HA-A_1R/Myc-$P2Y_1R$-transfected cell membranes (data not shown). Interestingly, a potent $P2Y_1R$ agonist, ADPβS, was found to be quite active in displacing the ligands from the [³H]NECA binding site of co-transfected cell membranes with K_i values of 0.38 nM (high-affinity site) and 610 nM (low-affinity site). In contrast, ADPβS in the 10^{-6} M range only slightly inhibited [³H]NECA binding of cell membranes expressing HA-A_1R alone (Figure 9.1a, right), indicating an alteration of ligand-binding specificity induced by co-transfection.

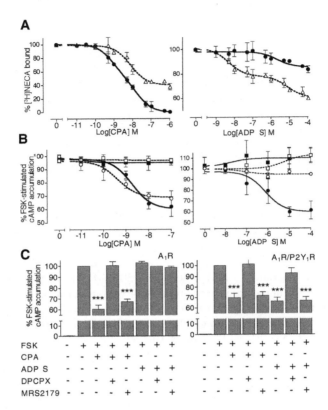

FIGURE 9.1 Co-expression with P2Y$_1$R modulates A$_1$R binding pharmacology and generates P2Y$_1$R agonist-sensitive adenylyl cyclase inhibition of A$_1$R. (a) Displacement of [^{3}H]NECA (40 nM) binding with transfected cell membranes by CPA (left) and ADPβS (right). Membranes from HA-A$_1$R-transfected (closed circle) or HA-A$_1$R/P2Y$_1$R-transfected (open triangle) cells were incubated with indicated concentrations of each ligand. [^{3}H]NECA concentrations were selected to ensure maximal saturation binding. Data represent the means ± SEM of the percentage of [^{3}H]NECA specifically bound values. Results from three independent experiments performed in duplicate are shown. (b) Concentration-dependent reduction of maximal forskolin (FSK; 10 μM)-stimulated intracellular cAMP accumulation by CPA (left) or ADPβS (right) in A$_1$R/P2Y$_1$R-transfected cells. Dotted line, cells expressing HA-A$_1$R alone; solid line, cells co-expressing HA-A$_1$R and Myc-P2Y$_1$R; circle, nontreated cells; square, PTX-pretreated cells (100 ng/ml, 16 h). The 100% values of cAMP for cells transfected with HA-A$_1$R, and HA-A$_1$R plus Myc-P2Y$_1$R were 72 ± 14 and 67 ± 19 pmol/10^5 cells, respectively (mean ± SEM, $n = 5$). (c) Pretreatment of cells with A$_1$R antagonist DPCPX, but not P2Y$_1$R antagonist MRS2179, significantly inhibited maximal ADPβS-induced adenylyl cyclase attenuation in A$_1$R/P2Y$_1$R-transfected cells. Left, HA-A$_1$R transfected cells; right, HA-A$_1$R/Myc-P2Y$_1$R co-transfected cells. The 100% values of cAMP for the cells transfected with HA-A$_1$R, and HA-A$_1$R/Myc-P2Y$_1$R were 70 ± 12 and 71 ± 17 pmol/10^5 cells, respectively (mean ± SEM, $n = 5$). Data represent the means ± SEM of the percentage of FSK-induced cAMP accumulation values. Results from three to five independent experiments performed in duplicate are shown. *** $p < 0.01$, student's t-test. (Part of this figure is reprinted with permission from the National Academy of Sciences USA, from Yoshioka, K., Saitoh, O., and Nakata, H., *Proc. Natl. Acad. Sci. USA*, 98, 7617, 2001.)

PROTOCOL 9-2:   MEMBRANE PREPARATION

1. Transfect approximately 2×10^6 HEK293T cells in 100 mm-diameter tissue culture dishes using Effectene transfection reagent (Quiagen) as described above.
2. At 48 h after transfection, wash the dishes quickly and gently with 10 ml of ice-cold Ca^{2+}- and Mg^{2+}-free Dulbecco's phosphate-buffered saline [PBS(-)].
3. Add 10 ml of ice-cold hypotonic lysis buffer containing 50 mM Tris-acetate buffer (TAB), pH 7.4, with a protease-inhibitor cocktail (one tablet of Complete/50 ml, Roche, Basel, Switzerland). Scrape cells off dishes with rubber policeman and transfer to 15 ml conical centrifuge tubes. Lyse cells by sonication on ice. All subsequent steps in this procedure are performed on ice.
4. Centrifuge the mixture at $750 \times g$ for 5 min to remove organelles and nuclei. The resulting supernatant is subjected to centrifugation at 30,000 $\times g$ for 20 min, and precipitated cell membranes are collected, washed twice by suspension followed by centrifugation, resuspended in a small volume (0.5 to 1.0 ml) of the ice-cold lysis buffer, and stored at 80°C. Retain a small aliquot (10 to 15 µl) to determine the protein concentration of the crude membrane preparation. The protein concentration can be determined by the Bradford method using a Bio-Rad protein assay kit (Bio-Rad, Hercules, CA).
5. Examine expression of the transfected receptors by immunoblot, receptor–ligand binding, or functional assays, as described below.

9.2.1.3 Adenylyl Cyclase Coupling in Co-Transfected Cells

Cells expressing A_1R alone reveal an inhibition of forskolin (FSK)-stimulated cAMP accumulation by CPA, a specific A_1R agonist, in a dose-dependent manner, with the estimated concentration for half-maximal response (IC_{50}) of 0.42 nM to a maximum inhibition of 70%. This activity is completely abolished by pretreatment of the cells with pertussis toxin (PTX). CPA-induced inhibition of FSK-stimulated adenylyl cyclase activity is also detected with the estimated IC_{50} value of 1.0 nM in cells co-expressing $A_1R/P2Y_1R$ (Figure 9.1b, left). This activity is also abolished by PTX treatment. However, the potency of adenylyl cyclase attenuation by CPA is reduced significantly in co-expressing cells compared with cells expressing A_1R alone ($p <$ 0.05, student t-test). Interestingly, in cells co-expressed with A_1R and $P2Y_1R$, ADPβS markedly reduced FSK-evoked adenylyl cyclase activity in a concentration-dependent manner, with the estimated IC_{50} value of 730 nM, to a maximum inhibition of 62% (Figure 9.1b, right). This $P2Y_1R$ agonist-dependent attenuation is PTX-sensitive, suggesting the involvement of a PTX-sensitive $G_{i/o}$ protein. As the ADPβS-induced adenylyl cyclase inhibition in co-expressed cells is blocked in the presence of A_1R antagonist DPCPX, whereas MRS2179, a specific $P2Y_1R$ antagonist, shows no effect on the ADPβS-evoked adenylyl cyclase inhibition, it is likely that ADPβS exerts adenylyl cyclase inhibitory activity through xanthine-sensitive ligand-binding

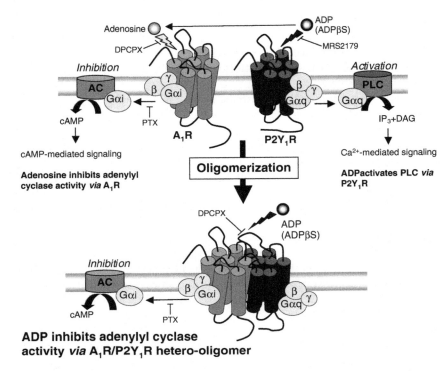

FIGURE 9.2 Presumed signaling pathways induced by the heteromeric oligomerization of A_1R and $P2Y_1R$ in HEK293T cells. After the heteromeric oligomerization of A_1R and $P2Y_1R$, a potent $P2Y_1R$ agonist, ADPβS, can bind with an A_1R binding pocket that is xanthine sensitive and inhibits adenylyl cyclase activity via the $G_{i/o}$ protein-linked effector system. AC, adenylyl cyclase; IP_3, inositol 1,4,5-trisphosphate; DAG, diacylglycerol; PTX, pertussis toxin; DPCPX, 8-cyclopentyl-1,3-dipropylxanthine (A_1R antagonist); MRS2179, N^6-methyl-2-deoxyadenosine-3,5-bisphosphate ($P2Y_1R$ antagonist). (From Yoshioka, K. and Nakata, H., *J. Pharmacol. Sci.*, 94, 88, 2004, with some modifications. With permission from The Japanese Pharmacological Society.)

sites of A_1R via a $G_{i/o}$ protein-linked effector system. These results are summarized in Figure 9.1c. The presumed signal transduction in the co-transfected cells is illustrated in Figure 9.2.

PROTOCOL 9-3: cAMP DETERMINATION

1. Culture the transfected cells in 24-well plates (1×10^5 cells /well), and then aspirate the media from the cells. Add 0.8 ml of assay medium (serum-free DMEM) containing 2 units/ml adenosine deaminase (ADA). Incubate for 60 min at 37°C.
2. Add 100 μl of assay medium containing 1 mM Ro201724 (final 100 μM) to inhibit phophodiesterase activities. (A stock solution of 100 mM

Ro201724 in dimethyl sulfoxide (DMSO) is made up fresh on the day of the assay.) Incubate for 15 min at 37°C.

3. Add 100 μl of various concentrations of receptor agonists, and incubate for 10 min in the presence of 10 μM FSK at 37°C.

4. Aspirate and add 300 μl of ice-cold 0.1 N HCl to lyse cells. Place dishes on ice for 10 min. Neutralize with 30 μl of 1 N NaOH.

5. Add 650 μl of ice-cold absolute ethanol (final ~65%). Incubate for 10 min to extract cAMP from the cells.

6. Use aliquots of extracted solution as samples for the cAMP assay system. Cyclic AMP extracted from cells was quantified by a cAMP EIA kit as described in the manufacturer's manual (Bio-Trak, Amersham Pharmacia).

9.2.2 IMMUNOPRECIPITATION

9.2.2.1 Co-Immunoprecipitation of Epitope-Tagged A_1R and $P2Y_1R$ in Co-Expressed HEK293T Cells

Immunoprecipitation of detergent-solubilized receptors followed by Western blotting has often been used as a useful technique for the possible dimerization of many GPCRs. In particular, differentially epitope-tagged receptors such as HA- or Myc- are often employed in co-immunoprecipitation experiments because of the availability of efficient antibodies against these tags.

Transient co-expression of the HA-tagged A_1 receptor (HA-A_1R) and Myc-tagged $P2Y_1$ receptor (Myc-$P2Y_1R$) in HEK293T cells revealed that they associate with each other as a heteromeric complex.[8] Immunoprecipitation of cell membrane lysates by an anti-HA antibody precipitated both Myc-$P2Y_1R$ and HA-A_1R, as shown in Figure 9.3. Conversely, the anti-Myc antibody immunoprecipitated HA-A_1R along with Myc-$P2Y_1R$ (data not shown), indicating that A_1R is able to form heteromeric complexes with $P2Y_1R$ when transfected simultaneously in HEK293T cells. The Myc-tagged $P2Y_2$ receptor ($P2Y_2R$), another subtype GPCR of the P2 receptor, was found to co-immunoprecipitate with HA-A_1R, but the myc-tagged dopamine D_2 receptor (D_2R) was not (Figure 9.3).

PROTOCOL 9-4: IMMUNOPRECIPITATION OF HA-A_1R/MYC-$P2Y_1$R EXPRESSED IN HEK293T CELLS

---

1. For a HEK293T cell membrane preparation, wash cells (~2 × 10^7 cells), expressing single receptors or combinations of receptors, HA-A_1R and Myc-$P2Y_1R$, twice with ice-cold phosphate-buffered saline (PBS), and scrape with a rubber policeman in ice-cold hypotonic lysis buffer containing 50 mM Tris-acetate buffer, pH 7.4, with a protease-inhibitor cocktail (Roche) on ice.

2. Sonicate the mixture on ice, and subject to low-speed centrifugation to remove organelles and nuclei. Subject the resulting supernatant to centrifugation at 30,000 × g for 20 min. Precipitated cell membranes are collected, washed twice, resuspended in lysis buffer, and stored at 80°C.

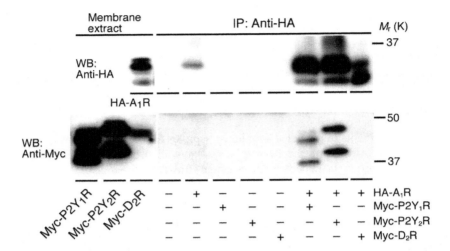

FIGURE 9.3 Co-immunoprecipitation of A_1R and $P2Y_1R$ from HA-A_1R/Myc-$P2Y_1$R-transfected HEK293T cells. Co-immunoprecipitation of cell lysates by anti-HA antibody was performed followed by Western blotting with anti-HA (upper) and anti-Myc (lower) antibodies. In addition to HA-A_1R (35 and 31 kDa), anti-HA antibody co-immunoprecipitated Myc-$P2Y_1$R (42 and 37 kDa) from the cell membrane lysates co-expressing HA-A_1R/Myc-$P2Y_1$R (lower, lane 9 from the left). Myc-$P2Y_2$R (39 and 45 kDa) also was co-immunoprecipitated by anti-HA antibody along with HA-A_1R from cell lysates co-expressing HA-A_1R/Myc-$P2Y_2$R (lower, lane 10 from the left). In contrast, Myc-D_2R was not immunoprecipitated from cell lysates co-expressing HA-A_1R/Myc-D_2R (lower, lane 11 from the left) by anti-HA antibody. (Part of this figure was reprinted with permission from National Academy of Sciences USA, from Yoshioka, K., Saitoh, O., and Nakata, H., *Proc. Natl. Acad. Sci. USA*, 98, 7617, 2001.)

3. For solubilization, incubate the crude membrane preparation with Tx-buffer (50 mM Tris-HCl buffer, pH 7.4, containing 1% Triton X-100, 300 mM NaCl, 100 mM idoacetamide and a protease-inhibitor cocktail) for 60 min at 4°C on a rotator. Centrifuge the mixture at 18,500 × g for 20 min, and collect the supernatant as the cell membrane lysate. An aliquot of the cell membrane lysate (~500 µg protein) is precleared with 30 µl of protein G-agarose (50% suspension in PBS) at 4°C for 30 min on a rotator. The protein G-agarose is then removed by microcentrifuging the lysate at maximum speed (18,500 × g) for 5 min at 4°C.

4. Incubate the precleared cell membrane lysate with 1 µg of anti-Myc 9E10 mAb (Roche) or anti-HA 3F10 mAb (Roche) for 60 min at 4°C on a rotator. Then add 50 µl of protein G-agarose to the mixture. Continue incubation for an additional 120 min at 4°C. Antibody suppliers routinely provide guidelines for the amount of antibody to use in immunoprecipitation procedures.

5. Wash three times with ice-cold Tx-buffer, and subsequently elute from protein G-agarose by the addition of 50 µl of the 2X Laemmli sample

buffer (125 mM Tris-HCl pH 6.8, 200 mM DTT, 2% SDS, 20% glycerol) used for sodium dodecyl sulfate-polyacrylamide gel electrophoresis (SDS-PAGE). Vortex and incubate the mixture at room temperature for 30 min to avoid aggregating receptor proteins. An appropriate amount of immunoprecipitated protein is subjected to SDS-PAGE, after which the protein on the gel is electrotransferred to a nitrocellulose membrane. Running a lane containing either the immunoprecipitated protein of interest or a cell lysate known to express the protein of interest is appropriate as a positive control for Western blotting.

6. After blocking with 5% nonfat skim milk dissolved in washing buffer (0.1% Tween 20 in Tris-HCl-buffered saline; TBS-T), detect HA-A_1R, Myc-$P2Y_1$R, Myc-$P2Y_2$R, or Myc-D_2R on the blot using anti-HA 3F10 mAb (50 ng/ml) or anti-Myc PL14 mAb (100 ng/ml, MBL) in TBS-T containing 2% nonfat skim milk for 1 h, shaking gently on an orbital shaker (or at 4°C overnight if required). Wash the nitrocellulose membrane three times with TBS-T, once for 15 min and twice for 10 min, shaking gently on an orbital shaker. The dilution may need to be optimized depending on the antiserum or purified antibody used.

7. Incubate the nitrocellulose membrane with the secondary antibody (horseradish peroxidase-conjugated goat anti-rat IgG antibody (for anti-HA mAb), goat anti-mouse IgG antibody (for anti-Myc mAb or anti-A_1R mAb), diluted 1:2000 in TBS-T containing 5% nonfat skim milk. Dilutions of the secondary antibody may need to be optimized if the background is high. Wash the nitrocellulose membrane three times with TBS-T, once for 30 min and twice for 15 min, shaking gently. Insufficient washing at this step will increase the background considerably.

8. Visualize the reactive bands by using the enhanced chemiluminescence system (ECL Western Blotting System, Amersham Pharmacia) according to the manufacturer's instructions. Briefly, mix an equal volume of detection solution 1 with detection solution 2 (a total volume of 5 ml is sufficient for a 40 cm^2 membrane), and pour onto the washed membrane. After incubation for 1 min at room temperature, wrap the membranes with a thin sheet, such as Saran Wrap, and place in an x-ray film cassette. Using a sheet of autoradiography film (for example, Hyperfilm ECL), expose for 1 to 60 min according to its appearance. A similar detection kit from Pierce (SuperSignal Western Blotting Kits, Pierce, Rockford, IL) is also applicable with a more sensitive result.

9.2.2.2 Co-Immunoprecipitation of A_1R and $P2Y_1$R from Rat Brain

It is important to detect oligomerization when receptors are expressed at physiological levels in order to address the physiological role of oligomerization. Heterooligomer formation between distinct GPCRs, such as AT1 and bradykinin B2 receptors[15] and metabotropic glutamate 1 and 5 receptors[16] has been shown *in vivo* using co-immunoprecipitation. As described below, the existence of A_1R/$P2Y_1$R heteromeric complexes *in vivo* was also demonstrated by co-immunoprecipitation experiments using a soluble extract from the rat cortex, hippocampus, and cerebellum membranes.[17]

The existence of $A_1R/P2Y_1R$ heteromeric complexes *in vivo* can be demonstrated by co-immunoprecipitation experiments using a soluble extract from rat cortex, hippocampus, and cerebellum membranes, where these two receptors are known to be co-localized.[17] As shown in the protocol below, co-immunoprecipitation followed by immunoblotting should be carried out using antibodies specific to receptors, i.e., anti-A_1R and anti-$P2Y_1R$ antibodies. It should be emphasized that the selection of antibodies is important to obtain reliable results, because it is common to find antibodies that are only applicable in Western blotting. Therefore, detailed characterization of receptor-specific antibodies is necessary before performing the full immunoprecipitation experiment. When brain extracts from three regions were similarly immunoprecipitated by anti-A_1R antibodies, a $P2Y_1R$ band (62 kDa) was clearly detected in addition to the A_1R bands (33, 39 kDa) in all immunoprecipitates (Figure 9.4, lanes 4 to 6). These findings indicate that A_1R is able to interact with $P2Y_1R$ to form a heteromeric complex in rat cortex, hippocampus, and cerebellum.

PROTOCOL 9-5:   IMMUNOPRECIPITATION OF $A_1R/P2Y_1R$ IN RAT BRAINS

1. Dissect cortical, hippocampal, and cerebellar tissues from adult rat brains (Wistar, five males, 10 weeks old). Homogenize the tissues with a Polytron homogenizer for 5 s periods in 50 mM Tris-acetate, pH 7.4, containing a protease inhibitor cocktail (Complete, Roche Diagnostics), and centrifuge the cell suspensions at $30,000 \times g$ for 30 min at 4°C. Wash the precipitated membranes twice by repeating the suspension and centrifugation as above.
2. Suspend the washed membrane pellet with 10-vol. ice-cold lysis buffer (50 mM Tris-HCl, pH 7.4, 1% Triton X-100, 300 mM NaCl, and a protease inhibitor cocktail), and incubate for 60 min at 4°C on a rotator. Centrifuge the mixture at $18,500 \times g$ for 20 min at 4°C, and save the supernatant as a solubilized preparation.
3. Add protein G-agarose (50 μl, Roche Diagnostics) to the solubilized lysate (1 ml), and incubate for 60 min at 4°C followed by centrifugation to remove protein G-agarose. Subsequently, incubate the precleared lysate with rabbit polyclonal anti-A_1R antibody (1 μg/ml, Sigma) for 60 min at 4°C on a rotator, and then add protein G-agarose (50 μl) to the mixture. Continue the incubation for an additional 120 min.
4. Collect the protein G-agarose by centrifugation ($15,000 \times g$ for 5 min), and wash the protein G-agarose three times with the same lysis buffer. Elute the immunoprecipitate from the protein G-agarose by adding 50 μl of the 2X Laemmli sample buffer (125 mM Tris-HCl pH 6.8, 200 mM DTT, 2% SDS, 20% glycerol) used for SDS-PAGE. Vortex and incubate the mixture at room temperature for 30 min. Take the appropriate amount of eluted immunoprecipitated protein and apply to 12% SDS-PAGE. The

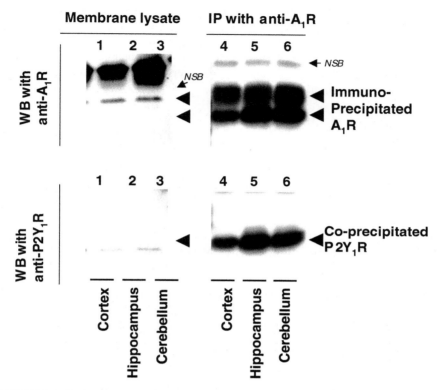

FIGURE 9.4 Co-immunoprecipitation of A_1R and $P2Y_1R$ from rat brains. Extracts from various regions of rat brains were immunoprecipitated with anti-A_1R antibodies, and analyzed by Western blotting with anti-A_1R (upper panels) or anti-$P2Y_1R$ (lower panels) antibodies. Anti-A_1R antibodies precipitated A_1R (upper panel, Mr = 33, 39 kDa) along with $P2Y_1R$ (lower panel, Mr = 62 kDa) from membrane lysates of rat cortex, hippocampus, and cerebellum. *NSB*, nonspecific band. (Part of this figure is from Yoshioka, K. et al., *FEBS Lett.*, 531, 299, 2002. With permission from Elsevier.)

resolved proteins are electrotransferred to a nitrocellulose membrane using semidry electrotransfer apparatus.

5. After blocking with 5% nonfat skim milk dissolved in washing buffer (0.1% Tween 20 in 25 mM Tris·HCl-buffered saline; TBS-T), incubate the blot membranes with rabbit polyclonal anti-A_1R antibody (0.5 μg/ml, Sigma) or anti-$P2Y_1R$ antibody (1:500 dilution, provided by Dr. Moore[18]), diluted in TBS-T containing 5% nonfat skim milk, shaking gently overnight at 4°C.

6. Wash the nitrocellulose membrane three times with TBS-T, once for 15 min and twice for 10 min, shaking gently on an orbital shaker. Then, incubate the membrane with the secondary antibody (horseradish peroxidase-conjugated goat anti-rabbit IgG antibody diluted 1:100,000 in TBS-T containing 5% nonfat skim milk) for 1 h at room temperature. Dilutions

of the secondary antibody may need to be optimized if the background is high. Wash the nitrocellulose membrane three times with TBS-T, once for 30 min and twice for 15 min, shaking gently.

7. Visualize the reactive bands using an enhanced chemiluminescence system (ECL Western Blotting System, Amersham Pharmacia) according to the manufacturer's instructions.

9.2.3 IMMUNOCYTOCHEMICAL ANALYSIS

Demonstration of the co-localization of the GPCRs of interest in heterologously expressed cultured cells or tissues by immunocytochemistry is of great importance to demonstrate the physiological relevancy of GPCR hetero-oligomerization. Again, as the selection of a highly specific antibody is the key factor to obtaining good results, the use of antibodies against tag peptides is preferable if tagged receptors are expressed.

9.2.3.1 Double Immunofluorescence Microscopy of Co-Expressed Purinergic Receptors

The subcellular distribution of HA-A_1R and Myc-$P2Y_1R$ in co-transfected cells can be examined immunocytochemically by confocal laser microscopy (Figure 9.5). When expressed in HEK293T cells individually, HA-A_1R and Myc-$P2Y_1R$ were localized near the plasma membranes (data not shown). Images taken at the microscopic level using an ×63 objective of co-transfected cells double labeled for HA-A_1R (red) and Myc-$P2Y_1R$ (green) are shown in Figure 9.5a and Figure 9.5b. Both receptors were expressed prominently near the plasma membranes. When the images were merged using confocal assistance software, there was a striking overlap (intense yellow spots) in the distribution of the two receptors (Figure 9.5c). The extent of overlapping pixels of the two signals was 35.4 ± 9.6% (n = 3). Immunostaining of unpermeabilized cells (without treatment with Triton X-100) was also performed with similar results (data not shown). As this co-localization occurred over plasma membranes, it supports the heteromeric association of A_1R and $P2Y_1R$. Similar co-localization of A_1R and $P2Y_1R$ has been demonstrated in rat brain sections.[17]

PROTOCOL 9-6:   IMMUNOCYTOCHEMICAL DETECTION OF A_1R/$P2Y_1R$ IN HEK293T CELLS

1. Culture HEK293T cells transfected with HA-A_1R alone, HA-A_1R/Myc-$P2Y_1R$, or Myc-$P2Y_1R$ alone on glass coverslips pretreated with 0.2% polyethyleneimine/0.15 M borate buffer, pH 8.4. Wash cells with PBS followed by fixing with 4% paraformaldehyde in PBS for 30 min, permeabilize with 0.25% Triton X-100 in PBS for 10 min, and incubate with the primary antibody against HA-tag (3F10, Roche, 100 ng/ml in 5% skim milk) or Myc-tag (9E10, Roche, 1μg/ml in 5% skim milk) for 90 min at room temperature.

2. Wash three times with PBS for 10 min.

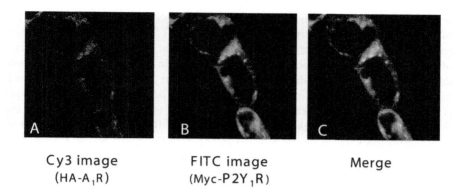

Cy3 image   FITC image   Merge
(HA-A$_1$R)   (Myc-P2Y$_1$R)

FIGURE 9.5 (See color insert following page 240) Confocal imaging of HEK293T cells co-expressing HA-A$_1$R/Myc-P2Y$_1$R. HA-A$_1$R (Cy3, red, Panel A) and Myc-P2Y$_1$R (FITC, green, Panel B) were detected using double fluorescent immunohistochemistry. The right panel is the product of merging the left two panels, showing the co-localization of HA-A$_1$R and Myc-P2Y$_1$R in co-transfected HEK293T cells (yellow, Panel C). (Part of this figure was reprinted with permission from National Academy of Sciences USA from Yoshioka, K., Saitoh, O., and Nakata, H., *Proc. Natl. Acad. Sci. USA*, 98, 7617, 2001.)

3. Incubate with the secondary antibody for 60 min at room temperature. Visualize rat anti-HA 3F10 mAb with Cy3-conjugated goat anti-rat IgG antibodies (Jackson ImmunoResearch, West Grove, PA, 1:1000 dilution in PBS). Use fluorescein isothiocyanate (FITC)-conjugated goat anti-mouse IgG antibody (Jackson ImmunoResearch, 1:1000 dilution in PBS) to detect mouse anti-Myc 9E10 mAb.
4. Note that the fluorescent images were obtained with a Zeiss LSM 410 confocal microscope. The extent of overlap of the two signals was determined using software for the Carl Zeiss LSM 4 Laser Scan Microscope.

9.2.4 BRET Analysis

The co-immunoprecipitation strategies described above seem limited because the solubilization of hydrophobic GPCRs can lead to aggregation that could be mistaken for dimerization, and the solubilization process with detergents can inhibit the association between GPCRs. To overcome these problems, biophysical assays based on light resonance energy transfer are of great value.

Bioluminescence resonance energy transfer (BRET) is a recently described biophysical method that represents a powerful tool with which to measure protein–protein interactions in live cells, in real time.[19,20] BRET results from nonradioactive energy transfer between a luminescent donor and fluorescent acceptor proteins. For example, in *Renilla reniformis*, the catalytic degradation of coelentrazine by luciferase (*Renilla* luciferase or Rluc) results in luminescence; this is, in turn, transferred to green fluorescent protein (GFP), which emits fluorescence. Two basic conditions are required for BRET to occur: first, the donor and the acceptor should be in close proximity (less than 100 Å as with two protein subunits in a dimer); and second, the emission spectrum of the donor and the excitation spectrum of the acceptor must overlap. The original BRET technology generally uses enhanced yellow fluorescent protein (EYFP) as an acceptor, a red-shifted variant of GFP that has an emission maximum at 530 nm. In contrast, the recently introduced BRET^2 uses a codon-humanized form of wild-type GFP, termed GFP^2, which has an emission maximum at 510 nm (Figure 9.6). The improved BRET technique (BRET^2),[21] which offers greatly improved separation of the emission spectra of the donor and acceptor moieties compared to traditional BRET,[22] was employed in an oligomerization study between A_1R and $P2Y_1R$,[23] as described below. BRET technology has been successfully used to show GPCR heterodimerization in living cells, including β_1 and β_2 adrenoceptors,[24] oxytocin and vasopressin receptors,[25] adenosine A_{2A} and dopamine D_2 receptors,[12] adenosine A_1 and $P2Y_1$ receptors,[23] thyrotropin-releasing hormone receptor 1 and 2,[26] and β_2 adrenoceptor and δ opioid receptors.[27] FRET (fluorescence resonance energy transfer) is another biophysical assay in which both the donor and acceptor are fluorescent. Other methods include photobleaching FRET and time-resolved FRET. These methods have been used to show the occurrence of dimerization in living cells for different GPCRs, including the δ-opioid receptor,[27,28] the thyrotropin-releasing hormone receptor,[28] and the SSTR5-somatostatin receptor.[29]

It is noted that the quantification of resonance energy transfer is much easier in BRET than in FRET. Reproducibility of the BRET ratio has been quite successful in my experience. Another related advantage of BRET is that the relative expression levels of the donor and acceptor partners can be quantified independently by measuring the luminescence of the donor and the fluorescence of the acceptor. In addition, a tenfold improvement in sensitivity could be attained by using BRET rather than FRET. As the BRET ratio is influenced by the relative expression of the donor and acceptor fusion proteins, i.e., increasing the concentration of the acceptor is likely to increase the probability of donor–acceptor interaction, it is important to control the protein expression when carrying out BRET assays. The relative expression of Rluc-tagged constructs can be quantified using a luminometer, and that of GFP variant-tagged constructs using fluorescence-activated cell sorting (FACS) or a fluorescence microtiter plate reader. BRET experiments are generally carried out with either a significantly greater concentration of acceptors than donors, such as with the 1:4 ratio,[28] or with equimolar donor/acceptor concentrations. The latter scenario was used when the percentage of the β_2-adrenergic receptor population existing as dimers was estimated. These calculations assumed a free equilibrium between the donor and acceptor constructs, thereby necessitating the 1:1 ratio.[24]

9.2.4.1   BRET of Co-Expressed Purinergic Receptors

We demonstrated that the co-expression of A_1R fused with GFP^2 and $P2Y_1R$ fused with Rluc in HEK293T cells resulted in a significant increase in the BRET signal upon the addition of Rluc substrates under basal conditions, indicating the protein–protein interaction between these receptors, as shown below.[23] The BRET signal for the co-expression of A_1R-GFP^2 and A_1R-Rluc was also significant, although the extent of heteromeric association was substantially greater than the homomeric

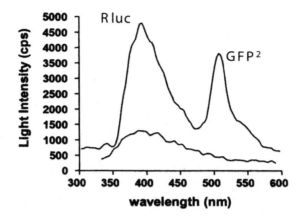

$$\text{BRET ratio} = \frac{E_m 515nm - Blank\,(E_m 515)}{E_m 410nm - Blank\,(E_m 410)}$$

FIGURE 9.6 BRET[2] emission spectra. BRET[2] uses advanced protein–protein interaction-assay technology, with *Renilla* luciferase (Rluc) as the donor and a modified green fluorescent protein (GFP[2]) as the acceptor molecule. It is based on energy transfer from a bioluminescent donor to a fluorescent acceptor protein. Rluc emits blue light in the presence of its substrate Deep Blue C and energy. If GFP[2] is in close proximity to Rluc, it absorbs blue light energy and re-emits green light. The BRET[2] signal, therefore, is measured by the amount of green light emitted by GFP[2] as compared to the blue light emitted by Rluc (BRET ratio). (Part of this figure was reprinted from Nakata, H., Yoshioka, K., and Saitoh, O., *Drug Dev. Res.*, 58, 340, 2003. With permission of Wiley InterScience.)

association of A_1R. A significant increase in the BRET signal was observed by combined treatment with A_1R and $P2Y_1R$ agonists.

9.2.4.2 Plasmid Constructs, Transfection, and BRET[2] Assay

The cDNA constructs of HA-A_1R-GFP[2] and Myc-$P2Y_1$R-Rluc were generated by amplification of the HA-tagged rat A_1R and Myc-tagged rat $P2Y_1R$ coding sequence,[8] without its stop codon, using sense and antisense primers containing distinct restriction enzyme sites at the 5' and 3'ends, respectively. The fragments were then subcloned in-frame into the appropriate sites of the codon-humanized pGFP[2]-N3 and pRluc-N3 expression vectors, respectively (GFP[2] fusion protein expression vector, #6310240, PerkinElmer). Rluc and GFP[2] were located at the C terminal end of the receptors. HEK293T cells were transfected with these plasmids transiently and were harvested 48 h later for BRET experiments.

To determine whether there is a constitutive association between A_1R and A_1R or A_1R and $P2Y_1R$ in living cells, BRET[2] was measured in HEK293T cells co-transfected with either HA-A_1R-GFP[2]/HA-A_1R-Rluc or HA-A_1R-GFP[2]/Myc-$P2Y_1$R-Rluc. As shown in Figure 9.7a, the co-expression of HA-A_1R-GFP[2]/HA-A_1R-Rluc

(BRET ratio = 0.062 ± 0.004, n = 15) or HA-A$_1$R-GFP^2/Myc-P2Y$_1$R-Rluc (BRET ratio = 0.07 ± 0.008, n = 20) on the addition of Rluc substrates resulted in a small but significant increase in the BRET ratio ($P < 0.05$ versus control cells) under basal conditions. Co-expression of the isolated Rluc along with HA-A$_1$R-GFP^2 resulted in weak energy transfer (BRET ratio = 0.045 ± 0.005, n = 6), indicating that there was no direct interaction between these two constructs (Figure 9.7a, lower line). Similarly, the co-expression of isolated GFP^2 with Myc-P2Y$_1$R-Rluc failed to produce a significant energy transfer signal (BRET ratio = 0.048 ± 0.006, n = 6) (data not shown). These results provide strong evidence of an association between either A$_1$R-GFP^2 and A$_1$R-Rluc or A$_1$R-GFP^2 and P2Y$_1$R-Rluc in intact cells. The extent of heteromeric association was substantially greater than that of the homomeric association of A$_1$R ($p < 0.05$). Incubation of HA-A$_1$R-GFP^2/Myc-P2Y$_1$R-Rluc-co-transfected cells with the agonists CPA and ADPβS increased the BRET ratio with a peak at 10 min (Figure 9.7a, upper line). The agonist-promoted BRET signal observed between HA-A$_1$R-GFP^2 and Myc-P2Y$_1$R-Rluc did not result from the nonspecific association between GFP^2 and Rluc proteins because no increase in signal intensity was detected in cells expressing either HA-A$_1$R-GFP^2/Rluc or HA-A$_1$R-GFP^2/HA-A$_1$R-Rluc, as shown above. It was also confirmed that the BRET signal did not strengthen in cells co-expressing GFP^2 and Rluc (0.045 ± 0.008, n = 6). Incubation of HA-A$_1$R-GFP^2/HA-A$_1$R-Rluc-expressing cells with agonists did not result in a significant increase in the BRET signal (Figure 9.7a, middle line).

To demonstrate the agonist-dependent increase in the BRET ratio, HA-A$_1$R-GFP^2/Myc-P2Y$_1$R-Rluc-transfected cells were incubated for 10 min in the presence of several ligands (Figure 9.7b). A significant increase in the ratio was again observed in the presence of both agonists, but not with either alone. This increase was significantly inhibited by pretreatment with MRS2179, a potent P2Y$_1$R antagonist, although the addition of MRS2179 alone had no effect on the BRET ratio.

This suggests that oligomerization between these purinergic receptors persists in living cells, and activation of the receptors may control the oligomerization process. It was noted however, that the detailed mechanism of BRET enhancement by receptor agonists should be further investigated because the possibility that enhanced BRET following agonist stimulation might be due to conformational changes in receptors that are already constitutively associated, rather than enhanced association between these receptors, has not been completely ruled out.

PROTOCOL 9-7:   BRET^2 ASSAY OF P2Y$_1$R-RLUC/A$_1$R-GFP^2 CO-EXPRESSING HEK293T CELLS

1. Transfect HEK293T cells transiently with HA-A$_1$R-GFP^2 and Myc-P2Y$_1$R-Rluc using Effectene transfection reagent (Qiagen) as described above. After 48 h culture, treat the cells with trypsin-EDTA (0.25%) solution for 1 to 3 min to detach cells from the dish. Collect the cells into DMEM-10% fetal bovine serum and centrifuge. Wash the cells with Dulbecco's PBS containing 0.1 g/l CaCl$_2$, 0.1 g/l MgCl$_2$, and 1 g/l D-glucose (assay buffer) two or three times. Suspend the cells in the assay buffer at 10^7 cells/ml.

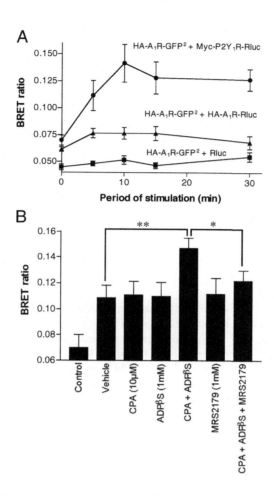

FIGURE 9.7 BRET^2 detection of constitutive and agonist-promoted oligomerization of HA-A$_1$R-GFP^2 and Myc-P2Y$_1$R-Rluc in living HEK293T cells. (a) Time-dependent BRET^2 signal in living HEK293T cells co-expressing HA-A$_1$R-GFP^2 and HA-A$_1$R-Rluc (homo-oligomer, triangle), HA-A$_1$R-GFP^2 and Myc-P2Y$_1$R-Rluc (hetero-oligomer, circle), or HA-A$_1$R-GFP^2 and Rluc (control, square). Cells were incubated with agonists of A$_1$R and P2Y$_1$R (1 μM CPA + 100 μM ADPβS) before the addition of Rluc substrates. The data shown represent the mean ± SEM of three independent experiments performed in triplicate for each time point. (b) BRET^2 ratio was measured in HEK293T cells co-transfected with HA-A$_1$R-GFP^2 and Myc-P2Y$_1$R-Rluc. Cells were incubated with either CPA (1 μM), ADPβS (100 μM), P2Y$_1$R antagonist MRS2179 (1 mM), or a combination thereof for 10 min before the addition of Rluc substrate. The data represent the mean ± SEM of three independent experiments, **$P < 0.01$ compared with vehicle treatment, *$P < 0.05$ compared with CPA and ADPβS treatment. (Part of this figure was reprinted from Yoshioka, K., Saitoh, O., and Nakata, H., *FEBS Lett.*, 523, 147, 2002. With permission from Elsevier.)

2. Prepare 1 mM DeepBlueC (PerkinElmer) stock solution, Rluc substrate, by adding absolute ethanol. Before the BRET assay, prepare a 20-fold (50 μM in the assay buffer) dilution of the stock DeepBlue C solution. The diluted DeepBlueC solution should be prepared fresh and discarded after the assay.

3. Dispense 50 μl of cells (~5 × 10⁵ cells/well) in a 96-well white-walled microplate (OptiPlate, PerkinElmer). The cells can be incubated with or without receptor ligands for specific periods at 37°C. Add 5 μl of freshly diluted 50 μM DeepBlue C solution to each well at a final concentration of 5 μM.

4. Determine the signal *immediately* by using a Fusion microplate analyzer (PerkinElmer) with 410 ± 40 nm and 515 ± 15 nm emission filters. The background was taken as the area of this region of the spectrum without transfectants. Data are represented as a BRET ratio defined as [(emission at 515 nm) (background emission at 515 nm)]/[(emission at 410 nm) (background emission at 410 nm)]. Although fluorescence decreases quickly during the assay, the BRET ratio remains constant for several minutes.

5. Note that the pBRET+ vector (#6310025, PerkinElmer) is useful as a positive control in a BRET² assay. This vector contains a fusion Rluc::GFP gene that once expressed in cells efficiently performs energy transfer on the addition of DeepBlueC. The fusion Rluc::GFP gene placed under the control of the cytomegalovirus (CMV) promoter thus shows high constitutive expression in a variety of cells, such as HEK293T cells. The BRET ratio obtained from HEK293T cells transfected with this vector is usually 0.3 to 0.4 in our experiments.

ACKNOWLEDGMENTS

This work was supported in part by grants for Scientific Research from the Ministry of Education, Culture, Sports, Science and Technology of Japan. It was also supported by the CREST program of the Japan Science and Technology Agency.

ABBREVIATIONS

BRET	bioluminescence resonance energy transfer
DMEM	Dulbecco's modified Eagle medium
GPCR	G protein-coupled receptor
HA	hemagglutinin
SDS-PAGE	sodium dodecyl sulfate-polyacrylamide gel electrophoresis

REFERENCES

1. Bouvier, M., Oligomerization of G-protein-coupled transmitter receptors, *Nat. Rev. Neurosci.*, 2, 274, 2001.

2. Angers, S., Salahpour,, A. and Bouvier, M., Dimerization: an emerging concept for G protein-coupled receptor ontogeny and function, *Annu. Rev. Pharmacol. Toxicol.*, 42, 409, 2002.

3. George, S.R., O'Dowd, B.F., and Lee, S.P., G-protein-coupled receptor oligomerization and its potential for drug discovery, *Nat. Rev. Drug Discov.*, 1, 808, 2002.

4. Kroeger, K.M., Pfleger, K.D., and Eidne, K.A., G-protein coupled receptor oligomerization in neuroendocrine pathways, *Front Neuroendocrinol.*, 24, 254, 2003.

5. Milligan, G., G protein-coupled receptor dimerization: function and ligand pharmacology, *Mol. Pharmacol.*, 66, 1, 2004.

6. Salim, K. et al., Oligomerization of G-protein-coupled receptors shown by selective co-immunoprecipitation, *J. Biol. Chem.*, 277, 15,482, 2002.

7. Ralevic, V. and Burnstock, G., Receptors for purines and pyrimidines, *Pharmacol. Rev.*, 50, 413, 1998.

8. Yoshioka, K., Saitoh, O., and Nakata, H., Heteromeric association creates a P2Y-like adenosine receptor, *Proc. Natl. Acad. Sci. USA*, 98, 7617, 2001.

9. Gines, S. et al., Dopamine D_1 and adenosine A_1 receptors form functionally interacting heteromeric complexes, *Proc. Natl. Acad. Sci. USA*, 97, 8606, 2000.

10. Ciruela, F. et al.. Metabotropic glutamate 1alpha and adenosine A_1 receptors assemble into functionally interacting complexes, *J. Biol. Chem.*, 276, 18,345, 2001.

11. Ferre, S. et al., Synergistic interaction between adenosine A_{2A} and glutamate mGlu5 receptors: implications for striatal neuronal function, *Proc. Natl. Acad. Sci. USA*, 99, 11,940, 2002.

12. Kamiya, T. et al., Oligomerization of adenosine A_{2A} and dopamine D_2 receptors in living cells, *Biochem. Biophys. Res. Commun.*, 306, 544, 2003.

13. Hillion, J. et al., Coaggregation, cointernalization, and codesensitization of adenosine A_{2A} receptors and dopamine D_2 receptors, *J. Biol. Chem.*, 277, 18,091, 2002.

14. Fuxe, K. et al., Receptor heteromerization in adenosine A_{2A} receptor signaling: relevance for striatal function and Parkinson's disease, *Neurology*, 61, S19, 2003.

15. AbdAlla, S., Lother, H., and Quitterer, U., AT_1-receptor heterodimers show enhanced G-protein activation and altered receptor sequestration, *Nature*, 407, 94, 2000.

16. Gama, L., Wilt, S.G., and Breitwieser, G.E., Heterodimerization of calcium sensing receptors with metabotropic glutamate receptors in neurons, *J. Biol. Chem.*, 276, 39,053, 2001.

17. Yoshioka, K. et al., Hetero-oligomerization of adenosine A_1 receptors with $P2Y_1$ receptors in rat brains, *FEBS Lett.*, 531, 299, 2002.

18. Moore, D. et al., Immunohistochemical localization of the $P2Y_1$ purinergic receptor in Alzheimer's disease, *Neuroreport*, 11, 3799, 2000.

19. Eidne, K.A., Kroeger, K.M., and Hanyaloglu, A.C., Applications of novel resonance energy transfer techniques to study dynamic hormone receptor interactions in living cells, *Trends Endocrinol. Metab.*, 13, 415, 2002.

20. Pfleger, K.D. and Eidne, K.A., New technologies: bioluminescence resonance energy transfer (BRET) for the detection of real time interactions involving G-protein coupled receptors, *Pituitary*, 6, 141, 2003.

21. Joly, E. et al., Bioluminescence resonance energy transfer (BRET2). Principles, applications, and products, in *Application Note # BRT-001*, PerkinElmer Inc., Wellesley, MA, 2002.

22. Xu, Y., Piston, D.W., and Johnson, C.H., A bioluminescence resonance energy transfer (BRET) system: application to interacting circadian clock proteins, *Proc. Natl. Acad. Sci. USA*, 96, 151, 1999.

23. Yoshioka, K., Saitoh, O., and Nakata, H., Agonist-promoted heteromeric oligomerization between adenosine A_1 and $P2Y_1$ receptors in living cells, *FEBS Lett.*, 523, 147, 2002.

24. Mercier, J.F. et al., Quantitative assessment of β_1- and β_2-adrenergic receptor homo- and heterodimerization by bioluminescence resonance energy transfer, *J. Biol. Chem.*, 277, 44,925, 2002.

25. Terrillon, S. et al., Oxytocin and vasopressin V1a and V2 receptors form constitutive homo- and heterodimers during biosynthesis, *Mol. Endocrinol.*, 17, 677, 2003.

26. Hanyaloglu, A.C. et al., Homo- and hetero-oligomerization of thyrotropin-releasing hormone (TRH) receptor subtypes. Differential regulation of beta-arrestins 1 and 2, *J. Biol. Chem.*, 277, 50,422, 2002.

27. McVey, M. et al., Monitoring receptor oligomerization using time-resolved fluorescence resonance energy transfer and bioluminescence resonance energy transfer. The human delta-opioid receptor displays constitutive oligomerization at the cell surface, which is not regulated by receptor occupancy, *J. Biol. Chem.*, 276, 14,092, 2001.

28. Kroeger, K.M. et al., Constitutive and agonist-dependent homo-oligomerization of the thyrotropin-releasing hormone receptor. Detection in living cells using bioluminescence resonance energy transfer, *J. Biol. Chem.*, 276, 12,736, 2001.

29. Rocheville, M. et al., Subtypes of the somatostatin receptor assemble as functional homo- and heterodimers, *J. Biol. Chem.*, 275, 7862, 2000.

30. Ciruela, F. et al., Immunological identification of A_1 adenosine receptors in brain cortex, *J. Neurosci. Res.*, 42, 818, 1995.

31. Canals, M. et al., Homodimerization of adenosine A_{2A} receptors: qualitative and quantitative assessment by fluorescence and bioluminescence energy transfer, *J. Neurochem.*, 88, 726, 2004.

32. Yoshioka, K. and Nakata, H., ATP- and adenosine-mediated signaling in the central nervous system: purinergic receptor complex: generating adenine nucleotide-sensitive adenosine receptors, *J. Pharmacol. Sci.*, 94, 88, 2004.

33. Nakata, H., Yoshioka, K., and Saitoh, O., Hetero-oligomerization between adenosine A_1 and $P2Y_1$ receptors in living cells: formation of ATP-sensitive adenosine receptors, *Drug Dev. Res.*, 58, 340, 2003.

Part III

Tertiary Structure of GPCRs and Their Ligands

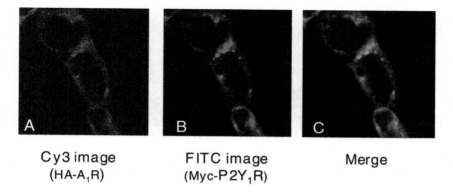

Cy3 image (HA-A₁R)	FITC image (Myc-P2Y₁R)	Merge

COLOR FIGURE 9.5 Confocal imaging of HEK293T cells co-expressing HA-A$_1$R/Myc-P2Y$_1$R. HA-A$_1$R (Cy3, red, Panel A) and Myc-P2Y$_1$R (FITC, green, Panel B) were detected using double fluorescent immunohistochemistry. The right panel is the product of merging the left two panels, showing the co-localization of HA-A$_1$R and Myc-P2Y$_1$R in co-transfected HEK293T cells (yellow, Panel C). (Part of this figure was reprinted with permission from National Academy of Sciences USA from Yoshioka, K., Saitoh, O., and Nakata, H., *Proc. Natl. Acad. Sci. USA*, 98, 7617, 2001.

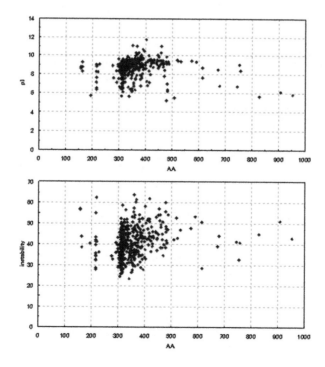

COLOR FIGURE 10.1 Properties of human class A GPCRs. Examined were 648 receptors included in the Swiss-Prot database. (Upper) Distribution of pI against the polypeptide length. (Lower) Distribution of instability against the polypeptide length.

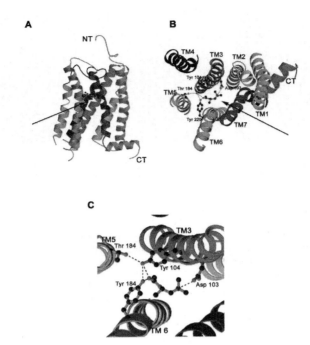

COLOR FIGURE 11.7 M_2 receptor model in the active form bound to (S)-methacholine viewed from the side (a), and from the top of the N-terminus (NT) (b). The seven-transmembrane segments (TM) are colored in aquamarine (TM1), gold (TM2), red (TM3), navy blue (TM4), orange (TM5), gray (TM6), and dark green (TM7); the C-terminal (CT) helix is colored in light purple. Arrows in (a) and (b) point to the (S)-methacholine binding site. The closer view of the (S)-methacholine binding site (c) implies the possible electrostatic interaction between the quaternary amine and Asp103 side chain, and hydrogen bonds involving the acetyl group and residues such as Tyr 104, Thr 184, and Tyr 184. The O-C2-C1-N dihedral angle of (S)-methacholine in this energy minimized model is +55.5°.

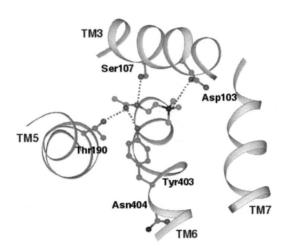

COLOR FIGURE 12.8 Complex model of acetylcholine at the binding cleft of the fully active form of the M_2 receptor models. (View from the extracellular site.) Transmembrane helical regions (TM) at the binding clefts are shown with gray ribbon. Hydrogen bonds are indicated with dotted lines. Oxygen and nitrogen atoms are black; carbon atoms are gray.

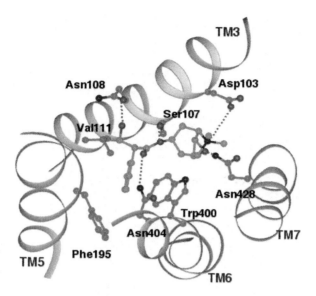

COLOR FIGURE 12.9 Complex model of N-methylscopolamine at the binding cleft of the physiologically inactive form of the M$_2$ receptor models.

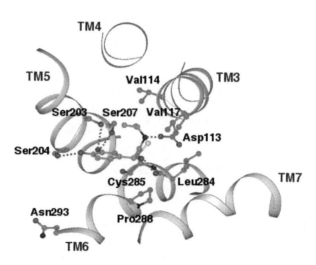

COLOR FIGURE 12.10 Complex model of *R*-isoproterenol at the binding cleft of the fully active form of the β$_2$ receptor models.

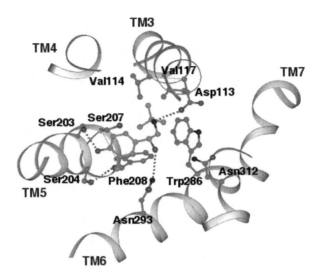

COLOR FIGURE 12.11 Complex model of salbutamol at the binding cleft of the partially active form of the β_2 receptor models.

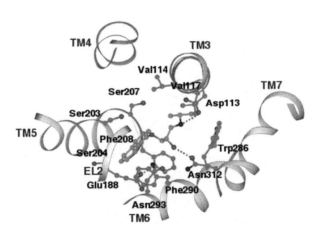

COLOR FIGURE 12.12 Complex model of propranolol at the binding cleft of the fully inactive form of the β_2 receptor models.

10 Methods and Results in X-Ray Crystallography of Bovine Rhodopsin

Tetsuji Okada, Rumi Tsujimoto, Miho Muraoka, and Chihiro Funamoto

CONTENTS

10.1 INTRODUCTION

Retinal photoreceptor proteins for visual function comprise one of the large subfamilies in class A G protein-coupled receptors (GPCRs). Rhodopsin in the rod cells is the best studied and represents a paradigm for understanding the structure and function of GPCRs. The heptahelical transmembrane motif shared by all GPCRs was first demonstrated by electron cryomicroscopy of bovine rhodopsin in two-dimensional crystals.[1] Recent x-ray crystallographic studies[2–4] revealed further details of the three-dimensional structure and provided a template model for hundreds of GPCRs having similar amino acid residues to rhodopsin. Despite considerable efforts made to obtain the structural information of another member in the GPCR superfamily, no atomic coordinates other than bovine rhodopsin is available at this moment.

The success of our x-ray crystallographic structure determination obviously took the advantages that rhodopsin could be obtained in milligram quantity from bovine retina and purified with a simple purification method, which includes membrane manipulation and selective solubilization.[5] The remarkable stability of bovine rhodopsin, having covalently bound intrinsic inverse-agonist 11-*cis*-retinal, must also aid other various aspects in sample preparation and crystallization. In fact, removal of the chromophore leaves much unstable apoprotein called opsin, which was not yet successfully crystallized.

GPCRs frequently have some posttranslational modifications, such as glycosylation and acylation at the extracellular and intracellular domains, respectively. In bovine rhodopsin, Asn2 and Asn15 in the N-terminal tail are known to link hexasaccharides,[6] and Cys322 and Cys323 in the C-terminal tail are palmitoylated.[7] It was also shown that the glycosylation contained some heterogeneous sugar components. Including these modifications and the retinal, total molecular weight is about 42,000 Da. For the crystallographic studies, it was anticipated that these nonprotein moieties would be unfavorable and could negatively affect the crystallization efforts of other GPCRs.

The calculated theoretical pI of bovine rhodopsin is 5.9, which is quite lower than most of the class A GPCRs (Figure 10.1). The stability of mammalian opsin is also recognized from the calculation of the instability index,[8] which is found to be usually small in the case of membrane proteins with structures that were successfully solved at atomic resolution. From these examinations, it should be noted that bovine rhodopsin has somewhat peculiar properties among the class A GPCRs. However, it is also true that some structurally related membrane proteins have been crystallized under similar conditions.

In this chapter, the experimental details in the x-ray crystallography of bovine rhodopsin in the ground state and some structural features that appear to be of general significance are described.

10.2 PURIFICATION AND CRYSTALLIZATION

10.2.1 Isolation of Membranes from Retina

We used bovine retina purchased from Lawson Co. (Lincoln, NE) in frozen, dark-adapted conditions for obtaining the rhodopsin sample. Although we have not yet figured out the reason, the batch of retina appears to be one of the critical factors determining the final diffraction quality of the crystals. It appears to also be true that it is the method of membrane preparation and not the absolute purity of the membrane that significantly affects the quality of the crystals.

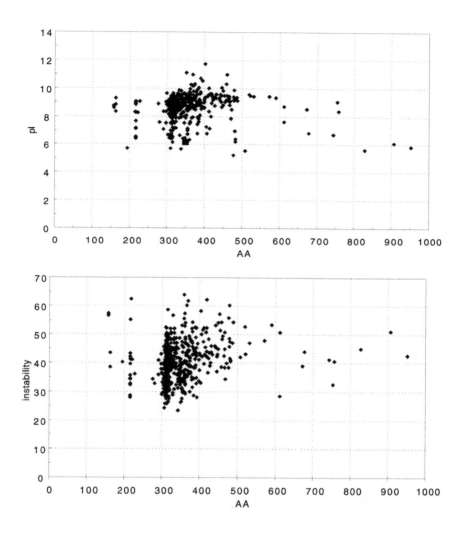

FIGURE 10.1 (**See color insert following page 240**) Properties of human class A GPCRs. Examined were 648 receptors included in the Swiss-Prot database. (Upper) Distribution of pI against the polypeptide length. (Lower) Distribution of instability against the polypeptide length.

In vertebrate retina, rhodopsin is exclusively localized in the outer segment of the rod cells that is responsible for the scotopic visual function. Isolation of the rod outer segment (ROS) membranes can be done by the sucrose flotation/density gradient method outlined in Protocol 10-1. All the procedures should be done on ice and under dim red light (>640 nm) unless otherwise stated.

PROTOCOL 10-1: ISOLATION OF MEMBRANES FROM RETINA

1. Materials required:
   a. 200 bovine retinas, frozen, dark-adapted (wrapped with aluminum foil)
   b. ROS buffer [10 mM MOPS/NaOH (pH 7.5), 30 mM NaCl, 60 mM KCl, 2 mM $MgCl_2$, 1 mM DTT, 0.1 mM PMSF, 4 mg/ml leupeptin, 50 KIU/ml aprotinin]
   c. 400 ml 40% (w/v) sucrose in ROS buffer
   d. 100 ml 29% (w/v) sucrose in ROS buffer
   e. 100 ml 34% (w/v) sucrose in ROS buffer
2. Thaw the retinas quickly at about 30°C, and put them into a 500 ml centrifuge bottle.
3. Add 200 ml 40% sucrose, and shake the bottle for 1 min × two times.
4. Centrifuge at 10,000 rpm for 30 min, and collect the supernatant, including some membranes adhered to the bottle wall.
5. Repeat steps 3 and 4 once more, mix and divide all the supernatants into four 500 ml bottles, and add roughly an equal volume of the ROS buffer to each of the bottles.
6. Centrifuge at 10,000 rpm for 60 min, and discard the supernatants.
7. Suspend the pellets with a small volume of ROS buffer to give a total volume of about 50 ml.
8. Prepare six swinging rotor centrifuge tubes containing a step gradient of 14 ml 34% and 8 ml 29% sucrose, then put the membrane suspension onto the top of each tube.
9. Centrifuge at 25,000 rpm for 90 min, and collect the membranes at the interface between the 29% and 34% sucrose layers.
10. Dilute the suspension with more than an equal volume of ROS buffer, centrifuge at 15,000 rpm for 30 min, and discard the supernatant.

10.2.2 PURIFICATION OF MEMBRANES

The ROS membranes prepared as described in Protocol 10-1 are already pure but must be washed once with the ROS buffer and then at least four times with distilled water to remove sucrose, ions, and the proteins peripherally associated with the membranes. Usually, we divide the total ROS membranes from 200 retinas into four parts for further washing. One of them, containing optimally about 50 mg rhodopsin, is homogenized with 15 ml of deionized water containing 1 mM DTT and 0.01% NaN_3, centrifuged at 20,000 to 30,000 rpm for 30 to 40 min, and then the supernatant is discarded. After the second wash, the pellet is subjected to a freeze–thaw step to increase the efficiency of the removal of the peripheral proteins. It also works to tighten the packing of the pellet, which gradually tends to swell under the hypotonic condition. The washed membranes are stored at 80°C until further use. Because membrane proteins are most stable in the lipid bilayer environment, purification of membranes should not be neglected, even if it is somewhat time consuming.

10.2.3 Selective Solubilization

Before solubilization, the washed ROS membranes are concentrated to about 12 mg rhodopsin/ml by centrifugation at 70,000 rpm for 30 min. We previously found that rhodopsin can be selectively extracted from bovine ROS membranes by utilizing a combination of alkyl(thio)glucoside and divalent cations.[5] A possible mechanism for this reaction was proposed recently.[9] The currently optimized procedure is summarized in Protocol 10-2. This is modified from the original works[5,10] in that the mixed micelle composed of nonylglucoside and 1,2,3-heptanetriol is replaced with heptylthioglucoside.

Protocol 10-2: Selective Solubilization of Rhodopsin

1. Materials required:
   a. 15% (w/w) heptylthioglucoside
   b. 0.5 M MES/NaOH (pH 6.5)
   c. 1 M zinc acetate
   d. Portable centrifuge (6400 rpm and 12,000 rpm)
2. Mix 7 µl of MES, 17 µl zinc acetate, and 21.5 µl of heptylthioglucoside in an Eppendorf tube at room temperature.
3. Add 110 µl of completely thawed ROS membrane suspension to the tube, and mix quickly for at least 1 min.
4. Spin the tube with a portable low-speed centrifuge at 6400 rpm for 1 min.
5. If the separation of supernatant and pellet is clear, remix the solution and leave it for at least 3 h at room temperature.
6. Centrifuge the tube at 12,000 rpm for 3 min and collect the supernatant.

A possible problem in following this protocol may occur at step 5. Sometimes the separation of supernatant and pellet does not appear to be so good, and this is usually cleared by slightly adjusting the volume ratio between the membrane suspension and the detergent. For sure, it is better to make a set of tubes differing slightly in the ratio for further crystallization trials. A number of previous experiments established that the solubilization solutions at step 3 should exhibit considerable turbidity immediately upon mixing, so that just a 1 min centrifugation with a portable Eppendorf tube spinner is enough to give a clear supernatant containing rhodopsin and an almost colorless pellet. If the amount of the detergent is either not enough or too much, the separation becomes considerably worse.

It is better for the collected rhodopsin solution to be incubated on ice for at least 3 days before use to remove additional amorphous material (which occasionally appears and sediment during this period). Leaving the sample at 5°C for more than a month does not significantly affect the results of crystallization.

10.2.4 Crystallization

Three-dimensional crystallization of bovine rhodopsin is carried out by hanging drop vapor diffusion with ammonium sulfate as a precipitant. Before the optimal purification procedure was found, a variety of crystals were obtained using more contaminated and less stable samples. In those days, the appearance of the crystals was rather nonspecific against the pH. It was gradually converged around 6, which is fairly close to the pI (5.9) of bovine rhodopsin, throughout the optimization process. One of the keys for getting the first crystal might be the screening of a wide range of the precipitant concentrations for both the initial and the final vapor diffusion equilibrium. The current procedure is outlined in Protocol 10-3.

PROTOCOL 10-3: THREE-DIMENSIONAL CRYSTALLIZATION OF RHODOPSIN

1. Materials required:
   a. 0.5 M 2-mercaptoethanol
   b. 15% (w/w) heptylthioglucoside
   c. 3.5 M and 3.99 M ammonium sulfate
   d. 1 M MES/NaOH (pH 6.0)
   e. 20% PEG600
   f. 18 mm siliconized round coverslips
   g. 24-well culture plate
2. Prepare reservoir solutions [2.7 to 3.2 M ammonium sulfate, 40 mM MES (pH 6.0)] in one row of a culture plate, and put grease or liquid pallafin on the lids.
3. Mix 0.4 µl of 2-mercaptoethanol, 0.3 µl of heptylthioglucoside, 4.5 µl of ammonium sulfate, 0.8 µl of PEG600, and 21 µl of purified rhodopsin solution in an Eppendorf tube.
4. Mix 1 µl of acetic acid, 16 µl of water, and 1 µl of sodium silicate quickly, take 3 µl from the mixture, add it to the solution of step 2, and mix quickly again.
5. Put 5 µl of the mixed sample on a coverslip for a total of six, and fix them to each of the lids of the plate.
6. Put the plate in an incubator at 10°C.

This protocol includes two modifications from the previous one.[10] First, we applied a unique use of sodium silicate that builds up a gel upon neutralization at a high concentration.[11] By decreasing the concentration substantially, it can be mixed homogeneously with membrane protein solution containing some amount of detergent. As shown in Figure 10.2, the presence of silicate suppresses the formation of amorphous aggregate and helps in the crystallization of rhodopsin. We speculate that a sort of silica network formation might work as the mechanism of this favorable effect.

Second, the use of a low concentration of PEG600 as an additive is found to improve the diffraction quality of rhodopsin crystals. Low-molecular-weight PEGs have been used successfully as a precipitant for the crystallization of membrane proteins, including most of the prokaryotic channel proteins solved by x-ray crystallography. In the presence of PEG600, the crystallization drop of rhodopsin tends to cause weak phase separation, giving rise to apparent defects in the crystals. It

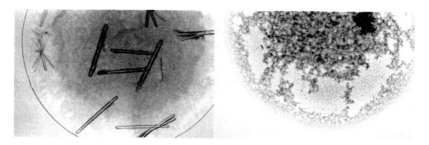

FIGURE 10.2 The effect of silicate on the three-dimensional crystallization of bovine rhodopsin. The sample prepared according to Protocols 10-1 and 10-2 was crystallized with (left) and without (right) silicate, as described in Protocol 10-3. To demonstrate the effect clearly, crystallization was conducted at 20°C. The images shown were taken after 24 days of vapor diffusion.

should be noted, however, that such defects do not negatively affect the diffraction quality.

As a result of these modifications, the diffraction limit extended roughly from 2.5 to 2.0 Å resolution.

10.2.5 Application to Rhodopsin Analogue

A number of studies on rhodopsin containing some artificial retinal have provided valuable information about its structure and function.[12] Such analogue pigment can be formed by first bleaching rhodopsin in the presence of hydroxylamine to remove the original chromophore and then adding an excess amount of the isomerically different or chemically modified retinal. Here, an example applying the procedures described in the previous sections is presented, using the making of the crystal of 9-*cis*-rhodopsin as a model.

Protocol 10-4: Preparation and Crystallization of 9-*cis*-Rhodopsin

1. Materials required:
   a. Purified ROS membranes prepared according to the steps in Sections 10.2.1 and 10.2.2
   b. 9-*cis*-retinal in ethanol
   c. 1 M hydroxylamine (pH 7.0)
   d. Regeneration buffer [25 mM MES (pH 6.4), 200 mM NaCl]
   e. Meterials listed in Protocols 10-2 and 10-3
2. Add 1 M hydroxylamine to the purified membrane suspension to give the final concentration of 10 mM.
3. Illuminate the mixture to completely bleach rhodopsin with intense light of >520 nm.
4. Wash the bleached membranes with the regeneration buffer to decrease the concentration of hydroxylamine to as low as 10 μM.

5. Add an excess amount of 9-*cis*-retinal to the membrane suspension in a volume less than 1% of the total and allow it to combine with opsin at room temperature for at least 2 h.
6. Add 1 M hydroxylamine, giving the final concentration of 1 mM, to convert the excess retinal to the oxime.
7. Wash the regenerated membranes again with distilled water to reduce the concentration of both NaCl and hydroxylamine to less than 2 mM and 10 μM, respectively.
8. Solubilize the membrane according to the procedure in Protocol 10-2.
9. Crystallize 9-*cis*-rhodopsin according to the procedure in Protocol 10-3.

We obtained the crystals of 9-*cis*-rhodopsin with simillar appearance to rhodopsin, and their x-ray diffraction data set to 2.9 Å resolution have been collected so far. Principally, this protocol can be applied to any other retinal analogues that bind to bovine opsin.

10.3 STRUCTURE DETERMINATION AND REFINEMENT

10.3.1 CHARACTERIZATION OF CRYSTALS

The crystals used for the structure determination and subsequent refinement are rod-shaped, despite the many changes made in the purification/crystallization procedures during the past several years. Even after finding the current optimal conditions, many of the crystals appear as a cluster, so careful isolation of the apparent single part is usually necessary for use. The dimension of each of the rods is typically $0.1 \times 0.1 \times 0.3$ mm, but that is variable in the longest axis.

In the process of the first structure determination,[2] a remarkable finding was that soaking the crystals in some mM mercury solution could significantly extend the diffraction limit from 3.5 Å to 2.5 Å. Because the cysteines to which mercury was found to bind are conserved in many of the other GPCRs in the class A GPCRs, this metal would also be useful for future structural studies on those receptors.

During the course of the diffraction data collections, it turned out that the crystals were merohedrally twinned in varying degrees. When the data obtained from the "native" (without mercury) crystals were processed, the space group appeared as $P4_122/P4_322$ due to the addition of symmetry originating from the perfect twinning. Thus, derivatization with mercury could partially lower the twinning ratio, and this helped us to determine the correct space group as $P4_1$. Recent modifications of the crystallization conditions further reduced the probability of twinning, so that we frequently find crystals with less than 5% in twinning fraction.

The unit cell dimension containing 2×4 molecules of rhodopsin is 97 Å × 97 Å × 150 Å on average, giving about 70% solvent content. It is noteworthy that recent improvement in the diffraction limit from 2.5 to 2.2 Å tends to show slightly longer lengths of the *c*-axis than before, which is an opposite trend to that of most cases, where smaller solvent content tends to exhibit higher x-ray diffraction. Because the crystal contacts between the asymmetric units (artificially flipped dimer) along the *c*-axis are mediated by the extracellular and cytoplasmic surfaces, better-ordered structures of some of these parts might require much larger space in the lattice.

10.3.2 SUMMARY OF STRUCTURE DETERMINATION

The initial phase information for the structure determination of bovine rhodopsin was obtained at 3.8 Å using a multiwavelength anomalous diffraction (MAD) data set collected at BL1-5 in the Stanford synchrotron radiation laboratory (SSRL). Then, the mercury binding sites were confirmed, and the phases for the model building were obtained using a MAD data set collected at BL45XU in SPring8 on a crystal with the twin fraction of 10%. The refinement of the model was proceeded to 2.8 Å with a data set obtained at 19-ID in the Advanced Photon Source (APS). Two coordinates were deposited using this data,[2,3] 1F88 and 1HZX, the latter of which contained some nonprotein components, such as palmitoyl chains and additional sugar moieties. A part of the third cytoplasmic loop and the C-terminal tail could not be constructed because of insufficient electron densities. All of the data were collected under cryogenic conditions, for which 15% sucrose was used as a cryoprotectant.

10.3.3 STRUCTURE REFINEMENTS AT HIGHER RESOLUTION

The initial studies on the crystal structure of bovine rhodopsin were followed by further refinement using the data to 2.6 Å[4] and then to 2.2 Å resolution. With the data to 2.6 Å, the major focus of refinement was to unequivocally reveal the distribution of internal water molecules in the transmembrane helical region. Seven sites were identified for each of the two rhodopsins in the asymmetric unit, and their functional roles are described in the next section. The latest refinement to 2.2 Å resolution completed the whole polypeptide chain of rhodopsin for the first time. It also revealed the details of the chromophore structure and water distribution around the second extracellular loop.

10.4 CRYSTAL STRUCTURE

10.4.1 CRYSTAL LATTICE

The building block of the crystal lattice is an artificial dimer, in which two rhodopsin molecules are associated in nearly an upside-down fashion. The hydrophobic interaction within a dimer involves the two transmembrane helices I and the four palmitoyl acyl chains that are attached to Cys322 and Cys323 in the C-terminal tail. Such a lattice composed of nonphysiological dimers appears to be exceptional among the previously determined crystal structures of transmembrane proteins. It is also common, in many cases, that the physiological multimeric form of membrane protein is retained even in the crystal lattice.

It appears to be reasonable to suppose that this unusual dimerization is critical for the crystallization process of rhodopsin, and it might also be a rate-limiting step. Like many transmembrane proteins, including GPCRs, bovine rhodopsin exhibits a dipolar charge distribution, with the positive charge on the cytoplasmic side and the negative charge on the extracellular side. Thus, this kind of flipped-dimerization may occur in the future crystallization process of GPCRs.

The regions close to the glycosylation sites are involved in the molecular packing between the adjacent dimers. This finding demonstrates that a large moiety of the posttranslational modifications could be accommodated in the case of typical "type II" packing[13] in the crystal of a membrane protein.

Another interesting finding in the crystal lattice is that there are continuous solvent tunnels that penetrate along the fourfold axis of a crystal. Because these regions are likely to be occupied by the detergent to some extent, this might support the packing with its interconnected structure.

10.4.2 Overall Structure

Each of the extracellular and intracellular regions of rhodopsin consists of three interhelical loops and a terminal (COOH- or NH2-) tail. Although the mass distribution to these two regions is comparable, the three-dimensional structure demonstrates a clear contrast: the four extracellular domains associate significantly with each other, while only a few interactions are observed among the cytoplasmic domains.

The center of the organized extracellular structure was occupied by the second loop (E-II). The latest refinement revealed that this loop is associated with many water molecules, which appears to mediate the interactions with the other parts in this domain. The E-II loop is connected to helix III via a disulfide bridge, a common feature among the hundreds of GPCRs. Whereas the E-II of rhodopsin fits nicely into a limited space inside the bundle of seven helices and comprises a substantial part of chromophore binding pocket in the ground state, it would be possible to rearrange during either photoactivation or passing of the retinal.

The fourth cytoplasmic loop, which is formed by anchoring the C-terminal tail to the membrane via two Cys residues carrying palmitoyl chain, was unexpectedly found to form a short helical structure (helix VIII) lying nearly parallel to the membrane surface. There is increasing evidence that a structure like helix VIII in rhodopsin must exist in some other class A GPCRs.

10.4.3 Transmembrane Region

The seven transmembrane helices of rhodopsin vary in length, in the degree of bending around Gly/Pro residues, and also in the tilt angles to the expected membrane surface. Because of the scattered distribution of some highly conserved residues in this helical domain, the overall arrangement of the heptahelical bundle is likely to be shared by many of the class A GPCRs. Many of such residues are, however, found in the cytoplasmic half of the helical bundle, because the extracellular side has to be registered to vary for the ligand-binding function.

The highly tilted helix III, first suggested by electron microscopy,[1] contains some key residues for the activation of rhodopsin (Figure 10.4). Cys110 at the extracellular end participates in the disulfide bond to Cys187 in the E-II. Glu113 (3.28, numbering for GPCRs according to Ballesteros and Weinstein[14]) is a counterion to the protonated

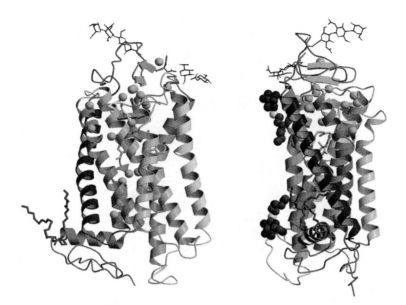

FIGURE 10.3 The complete crystal structure model of bovine rhodopsin at 2.2 Å resolution. The two images are drawn by rotating approximately 90 degrees around the vertical axis. Small spheres represent water molecules identified consistently in the two rhodopsin of the asymmetric unit. In the right figure, two heptylthioglucoside molecules found in the electron density map are included to show the tentative limit of the transmembrane region. The extracellular domains are shown in the upper side.

Schiff base of the retinal chromophore. In some GPCRs for cationic amines, an acidic amino acid that is responsible for the ligand binding exists in the position (3.32) shifted one turn of the helix to the cytoplasmic side of Glu113. Rhodopsin and those GPCRs are thought to share a similar mechanism of activation involving neutralization of the acidic amino acid side chain. One of the most important findings in the crystal structure of ground state rhodopsin is the arrangement around the so-called D(E)RY(W) sequence in the cytoplasmic end of helix III. The positively charged side chain of highly conserved Arg135 (3.50) appears to form an ion pair with Glu134 (3.49), both of which are surrounded mostly by the hydrophobic residues in helices II, III, IV, V, and VI, with the exception of two nearby polar residues in helix VI, Glu247 (6.30) and Thr251 (6.34). Rearrangement of these cytoplasmic surface residues must be critical for the activation of G protein.

Helix VI also contains some highly conserved amino acids that are supposed to determine the activity of GPCRs, such as Phe261 (6.44) and Trp265 (6.48) in rhodopsin. Another significant feature of this helix is a strong distortion by Pro267 (6.50), one of the most conserved residues among GPCRs. It is likely that activation of GPCRs requires some mechanism that allows these residues to rearrange by removing the interactions with the other helices upon ligand binding or photoisomerization in rhodopsin.

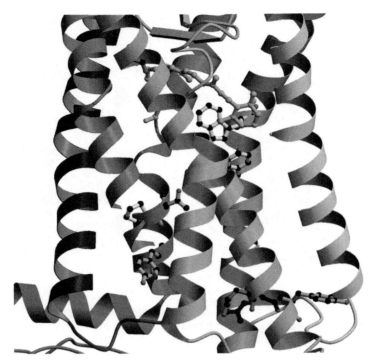

FIGURE 10.4 An expanded view of the transmembrane helical bundle of bovine rhodopsin, indicating some highly conserved amino acid residues.

In some interhelical spaces of rhodopsin, we find either ions or water molecules. The most outstanding site is surrounded by helices I, II, III, VI, and VII, and we identified a cluster of four water molecules there at 2.2 Å resolution (Figure 10.5). The hydrogen bonds with the water involve some highly conserved residues, such as Asn55 (1.50) and Asp83 (2.50), both of which are referred to as the N–D pair in class A GPCRs, and Asn302 (7.49) at the initial position of the so-called NPxxY motif. It is possible that a putative cation-binding site in some GPCRs coincides with this water cluster region in rhodopsin. Therefore, the hydrogen-bonded network among helices I, II, III, VI, and VII may vary and change upon ligand binding for a distinct class of receptors. Such flexible structural rearrangement would partly explain substrates of GPCRs exhibiting distinct affinity for a ligand and a target G protein.

Probably due to the high concentration included in our purification procedure, a zinc ion is found to bind in the transmembrane region of rhodopsin. It is coordinated by the major ligand His211 (5.46) and is also surrounded by Glu122 (3.37), Trp126 (3.41), Met163 (4.52), and Cys167 (4.56), some of which form a part of the retinal binding pocket. The interactions among helices III, IV, and V are supposed to change directly upon binding of a ligand in some class of GPCRs, because the binding sites were identified in the positions close to the His211 of rhodopsin. Therefore, disrup-

FIGURE 10.5 A projection slab view of the transmembrane helical bundle of bovine rhodopsin around the water site 1 containing a cluster of four water molecules and some highly conserved amino acid residues. The three water molecules identified at 2.6 Å resolution and the additional one at 2.2 Å resolution are shown .

tion of this interhelical restraint would trigger the subsequent activation in many class A GPCRs.

10.4.4 Constitutive Activity

The hypothesis that rhodopsin and other GPCRs for diffusible ligands share a common mechanism of activation is supported by the studies on a number of constitutively active mutants (CAMs). Table 10.1 and Figure 10.6 show some positions of amino acids exhibiting such activity. Although the first CAMs found for an adrenergic receptor were supposed to reside in the third cytoplasmic loop,[15] the crystal structure of bovine rhodopsin strongly suggests that those residues are included in helix VI. One of them corresponds to Thr251 (6.34) in bovine rhodopsin, which is in the proximity of Arg135 (3.50) in the ERY motif. Recent studies on an opioid receptor demonstrated clear correlation between the constitutive activity and the charged state at 6.34, supporting the direct interaction between 3.50 and 6.34 in this receptor.[16]

Even in the interior region of the transmembrane helical bundle, a number of CAM sites were identified, such as 3.36, 6.40, 6.44 in bovine rhodopsin,[17,18] 3.43 in M_1 muscarinic receptor,[19] 6.40 and 6.44 in M_5 muscarinic receptor,[20] 3.43 in FSH receptor,[21] and 3.36 in TSH receptor.[22] This experimental evidence supports the idea of the existence of an activation mechanism involving interaction changes between helices III and VI in many of the class A GPCRs. Disruption of the salt bridge between Glu113 and the protonated Schiff base can be mimicked by mutating either Glu113 or Lys296, resulting in activation of rhodopsin in the dark. Similar positions

TABLE 10.1
Amino Acid Residues in the Transmembrane Region Known to Evoke Constitutive Activity upon Mutation

TM[a]	Rho.[b]	Property	Number[f]	Receptors[i]	Ligand	Property	Gi
III	Glu113	CAM[c]	3.28 (3.32)[g]	α_{1B}, δ	Amine, peptide	CAM	G_q, $G_{i/o}$
III	Gly120	Wat1[d]	3.35	α_{1B}, AT1A, B$_1$, PAF	Amine, Peptide, lipid	CAM	G_q, G_i
III	Gly121	CAM	3.36	TSH	Glycoprotein	CAM	G_s
III	Ala124	Wat1	3.39	D$_2$	Amine	Na$^+$	$G_{i/o}$
III	Leu128	N.A.[e]	3.43	M$_1$, FSH	Amine, glycoprotein	CAM	G_q, G_s
III	Glu134	CAM	3.49	α_{1B}, β_2, H$_2$, V$_2$	Amine, peptide	CAM	G_q, G_s
III	Arg135	CAM	3.50	β_2	Amine	CAM	G_s
VI	Glu247	N.A.	6.30	β_2, LH, TSH	Amine, glycoprotein	CAM	G_s
VI	Thr251	N.A.	6.34	α_{1B}, β_2, μ, LH, TSH	Amine, peptide, glycoprotein	CAM	G_q, G_s, $G_{i/o}$
VI	Met257	CAM	6.40	M$_5$, TSH	Amine, glycoprotein	CAM	G_q, G_s
VI	Phe261	CAM, Wat1	6.44	M$_5$, LH, TSH	Amine, glycoprotein	CAM	G_q, G_s
VII	Lys296	CAM	7.43 (7.36)[h]	α_{1B}, δ	Amine	CAM	G_q, $G_{i/o}$
VII	Asn302	Wat1	7.49	TSH	Glycoprotein	CAM	G_s
VII	Pro303	Bending	7.50	5-HT-2A	Amine	Bending	$G_{i/o}$

[a] Number of transmembrane helix; [b] Number in bovine rhodopsin; [c] Constitutively active mutant; [d] Involved in water site 1; [e] Not available; [f] Numbering for GPCRs according to Ballesteros, J.A. and Weinstein, H., *Methods Neurosci.*, 25, 366, 1995; [g] Shifted four residues to the cytoplasmic side in the receptors; [h] Shifted seven residues to the extracellular side in α_{1B} receptor; [i] Receptors in the class A GPCRs known to exhibit constitutive activity in this site; α_{1B}, β_2: adrenergic receptors; AT1A: angiotensin receptor; B$_1$: bradykinin receptor; D$_2$: dopamine receptor; M$_1$, M$_5$: muscarinic receptors; H$_2$: histamine receptor; V$_2$: vasopressin receptor; δ, μ: opioid receptors; FSH: follicle-stimulating hormone receptor; LH: luteinizing hormone receptor; PAF: platelet-activating factor receptor; TSH: thyroid-stimulating hormone receptor; 5-HT-2A: serotonin receptor; [j] Major target G protein subtypes.

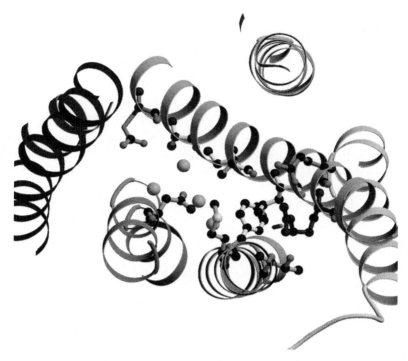

FIGURE 10.6 A projection view of the transmembrane helical bundle of bovine rhodopsin with some of the residue positions that are known to be the CAM sites.

in the adrenergic receptor to these residues could also evoke constitutive activity upon mutation,[23] suggesting that these changes at the extracellular side result in similar structural consequences in rhodopsin and in the adrenergic receptor.

As described above, only partly to the known CAM sites, it is now clear that a variety of receptors exhibit constitutive activity when some structural perturbations are given, regardless of their distance from the cytoplasmic surface, where the binding of G proteins occurs. This was particularly demonstrated for helix III, as shown in Figure 10.6.

10.5   REMARKS

Crystallization and structure determination of bovine rhodopsin, besides its biological and physiological implications, demonstrate the possibility of challenging the structure determination of a number of membrane proteins in the rhodopsin family of GPCRs. Many of the details in the procedures described above might not be so helpful if trying to determine the crystal structure of other GPCRs for diffusible molecules. It is certainly true that other visual pigments and GPCRs require much care to keep them in stable forms for three-dimensional crystallization. To fill the gap between the rhodopsin system and others, even partially, we have been extending the methodology to the membranes of mammalian cells overexpressing rhodopsin and its mutants. On the other hand, the crystal structure model of rhodopsin in its inactive form has proven to be valuable for structural and functional studies of a number of GPCRs in recent years. Further x-ray crystallographic studies of the photoreaction intermediates of rhodopsin will help in our understanding of the mechanism of activation.

ACKNOWLEDGMENTS

We are grateful to the collaborators and the synchrotron people for their contributions during the early stages of this project. Excellent support by Drs. H. Sakai and M. Kawamoto at BL41XU of SPring-8 is also gratefully acknowledged. This work was supported by the Japanese Ministry of Education, Culture, Sports, Science and Technology (MEXT) and NEDO.

REFERENCES

1. Schertler, G.F., Villa, C., and Henderson, R., Projection structure of rhodopsin, *Nature*, 362, 770, 1993.
2. Palczewski, K. et al., Crystal structure of rhodopsin: a G protein-coupled receptor, *Science*, 289, 739, 2000.
3. Teller, D.C. et al., Advances in determination of a high-resolution three-dimensional structure of rhodopsin, a model of G-protein-coupled receptors (GPCRs), *Biochemistry*, 40, 7761, 2001.
4. Okada, T. et al., Functional role of internal water molecules in rhodopsin revealed by x-ray crystallography, *Natl. Acad. Sci. USA*, 99, 5982, 2002.
5. Okada, T., Takeda, K., and Kouyama, T., Highly selective separation of rhodopsin from bovine rod outer segment membranes using combination of divalent cation and alkyl(thio)glucoside, *Photochem. Photobiol.*, 67, 495, 1998.
6. Fukuda, M.N., Papermaster, D.S., and Hargrave, P.A., Rhodopsin carbohydrate. Structure of small oligosaccharides attached at two sites near the NH2 terminus, *J. Biol. Chem.*, 254, 8201, 1979.
7. Ovchinnikov, Y.A., Abdulaev, N.G., and Bogachuk, A.S., Two adjacent cysteine residues in the C-terminal cytoplasmic fragment of bovine rhodopsin are palmitylated, *FEBS Lett.*, 230, 1, 1988.
8. Guruprasad, K., Reddy, B.V.B., and Pandit, M.W., Correlation between stability of a protein and its dipeptide composition: a novel approach for predicting *in vivo* stability of a protein from its primary sequence, *Protein Eng.*, 4, 155, 1990.
9. Okada, T., Crystallization of bovine rhodopsin, a G protein-coupled receptor, in *Methods and Results in Crystallization of Membrane Proteins*, Iwata, S., Ed., International University Line, La Jolla, CA, 2003.
10. Okada, T. et al., X-ray diffraction analysis of three-dimensional crystals of bovine rhodopsin obtained from mixed micelles, *J. Struct. Biol.*, 130, 73, 2000.
11. Cudney, B., Patel, S., and McPherson, A., Crystallization of macromolecules in silica-gels, *Acta Crystallogr. D*, 50, 479, 1994.
12. Crouch, R. K. et al., Use of retinal analogues for the study of visual pigment function, *Methods Enzymol.*, 343, 29, 2002.

13. Michel, H., General and practical aspects of membrane protein crystallization, In *Crystallization of Membrane Proteins*, Michel, H., Ed., CRC Press, Boca Raton, FL, 1991, p. 74.

14. Ballesteros, J.A. and Weinstein, H., Integrated methods for the construction of three-dimensional models and computational probing of structure–function relations in G protein-coupled receptors, *Methods Neurosci.*, 25, 366, 1995.

15. Samama, P. et al., A mutation-induced activated state of the beta 2-adrenergic receptor. Extending the ternary complex model, *J. Biol. Chem.*, 268, 4625, 1993.

16. Huang, P. et al., Functional role of a conserved motif in TM6 of the rat mu opioid receptor: constitutively active and inactive receptors result from substitutions of Thr6.34(279) with Lys and Asp, *Biochemistry*, 40, 13,501, 2001.

17. Han, M. et al., Partial agonist activity of 11-*cis*-retinal in rhodopsin mutants, *J. Biol. Chem.*, 272, 23,081, 1997.

18. Han, M., Smith, S.O., and Sakmar, T.P., Constitutive activation of opsin by mutation of methionine 257 on transmembrane helix 6, *Biochemistry*, 37, 8253, 1998.

19. Lu, Z.L. and Hulme, E.C., The functional topography of transmembrane domain 3 of the M_1 muscarinic acetylcholine receptor, revealed by scanning mutagenesis, *J. Biol. Chem.*, 274, 7309, 1999.

20. Spalding, T.A. et al., Identification of a ligand-dependent switch within a muscarinic receptor, *J. Biol. Chem.*, 273, 21,563, 1998.

21. Tao, Y.X. et al., Constitutive activation of G protein-coupled receptors as a result of selective substitution of a conserved leucine residue in transmembrane helix III, *Mol. Endocrinol.*, 14, 1272, 2000.

22. Tonacchera, M. et al., Functional characteristics of three new germline mutations of the thyrotropin receptor gene causing autosomal dominant toxic thyroid hyperplasia, *J. Clin. Endocrinol. Metab.*, 81, 547, 1996.

23. Porter, J.E., Hwa, J., and Perez, D.M., Activation of the β_{1b}-adrenergic receptor is initiated by disruption of an interhelical salt bridge constraint, *J. Biol. Chem.*, 271, 28,318, 1996.

11 Determination of Steric Structure of Muscarinic Ligands Bound to Muscarinic Acetylcholine Receptors: Approaches by TRNOE (Transferred Nuclear Overhauser Effect)

Hiroyasu Furukawa, Toshiyuki Hamada, Hiroshi Hirota, Masaji Ishiguro, and Tatsuya Haga

CONTENTS

11.1 INTRODUCTION

Acetylcholine elicits a variety of cellular responses by acting on muscarinic acetylcholine receptors (mAChR), which belong to a family of seven-transmembrane G protein-coupled receptors (GPCRs). Signals of acetylcholine binding to five subtypes of mAChR[1-3] are transmitted via activation of heterotrimeric G proteins that, in turn, activate or regulate the function of their effectors, such as phospholipase C and adenyl cyclase.

GPCRs comprise the largest family of membrane protein receptors.[4] Because of their vast involvement in physiological activities, GPCRs account for a large portion of the targets for contemporary drugs. Despite much effort, the only atomic resolution structures available for GPCRs are of rhodopsin[5] and the extracellular ligand-binding domain of the metabotropic glutamate receptors.[6] The molecular mechanism for receptor-mediated G protein activation is still hypothetical at this point because of the lack of high-resolution structural information representing different states of receptors.

Even before the first GPCR was cloned, acetylcholine and its analogues were extensively studied by x-ray crystallography.[7-9] From a biological point of view, there has been a question of what conformation of acetylcholine is physiologically important. Acetylcholine is a molecule that contains two functional groups — a quaternary amine and an acetyl group — that are tethered by methylene carbons (Figure 11.1). After molecular cloning of mAChRs and extensive mutagenesis studies, it became clear that the conserved aspartic acid residue in the third transmembrane segment and tyrosine residue in the sixth transmembrane segment interact with the quaternary amine and the acetyl group, respectively.[10,11] Therefore, the positioning of the two groups is critical in receptor–acetylcholine interaction, and this positioning is governed, for the most part, by the dihedral angle of the freely rotating methylene group (Figure 11.1).

Using diastereomers of conformationally rigid acetylcholine analogues to measure activities such as ileum muscle contraction has been a popular approach to addressing the biologically active conformation of acetylcholine.[12-14] However, a clear picture of the physiologically relevant conformation of acetylcholine has yet to emerge because of ambiguities in degrees of conformational rigidity and the actual dihedral angles of such compounds. It is apparent that a more direct approach must be applied to solve this classical problem. The transferred nuclear Overhauser effect (TRNOE) method in nuclear magnetic resonance (NMR) is frequently used to extract structural information about small ligands bound to large molecules, such as a protein, and is, therefore, suitable for solving the above problem.[15,16] Here we discuss the method to prepare the receptor sample for study by NMR, synthesis and characterization of the ligand, and assessment of its conformation bound or unbound to mAChR M_2 subtype (M_2 receptors) by the TRNOE method.[17]

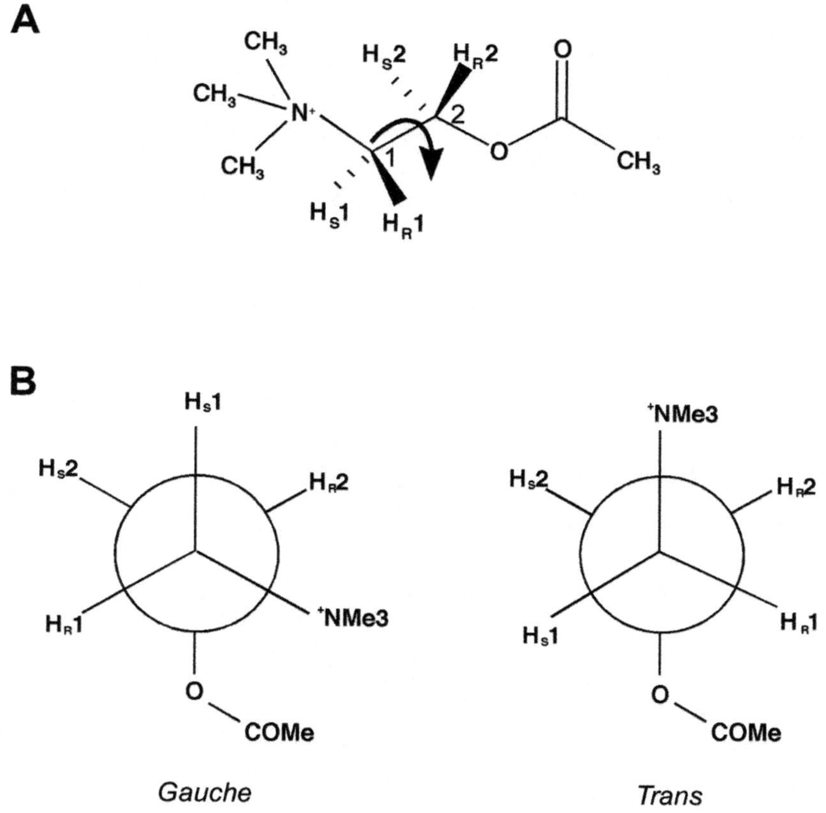

FIGURE 11.1 Acetylcholine. The N-C1-C2-O dihedral angle is controlled by the free rotation of C-C (a, arrow). This gives *gauche* and *trans* conformations regarding the position of two functional groups, a quaternary amine and an acetyl group, as represented by the Newman projection (b).

11.2 EXPRESSION AND PURIFICATION OF THE M_2 RECEPTORS

Biophysical studies of membrane proteins have lagged behind due to difficulties in obtaining large amounts of properly folded proteins. The NMR method used in this study requires milligram quantities of purified mAChRs. One approach to overcoming this difficulty is to find a prokaryotic homologue of the membrane proteins that can be overexpressed in bacterial cells. In fact, the discovery of the prokaryotic homologue, KcsA, led to the breakthrough in x-ray crystallography of potassium ion channels.[18] However, many physiologically important proteins, including GPCRs, are unique to eukaryotes. Therefore, a solid overexpression system to obtain milligram quantities of eukaryotic membrane proteins, such as the mAChR, has to be established to carry out biophysical studies.

Despite success in obtaining 10 to 15 pmol/mg of neurotensin receptors in bacterial cells through extensive studies on fusion partners as well as usage of promoters,[19] employing a similar method did not yield a sufficient amount of the M_2 receptors to allow for the NMR studies.[20] Among expression systems tested, the *Sporodoptera frugiperda* (*Sf*)9/baculovirus system has the highest expression level for the M_2 receptors. We cultured *Sf*9 cells in large scale (7 l) using a bioreactor, which tightly regulates conditions such as dissolved oxygen concentration, pH, and temperature, while allowing cells to grow to a high density.

To acquire a homogeneous population of receptors that retain ligand-binding activity, membrane fractions expressing the M_2 receptor are isolated, solubilized by a combination of digitonin and sodium cholate, and purified by a two-step purification using 3-(2'-aminobenzhydryloxy)tropane (ABT)-agarose affinity chromatography gel[21] and hydroxyl apatite.

11.2.1 METHODS FOR EXPRESSING THE M_2 MUTANT USING *Sf*9/BACULOVIRUS

In this section, we focus on the studies of the human M_2 receptors with the following genetic alterations (Figure 11.2):

1. Deletion of the central part of the protease-susceptible third intracellular loop (233-380)
2. Replacement of putative glycosylation residues Asn 2, 3, 6, and 9 with Asp for prevention of molecular heterogeneity
3. Addition of a hexa-histidine tag downstream of a thrombin cleavage site at the C-terminus for an additional purification option

Baculovirus harboring the above recombinant gene (M_2 mutant) is made using the Bac-to-Bac system (Invitrogen, Carlsbad, CA). Virus titer is measured by the endpoint dilution method.[22] Cells are cultured at 27 to 28°C in IPL-41 supplemented with 5% fetal bovine serum (FBS), although a serum-free medium, such as Sf900 II SFM (Invitrogen), may also be used. Polystyrene flasks (25 or 75 cm²) and spinner flasks (Bellco Glass, Vineland, NJ) are used for attached culture and suspension culture, respectively.

PROTOCOL 11-1: *Sf*9 CELL CULTURE

The following procedures are also illustrated in Figure 11.3:

1. Medium: IPL-41 medium (JRH Biosciences, Lenexa, KS) supplemented with 5% bovine serum (Cansera, Ontario, Canada), TC Yeastolate (Sigma-Aldrich, St. Louis, MO), tryptose phosphate broth (Sigma), Pluronic F68 (Sigma), Pennicilin/Streptomycine (Gibco/Invitrogen).
2. Thaw frozen *Sf*9 cells (10^7 cells) in 6 ml of IPL-41 medium and culture in 25-cm² flask for 2 to 3 days.

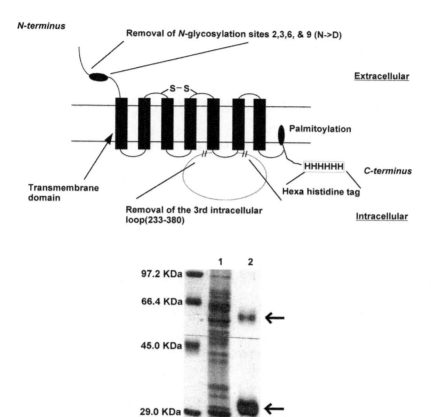

FIGURE 11.2 The M$_2$ mutant. This mutant of the M$_2$ receptor lacks four putative N-glyco-sylation sites at its N-terminus, has a deletion in the protease-susceptable third inner loop, and has a hexahistidine tag at its C-terminus as an option for purification (a). This M$_2$ mutant can be expressed in Sf9/baculovirus at the expression level of approximately 1 mg per liter culture and purified to homogeneity as indicated by the 12% SDS-PAGE gel stained with Coomassie brilliant blue (b). The arrows indicate bands for the monomeric and dimeric M$_2$ mutants.

3. Detach cells from the 25-cm^2 flasks by gentle pipetting, and transfer 1 ml of the cell suspension to a 75-cm^2 flask filled with 10 ml of medium. Prepare a total of six 75-cm^2 flasks.

4. Note that the 75-cm^2 flasks are confluent after 4 to 5 days. Detach the cells in the same manner as in step 2, and allocate them into two 300-ml spinner flasks. Adjust the total suspension volume to 100 ml. Spin the culture at 150 rpm.

5. Culture the cells for the next 3 to 4 days until the cell density is between 2–3 × 10^6 cells/ml. Then, scale up each 100-ml culture to 400 ml in a 1 l spinner flask.

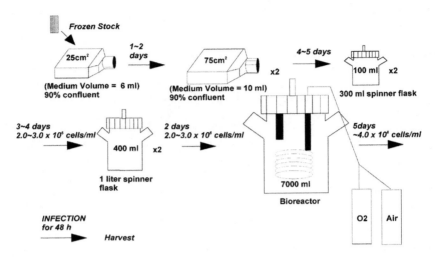

FIGURE 11.3 Scheme of the *Sf*9/baculovirus culturing. Cells are cultured at 27 to 28°C.

6. Note that after 2 days, the cell density should reach 2–3×10^6 cells/ml. Transfer the cells to the bioreactor filled with 6 l of air-saturated IPL-41 medium. At this point, the cell density has to be at least 0.2×10^6 cells/ml. Throughout the culturing process, keep the glucose concentration above 150 mg/dl for cell growth. Measure the glucose concentration using a blood glucose meter and adjust using 2× IPL-41 medium. Air is supplied by an air pump at the flow rate of 700 ml/min. Keep the dissolved oxygen (DO) concentration at approximately 20% by mixing air with oxygen as needed. Spin the culture at a speed of 50 rpm.

7. When the cell density reaches 4–5×10^6 cells/ml, infect the cells with baculovirus at multiplicity of infection (m.o.i.) = 5. The DO concentration decreases dramatically 1 to 2 h after the addition of the virus solution; therefore, pure oxygen is supplied at 700 ml/min to compensate for the loss.

8. After 48 h, harvest the cells by centrifugation at 5000 rpm for 20 min.

9. Resuspend cell pellet (from 6 l culture) in 3 l of phosphate buffer saline, and centrifuge at 5,000 rpm for 20 min.

10. Store the pellet at −80°C.

11.2.2 METHODS FOR MEMBRANE PREPARATION AND SOLUBILIZATION

The membrane fraction of the *Sf*9 cells expressing the M_2 mutant is isolated by cell lysis and a combination of high and low centrifugation. Total protein concentration of the membrane fraction is measured by BCA assay (Pierce, Rockford, IL). The membrane proteins are extracted by the use of detergent at the precise protein/detergent ratio. The solubilization condition has to be carefully chosen so that the M_2

mutant is efficiently extracted from the membrane fraction, while the ligand-binding activity is retained. No ligand should be added to the solubilization buffer (although the addition of ligand stabilizes the receptor) because it hinders efficient binding to the ABT-agarose. The detergent that satisfies the above requirement is a mixture of digitonin and sodium cholate.

PROTOCOL 11-2: MEMBRANE PREPARATION AND SOLUBILIZATION

Buffers

Buffer A: 20 mM HEPES-NaOH (pH 7.4), 5 mM EDTA, 2 mM $MgCl_2$, 5 µg/ml leupeptin, 5 µg/ml pepstatin A, 0.5 mM phenylmethane sulfonyl fluoride, and 5 mM benzamidine

Buffer B: 20 mM HEPES-NaOH (pH 7.4), 150 mM NaCl, 5 µg/ml leupeptin, 5 µg/ml pepstatin A, 0.5 mM phenylmethane sulfonyl fluoride, and 5 mM benzamidine

Preparation of Membrane Fraction

1. The frozen cell pellets are thawed quickly at 30°C and are resuspended in buffer A.
2. Once resuspended, the cells are placed into a N_2 cavitation instrument (Parr Cell Disruption Bomb 4635, Parr Instrument, Moline, IL). The chamber is filled with N_2 gas until the internal pressure reaches 10,000 psi. The cell suspension is stirred with a magnetic stir bar at 4°C for 30 min.
3. The pressure is released, and the resulting homogenetate is recovered.
4. The homogenate is centrifuged at 1,000 g for 10 min to remove nuclei and any high-ordered aggregates.
5. The supernatant is centrifuged at 150,000 g for 30 min.
6. The pellet is resuspended in 400 ml of buffer B and centrifuged again at 150,000 g for 30 min.
7. The pellet is resuspended with 200 ml of buffer B.
8. The total protein concentration is measured using a BCA assay kit (Pierce). Typically, the total protein concentration of the membrane fraction is 15 to 20 mg/ml.

Solubilization of the Membrane Fraction

1. Mix the digitonin powder in water at 4%, and boil until it is dissolved.
2. Leave the digitonin solution at 4°C and let impure materials precipitate.
3. Centrifuge the mixture at 150,000 g for 30 min. Recover the supernatant (regarded as 4% digitonin solution). Do the above procedure 1 to 2 days ahead of solubilization because long-term storage of this solution causes digitonin to precipitate out.

4. Add the appropriate amount of buffer B to the membrane fraction so that the total protein concentration after addition of detergent is 8 mg/ml.
5. To the membrane fraction suspension, gradually add digitonin and sodium cholate to 1% and 0.3%, respectively. Stir the mixture at 4°C for 1 h.
6. Centrifuge the mixture at 150,000 rpm for 30 min. Recover the supernatant. The typical recovery of the M_2 mutant is approximately 90% as assessed by comparison of [³H]N-methyl scopolamine (NMS) binding before and after solubilization.

11.2.3 Methods for Purification of the M_2 Mutant

The detergent solubilized M_2 mutant is purified to homogeneity by the two-step purification method, involving ABT-agarose column and hydroxyl apatite, as established previously.[21] The purification process is modified for NMR experiments. Specifically, the modification involves exchanging a portion of receptor-bound digitonin to sodium cholate, atropine to (S)-methacholine, and H_2O to D_2O. In this section, the protocol from 7 l insect cell culture is described.

Protocol 11-3: Purification of the M_2 Mutant

Buffers

Buffer A: 20 mM KPB (pH 7.0), 150 mM NaCl, and 0.1% digitonin
Buffer B: 20 mM KPB (pH 7.0), 150 mM NaCl, 0.1% digitonin, and 1 mM atropine sulfate
Buffer C: 10 mM KPB (pH 7.0), 0.1% digitonin, and 0.1 mM atropine sulfate
Buffer D: 10 mM KPB (pH 7.0), 0.3% sodium cholate, and 0.1 mM atropine sulfate
Buffer E: 1 M KPB (pH 7.0), 0.3% sodium cholate, and 0.1 mM atropine sulfate
Buffer F: 10 mM Tris(hydroxymethyl-$d3$)amino-$d2$-methane-deuterium chloride, and 0.2% sodium cholate in D_2O

Purification

All of the following, except for step 8, are done at 4°C:

1. Solubilize the membrane fraction from a 7 l culture as described above and load onto 400 ml of ABT-agarose gel at a flow rate of 1.5 ml/min.
2. Wash column with 5 column volume (CV) of buffer A at a flow rate of 1.5 ml/min.
3. Prepare hydroxyl apatite column (5 ml), equilibrated with buffer A, and connect it to the ABT-agarose column in tandem.
4. Elute the M_2 mutant from ABT-agarose with 5 CV of buffer B onto the hydroxyl apatite at the same flow rate.

5. Take out the hydroxyl apatite column, and wash it with 5 CV of buffer C by gravity flow.

6. Wash the hydroxyl apatite column with 12 CV of buffer D by gravity flow.

7. Elute the M_2 mutant with buffer E.

8. Equilibrate the PD10 column (Amersham Pharmacia Biotech, Piscataway, NJ) with 5 CV of buffer F, apply sample, and recover void volume of the column. Add (S)-methacholine to 1.5 mM.

9. Concentrate sample to 500 μl using Cenricon 30 (Amicon, Millipore, Billerica, MA) concentrator, add buffer F with 1.5 mM (S)-methacholine, and concentrate to 500 μl again.

10. Centrifuge the sample at 150,000 g for 20 min to remove possible protein aggregates.

11. Measure [³H]NMS binding activity, and adjust the concentration of the binding site to 50 μM.

11.3 SYNTHESIS AND CHARACTERIZATION OF (S)-METHACHOLINE

Acetylcholine is the physiological ligand for mAChRs. However, due to the presence of two pairs of chemically equivalent protons, conformation of acetylcholine cannot be determined by NMR (Figure 11.4). Therefore, to understand the mechanism of action of acetylcholine on mAChRs, structural studies on analogue compounds should be pursued. The analogue compound to be studied should fulfill the following criteria:

1. It should be similar to acetylcholine in its chemical structure.
2. It should have a similar effect on mAChRs.
3. It should be a compound with NMR spectra for each proton fully assigned but not averaged.

The most suitable compound satisfying the above is (S)-methacholine (Figure 11.4), which contains one methyl group at C2 in acetylcholine. The addition of the methyl group slows the free rotation between the C1–C2 bond (Figure 11.4), resulting in the separation of the NMR peaks for hydrogens attached to the methylene group. Because only a racemic mixture of methacholine is commercially available, (S)-methacholine has to be synthesized. In the first part of this section, the synthesis of (S)-methacholine is shown. The synthesized (S)-methacholine is later tested using one-dimensional ¹H-NMR spectroscopy to confirm purity.

In addition to suitability to the NMR experiment, similarity in biochemical properties of (S)-methacholine to acetylcholine has to be verified. Binding affinity (in K_d or K_i) is measured using tritium-labeled ligand binding to either membrane fraction expressing mAChRs or purified receptors. Receptor-mediated G protein activation is measured in the context of a muscarinic receptor (M_2)-$G_{i1}\alpha$ fusion

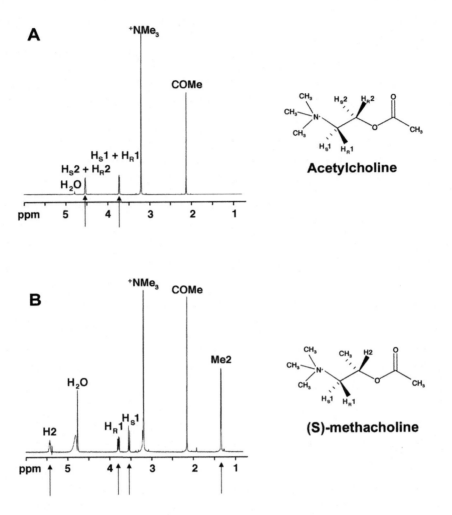

FIGURE 11.4 The one-dimensional ^{1}H-NMR spectra of acetylcholine (a) and (S)-methacholine (b). Note that all protons from the C1 and C2 methylene groups are distinguishable in (S)-methacholine, whereas they are averaged in acetylcholine as indicated by arrows. (Modified from Furukawa, H. et al., *Mol. Pharmacol.*, 62, 778, 2002.)

protein, in which G protein $G_{i1}\alpha$ unit is fused to the C-terminus of the M_2 mutant. The binding of nonhydrolysable GTP analogue, [^{35}S]guanosine 5'-3-O-(thio)triphosphate ([^{35}S]GTPγS), is measured in the presence of GDP and agonist. The M_2-$G_{i1}\alpha$ fusion protein is expressed in *Sf*9/baculovirus, as in Section 11.2.1. The expression level of this fusion protein is approximately 0.3 mg per liter culture.

11.3.1 Methods for Synthesis of (S)-Methacholine

Protocol 11-4: Synthesis of (S)-Methacholine Iodide

1. Take 5 g of ethyl L(-)-lactate and 30 ml of dimethylamine anhydrous and mix, heat at 80°C for 1 h in a sealed tube, and leave at room temperature for 24 h.
2. Remove unreacted dimethylamine by evaporation.
3. Dissolve the product in diethyl ether, and mix with 2 g of lithium aluminum hydride ($LiAlH_4$) for reduction.
4. Extract the product with diethyl ether, and dry over anhydrous sodium sulfate (Na_2SO_4). After evaporation of diethyl ether and distillation, the distillate is dissolved in ethanol.
5. Add 2 ml of methyl iodide, and mix well. Diethyl ether is added drop by drop until crystals of S(+)-β-methylcholine iodide can be seen.
6. Take 1 g of the S(+)-β-methylcholine iodide crystals, mix with acetic anhydride (12.5 ml), and stir for 30 min at room temperature. Unreacted acetic anhydride is removed under vacuum.
7. Dissolve the product in methanol, add diethyl ether, and crystallize S(+)-methacholine iodide [(S)-methacholine].

Protocol 11-5: Comparison of (S)-Methacholine to Acetylcholine in One-Dimensional Proton Spectroscopy

1. Dissolve (S)-methacholine or acetylcholine to yield a final concentration of 1 mM in D_2O (400 μl volume).
2. Place the sample to NMR tube (Shigemi, Allison Park, PA) and measure ^1H-NMR spectrum in Bruker Avance 600 at 296 K using the spectrum width 6127 Hz.

11.3.2 Methods for Activity Assays

Materials

[^3H]N-methyl scopolamine (NMS), [^{35}S]guanosine 5'-3-O-(thio)triphosphate (GTPγS) (NEN-Dupont), GF/B glass filter (Whatman), atropine sulfate (Sigma), and SephadexG50 fine (Amersham)

Protocol 11-6: [^3H]NMS Binding Assay for the Receptors in Membrane Fraction

The following is an example for a duplicate experiment. The typical [^3H]NMS displacement curve is shown in Figure 11.5a:

A

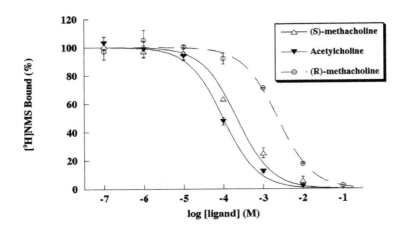

B

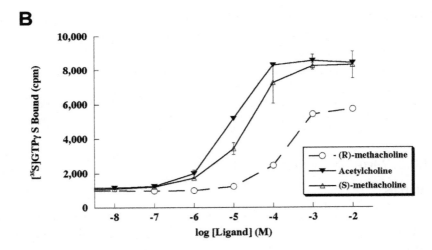

FIGURE 11.5 Displacement of [³H]NMS binding by agonists and antagonists (a) and dose–response binding of [³⁵S]GTPγS by agonists (b). The experiments are conducted using membrane fractions expressing the M_2 mutant (a) and the M_2-$G_{i1}\alpha$ (b). The K_i values are 19.5 μM (acetylcholine), 45.4 μM [(S)-methacholine], and 429 μM (R)-methacholine as calculated using the IC_{50} values determined in (a) and the known value of K_d for [³H]NMS [see Furukawa, H. and Haga, T., *J. Biochem. (Tokyo)*, 127, 151, 2000]. The EC_{50} values are 9.36 μM (acetylcholine), 23.8 μM [(S)-methacholine, and 1120 μM (R)-methacholine] as measured in (b).

1. Prepare binding assay buffer: 20 mM potassium phosphate buffer (KPB) (pH 7.4), 150 mM NaCl, and 1.5 nM [³H]NMS.
2. Put 10 µl of the membrane fraction (prepared as in the previous section) in four polypropylene tubes.
3. Add 10 µl of H_2O or various concentrations of agonists or antagonists to two tubes (Rxn1) and 10 µl of 10 mM atropine to the other two tubes (Rxn2).
4. Add 1 ml of the binding assay buffer, and mix by vortexing.
5. Incubate the samples at 30°C for 1 h.
6. Equilibrate the GF/B filter with an ice-cold buffer containing 20 mM KPB and 150 mM NaCl (wash buffer).
7. Apply samples to GF/B filter by vacuum filtration.
8. Wash the filter paper with ice-cold wash buffer (3 ml × 3).
9. Remove GF/B filters, and dry at approximately 70°C for 1 h.
10. Add scintillation liquid, and measure radioactivity. The specific [³H]NMS binding is defined as tritium count in Rxn1 minus Rxn2.

PROTOCOL 11-7: [³H]NMS BINDING ASSAY FOR THE DETERGENT-SOLUBILIZED RECEPTORS

1. Prepare binding assay buffer: 20 mM potassium phosphate buffer (KPB) (pH 7.4), 150 mM NaCl, 0.1% digitonin, and 1.5 nM [³H]NMS.
2. Put 10 µl of soluble receptors (solubilized fraction or purified receptors) in four polypropylene tubes.
3. Add 2 µl of H_2O to two tubes (Rxn1) and 2 µl of 10 mM atropine to the other two tubes (Rxn2).
4. Add 200 µl of the binding assay buffer, and mix by vortexing.
5. Incubate the samples at 30°C for 1 h.
6. Equilibrate SephadexG50 (2 ml bed volume) with a buffer containing 20 mM KPB, 150 mM NaCl, and 0.1% digitonin (wash buffer).
7. Apply 200 µl of the reaction mix carefully on top of the SephadexG50 gel surface. Let it run by gravity flow.
8. Apply 600 µl of wash buffer.
9. Apply 400 µl of wash buffer, and recover the eluent from the column.
10. Add 5 to 10 ml of scintilation liquid, mix well, and measure radioactivity.

PROTOCOL 11-8: [³⁵S]GTPγS BINDING ASSAY FOR THE M_2-$G_{i1}\alpha$ FUSION PROTEIN

In each experiment, the membrane fraction expressing the M_2-$G_{i1}\alpha$ is resuspended in 20 mM KPB so that the total protein concentration (as assessed by BCA assay) is 0.5 to 1 mg/ml. The typical dose–response graph from the experiment is shown in Figure 11.5b.

1. Take 20 μl of membrane fraction suspension and mix with 180 μl of buffer containing 20 mM KPB, 165 mM NaCl, 55 nM [^{35}S]GTPγS, 1.1 μM GDP, 11 mM MgCl$_2$, 1.1 mM dithiothreitol, and various concentrations of ligands [such as (S)-methacholine, acetylcholine, carbamylcholine, and atropine].
2. Incubate the samples at 30°C for 1 h.
3. Apply samples to the GF/B filter by vacuum filtration.
4. Wash the glass filter paper with ice-cold buffer containing 20 mM KPB, 150 mM NaCl, and 10 mM MgCl$_2$ (3 ml × 3).
5. Dry and measure radioactivity as in [^{3}H]NMS binding assay.

11.4   DETERMINATION OF THE (S)-METHACHOLINE CONFORMATION BY NOESY AND TRNOESY

The TRNOE is an extension of the nuclear Overhauser effect (NOE) and has been widely used to determine conformations of small ligands bound to large molecules, such as proteins, in an exchanging system.[15] The NOE experiment involves continuous saturation of the transition of one nucleus (A) and observation of change in the relaxation process (as represented by a NMR signal) of another nucleus (X) that are related to each other by a dipole–dipole interaction. In TRNOE, the pattern of the cross-relaxation between the two nuclei in the bound state is transferred to the free-state via chemical exchange. The NOE intensities change as a function of Larmor frequency (ω) and correlation time (τ$_c$). In small molecules where ωτ$_c$ < 1, proton–proton NOE signals are positive, while in large molecules where ωτ$_c$ > 1, NOE signals are negative. For ligands associated with large molecules, ωτ$_c$ is greater than 1, and a negative TRNOE signal is observed. Thus, the TRNOE method involves the measurement of the negative NOEs on the free ligand resonances following irradiation. The geometric information of the bound ligand is transferred to the free ligands as a result of a dissociation process in the chemical exchange.

The TRNOE is observed if the following conditions are met:

$$k_{off} \geq 10\rho_{Fi}, \text{ and 2) } |1 - a|\sigma^{BiBj} >> a\sigma^{FiFj},$$

where k_{off} is the chemical off rate between the free and bound ligands; ρ_{Fi} is the spin-lattice relaxation rate of proton i in the free state; a is the molar fraction of the free ligand; and σ^{BiBj} and σ^{FiFj} are the cross-relaxation rates between proton i and j in bound and free states, respectively. The binding constant values of ligands appropriate for TRNOE measurements are typically in the μM and mM range. The molecular size of the ligands has to be less than 1000 Da. There is virtually no limit to the size of the protein that the ligand binds. For example, the conformations of peptides bound to GroEL, a large homomeric protein complex composed of 14 identical 60 KDa subunits, were reported.[23]

The NOE or TRNOE intensity is inversely proportional to the distance between two protons raised to the sixth power. Therefore, NOE and TRNOE contain distance information between protons. Conformations of (S)-methacholine are estimated based on the intensities of NOEs or TRNOEs between protons within the molecule.

11.4.1 Methods for Determination of the Free (S)-Methacholine Conformation

The conformation of the (S)-methacholine in solution is determined, at first, by the use of two-dimensional (2D) nuclear Overhauser effect spectroscopy (NOESY) and the precise measurement of coupling constants in proton NMR. By doing so, one can compare the conformations of the free and the receptor-bound forms. All of the measurements are done in Bruker Avance 600 and 800.

Protocol 11-9: Two-Dimensional NOESY of (S)-Methacholine Conformation in Solution

1. Prepare 10 mM of (S)-methacholine solution in 500 μl of D_2O.
2. Set temperature of Bruker Avance 600 to 296 K.
3. Perform NOESY experiment with the following parameters: mixing time of 1.2 s, 64 scans/increment, raw data matrices (t_1 = 256, t_2 = 2 K), spectrum width = 6127 Hz, and a total relaxation delay of 5.0 s.
4. Make sure the spectrum is baseline corrected, multiplied by a π/2-shifted squared sine bell window function in F1 and F2 dimensions, Fourier transformed, and zero-filled to confer the final data matrices.
5. Measure the NOE cross-peak volume by xwinnmr (Bruker).
6. Estimate conformation qualitatively by cross-peak volumes. (The typical spectrum is illustrated in Figure 11.6.)

Protocol 11-10: Determination of Coupling Constant

1. Prepare 10 mM of (S)-methacholine solution in 500 μl of D_2O.
2. Set temperature of Bruker Avance 800 to 296K.
3. Perform 1D ^{1}H-NMR experiment with 32 scans and a spectrum width of 3600 Hz.
4. Multiply the free induction decay by the exponential or Gaussian function and then use Fourier transformation.
5. Measure the coupling constants for each proton by xwinnmr.
6. Calculate the O-C2-C1-N dihedral angle by using the modified Karplus equation:[24]

$$J = 13.89*\cos^2\phi - 0.98*\cos\phi + \Sigma\Delta\chi_I\{1.02 - 3.40 \cos^2(\xi_i*\phi + 14.9*|\Delta\chi_I|)\}$$

where φ is the degree of the dihedral angle, $\Delta\chi_I$ is the electronegativity difference between the substituents attached to the H-C-C-H system and the hydrogen, and ξ_I is the correction term (+1 in this case).

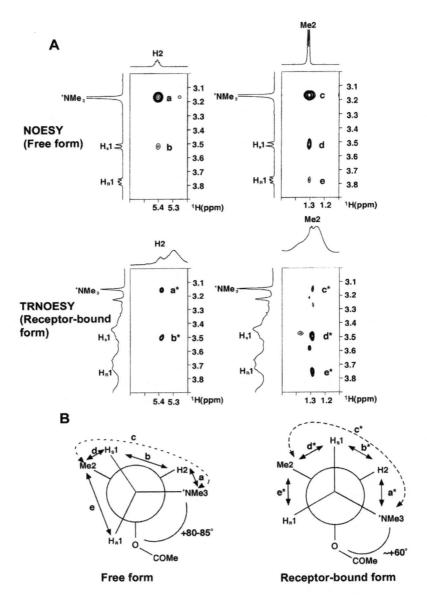

FIGURE 11.6 Typical spectra of NOESY (free form) and TRNOESY (receptor-bound form) (a) and the correlation of the NOE or TRNOE to the conformation of (S)-methacholine as shown in the Newman projections (b). The alphabetical labeling of NOE or TRNOE cross-peaks in (a) (a to e or a* to e*) correspond to the NOE or TRNOE correlation in (b). (Modified from Furukawa, H. et al., *Mol. Pharmacol.*, 62, 778, 2002.) The intensities of TRNOE cross-peaks, d* and e*, are always equal in experiments using various mixing times, indicating that Me2-H_s1 and Me2-H_R1 are essentially equidistant. This makes the O-C2-C1-N dihedral angle to be approximately +60°.

11.4.2 METHODS FOR DETERMINATION OF THE RECEPTOR-BOUND (S)-METHACHOLINE CONFORMATION

For precise measurement of TRNOE, the following factors are critical:

1. The sample buffer in the absence of protein should not produce any TRNOE signals.
2. The protein should retain activity during the experiment.
3. An appropriate protein–ligand ratio and mixing time should be chosen to maximize the TRNOE signals.

The use of more than 0.5% of digitonin causes TRNOE signals as a result of the interaction between (S)-methacholine and digitonin micelles. Digitonin molecules bind tightly to and cluster around the M_2 receptors (as assessed by thin-layer chromatography) in an unknown manner. Therefore, it is essential to substitute some of the digitonin molecules bound to the M_2 mutant by sodium cholate to minimize background TRNOE. The receptor-bound digitonin concentration is approximately 0.2% after the sample preparation protocol in the previous section, as assessed by thin-layer chromatography. The addition of soybean lipid to the sample before the experiment completely eliminates the background TRNOE. The final sample condition contains 10 mM Tris(hydroxymethyl-$d3$)amino-$d2$-methane-deuterium chloride, 0.2% sodium cholate, and 1.5 mM (S)-methacholine in D_2O. The M_2 mutants in this buffer condition were confirmed to retain ligand-binding activity at 23°C, but not at 30°C, for at least 3 days. The amount of (S)-methacholine has to be sufficiently high so that the ligand-binding site in the M_2 mutant is saturated with ligand, but it has to be within the extent that positive NOE (contributed by free ligands) does not interfere with the experiment. However, in theory, small ligands such as (S)-methacholine do not produce a strong NOE with the short mixing time used in TRNOE experiments. In this case, 1.5 mM (S)-methachioline has been confirmed to produce the largest interpretable TRNOE signals.

The actual measurements are taken using 2D ^1H-^1H transferred Overhauser effect spectroscopy (TRNOESY). Because detergent molecules contain a substantial number of protons, the spectrum becomes noisy. To eliminate the problem, differential spectra are obtained and analyzed. Specifically, the measurement is first made on the sample containing the purified M_2 mutant and (S)-methacholine, and later on the same sample containing 1 mM atropine sulfate (antagonist). The second spectrum is subtracted from the first one. By following this method, TRNOE signals can be unambiguously extracted from the spectra.

PROTOCOL 11-11: 2D TRNOESY EXPERIMENT FOR DETERMINATION OF THE RECEPTOR-BOUND (S)-METHACHOLINE CONFORMATION

1. Mix 200 μl of the purified M_2 mutant (50 μM) with methacholine (1.5 mM) prepared in Section 11.2.3 with 15 μl of 40 mg/ml crude soybean phosphatidylcholine (mixed micelles prepared in 0.2% sodium cholate).

2. Set temperature of Bruker Avance 600 to 296 K.
3. Acquire ^{1}H-NMR spectrum, and define the position of the water peak.
4. Perform TRNOESY experiment with the following parameters: mixing time randomized ±10% from 50 or 150 ms, 64 scans/increment, raw data matrices (t_1 = 256, t_2 = 2 K), spectrum width = 6127 Hz, and a total relaxation delay of 5.0 s. Water signal (with its peak position defined in ^{1}H-NMR spectrum) is presaturated (70 dB, 1.5 s) during relaxation delay.
5. Take out the sample tube, and add atropine sulfate to 1 mM.
6. Perform the same experiment as in step 4.
7. Note that the spectra obtained in steps 4 and 6 are baseline-corrected, multiplied by a $\pi/2$-shifted squared sine bell window function in F1 and F2 dimensions, Fourier transformed, and zero-filled to confer the final data matrices.
8. Subtract the processed spectrum from step 6 (plus atropine sulfate) from the one in step 4.
9. Measure the TRNOE cross-peak volume of the differential spectra by xwinnmr.
10. Estimate conformation qualitatively by cross-peak volumes. (The typical spectrum is illustrated in Figure 11.7.)

11.5 ASSESSMENT OF (S)-METHACHOLINE CONFORMATION BY DOCKING STUDIES

The relevance of the experimentally determined (S)-methacholine conformation is tested by docking study using the M_2 receptor molecular model. The M_2 receptor model is built based on the recent crystal structure of bovine rhodopsin[5] in combination with the information regarding the transmembrane domain movement by light activation from electron paramagnetic resonance studies.[25] The actual building process of the M_2 receptor model is only briefly discussed here.

PROTOCOL 11-12:   HOMOLOGY MODELING OF THE M_2 RECEPTOR AND
                  DOCKING OF (S)-METHACHOLINE

All of the following are carried out using modules in the Insight II package (Molecular Simulation Inc., San Diego, CA):

1. The model of photoactivated rhodopsin (metarhodopsin II) is constructed (adopting the rigid body movement of the transmembrane segments 3 through 6 to the crystal structure of rhodopsin in the dark state).
2. The transmembrane bundles are built by replacing the amino acid residues of the helices of rhodopsin with the sequence of the M_2 receptor by "Homology" module.
3. The loop structures are constructed by using the fragment library in the "Biopolymer" module.
4. The structure is energy-minimized by the use of the DISCOVER 3 force field.

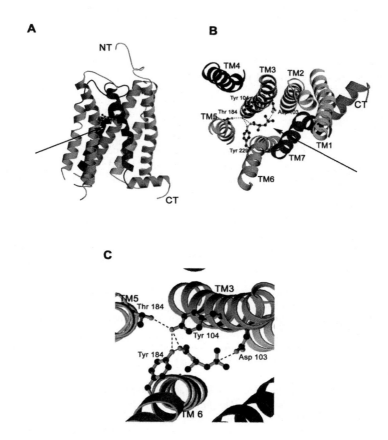

FIGURE 11.7 (**See color insert following page 240**) M$_2$ receptor model in the active form bound to (S)-methacholine viewed from the side (a), and from the top of the N-terminus (NT) (b). The seven-transmembrane segments (TM) are colored in aquamarine (TM1), gold (TM2), red (TM3), navy blue (TM4), orange (TM5), gray (TM6), and dark green (TM7); the C-terminal (CT) helix is colored in light purple. Arrows in (a) and (b) point to the (S)-methacholine binding site. The closer view of the (S)-methacholine binding site (c) implies the possible electrostatic interaction between the quaternary amine and Asp103 side chain, and hydrogen bonds involving the acetyl group and residues such as Tyr 104, Thr 184, and Tyr 184. The O-C2-C1-N dihedral angle of (S)-methacholine in this energy minimized model is +55.5°.

5. The model of (S)-methacholine is docked to the putative binding site of the M$_2$ mAChR model, guided by the ligand–receptor interactions between the quaternary amine-Asp 103, and the carbonyl oxygen-Tyr 403.

6. The (S)-methacholine bound model is subjected to molecular dynamics with simulated annealing in DISCOVER 3 force field involving the amino acid residues within 9 Å from the ligands.

ACKNOWLEDGMENT

Tomoaki Okada is greatly acknowledged for his technical assistance. We thank Dr. Kazuo Nagasawa for synthesis of (S)-methacholine. M. Rosconi and T. Kawate are thanked for critical reading of this manuscript.

REFERENCES

1. Kubo, T. et al., Cloning, sequencing and expression of complementary DNA encoding the muscarinic acetylcholine receptor, *Nature*, 323, 411, 1986.
2. Bonner, T.I. et al., Identification of a family of muscarinic acetylcholine receptor genes, *Science*, 237, 527, 1987.
3. Peralta, E.G. et al., Primary structure and biochemical properties of an M_2 muscarinic receptor, *Science*, 236, 600, 1987.
4. Pierce, K.L. et al., Seven-transmembrane receptors, *Nat. Rev. Mol. Cell Biol.*, 3, 639, 2002.
5. Palczewski, K. et al., Crystal structure of rhodopsin: a G protein-coupled receptor, *Science*, 289, 739, 2000.
6. Kunishima, N. et al., Structural basis of glutamate recognition by a dimeric metabotropic glutamate receptor, *Nature*, 407, 971, 2000.
7. Baker, R.W. et al., Structure and activity of muscarinic stimulants, *Nature*, 230, 439, 1971.
8. Chothia, C. and Pauling, P., Absolute configuration of cholinergic molecules; the crystal structure of (plus)-trans-2-aceoxy cyclopropyl trimethylammonium iodide, *Nature*, 226, 541, 1970.
9. Casy, A.F. et al., Conformation of some acetylcholine analogs as solutes in deuterium oxide and other solvents, *J. Pharm. Sci.*, 60, 67, 1971.
10. Wess, J. et al., Site-directed mutagenesis of the M_3 muscarinic receptor: identification of a series of threonine and tyrosine residues involved in agonist but not antagonist binding, *Embo. J.*, 10, 3729, 1991.
11. Ward, S.D. et al., Alanine-scanning mutagenesis of transmembrane domain 6 of the M(1) muscarinic acetylcholine receptor suggests that Tyr381 plays key roles in receptor function, *Mol. Pharmacol.*, 56, 1031, 1999.
12. Portoghese, P.S., Relationships between stereostructure and pharmacological activities, *Annu. Rev. Pharmacol.*, 10, 51, 1970.
13. Casy, A.F., Stereochemical aspects of parasympathomimetics and their antagonists: recent developments, *Prog. Med. Chem.*, 11, 1, 1975.
14. Lewis, N.J. et al., Diacetoxypiperidinium analogs of acetylcholine, *J. Med. Chem.*, 16, 156, 1973.
15. Clore, G.M. and Gronenborn, A.M., Theory and application of the transferred Overhauser effect to the study of the conformations of small ligands bound to proteins, *J. Magn. Reson.*, 48, 402, 1982.
16. Post, C.B., Exchange-transferred NOE spectroscopy and bound ligand structure determination, *Curr. Opin. Struct. Biol.*, 13, 581, 2003.
17. Furukawa, H. et al., Conformation of ligands bound to the muscarinic acetylcholine receptor, *Mol. Pharmacol.*, 62, 778, 2002.
18. Doyle, D.A. et al., The structure of the potassium channel: molecular basis of K+ conduction and selectivity, *Science*, 280, 69, 1998.

19. Tucker, J. and Grisshammer, R., Purification of a rat neurotensin receptor expressed in *Escherichia coli*, *Biochem. J.*, 317 (Pt 3), 891, 1996.
20. Furukawa, H. and Haga, T., Expression of functional M_2 muscarinic acetylcholine receptor in *Escherichia coli*, *J. Biochem. (Tokyo)*, 127, 151, 2000.
21. Haga, K. and Haga, T., Purification of the muscarinic acetylcholine receptor from porcine brain, *J. Biol. Chem.*,260, 7927, 1985.
22. O'Reilly, D.R. et al., *Baculovirus Expression Vectors: A Laboratory Manual*, Freeman, New York, 1992.
23. Wang, Z. et al., Basis of substrate binding by the chaperonin GroEL, *Biochemistry*, 38, 12,537, 1999.
24. Haasnoot, C.A.G. et al., The relationship between proton–proton NMR coupling constants and substituent electronegativities-I., *Tetrahedron*, 36, 2783, 1980.
25. Farrens, D.L. et al., Requirement of rigid-body motion of transmembrane helices for light activation of rhodopsin, *Science*, 274, 768, 1996.

12 Modeling of G Protein-Coupled Receptors for Drug Design

Masaji Ishiguro

CONTENTS

12.1 INTRODUCTION

G protein-coupled receptors (GPCRs) are heptahelical transmembrane-integrated proteins that transduce a large number of signals across the cell membrane by binding signaling molecules such as ions, odorants, biogenic amines, lipids, peptides, and proteins to the extracellular side of the membrane. A wide variety of intracellular biochemical events are initiated through interactions between the activated GPCR and heterotrimeric guanosinetriphosphate (GTP)-binding protein (G protein).

GPCRs in the rhodopsin family share seven hydrophobic transmembrane regions. The extracellular region of the transmembrane helices forms the ligand-binding pocket[1] for GPCR ligands, such as cationic biogenic amine ligands, while the intracellular loops mediate receptor–G protein coupling. Mutational analysis of the receptor functions[2–7] and observation of the rigid-body motion of the transmembrane segments (TM)[8,9] in photoactivation of rhodopsin suggest the presence of multiple structures in inactive and active GPCRs. Analysis of the structural changes in the fluorescence-labeled adrenergic receptor upon ligand binding suggested that

283

the partial agonist-bound receptor structure is distinct from that of the full agonist-bound receptor.[10] Plasmon waveguide resonance (PWR) measurements on structural alteration of opioid and β-adrenergic receptors further indicate the formation of different receptor structures in the binding of functionally different ligands.[11] Furthermore, a recent report on κ-opioid receptor ligands suggested that full agonist binding involves the rigid-body rotation of TM6.[12] Despite the progress in understanding pharmacological events, with the exception of rhodopsin, the structural basis for controlling the potency and selectivity of ligands and the efficacy of signal transduction at the atomic level remain unclear due to a lack of information on the three-dimensional structure of the receptors.[13–15]

Rhodopsin, an inactive form of GPCR, forms a protonated Schiff base (PSB) with the inverse agonist, 11-*cis*-retinal, at Lys296 of opsin, the protein moiety of rhodopsin. Rhodopsin can be photochemically converted to the activated form, metarhodopsin II (Meta II), by isomerization of the 11-*cis* retinylidene chromophore to the all-*trans* chromophore, a full agonist.[16,17] GPCRs share a few highly conserved residues with rhodopsin in each α-helical transmembrane segment. These highly conserved residues are thought to play important roles in the structural changes of the helical arrangement as well as in signal transduction. The roles of these residues have been investigated by modeling the photoactivated intermediate structures in the rhodopsin photocascade (Scheme 12.1).[18]

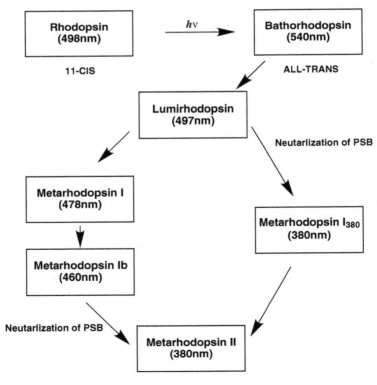

SCHEME 12.1

The photointermediates in the photocascade bind the G protein, transducin, activating the guanosinediphosphate (GDP)–GTP exchange in transducin. Metarhodopsin Ib (Meta Ib) is known to bind transducin without activation,[19,20] whereas an earlier intermediate in the photocascade, metarhodopsin I (Meta I), is unable to bind transducin. Opsin is known to weakly activate transducin under physiological conditions.[21] A mutant substituted at Gln for Glu113, the counterion of PSB, consistently shows increased activity with respect to opsin, yet is not fully active (partially active). Moreover, it exhibits full activity upon binding exogenous all-*trans* retinal.[3] Thus, the mutant is expected to have a structure analogous to a partial agonist-bound receptor.

The 11-*cis* retinylidene chromophore rapidly isomerizes to an all-*trans* form upon illumination.[16] A number of photointermediates are observed along the photocascade. An early photointermediate, bathorhodopsin (Batho), with a photoisomerized all-*trans* retinylidene chromophore, slowly decays and makes a conformational change to Meta I via lumirhodopsin (Lumi). Deprotonation of the Schiff base in the following thermal decay yields metarhodopsin II (Meta II), which activates transducin fully.[17,22] The *cis–trans* photoisomerization of the chromophore occurs within the vicinity of opsin, affording a highly strained conformation of the chromophore in Batho.[23] The flip of the modified β-ionone moiety has been suggested in the formation of Lumi, with a rearrangement of TM3 and TM4 to accommodate the modified β-ionone moiety.[24] The photoconversion process is dependent on two temperatures during the Lumi to Meta II transition. At physiological temperatures, Lumi rapidly equilibrates with metarhodopsin I_{380} (Meta I_{380}), and this is followed by the formation of Meta II.[24] In the photoconversion process at low temperatures, Meta I is a stable intermediate in the Lumi to Meta II transition. Time-resolved ultraviolet (UV) measurements detected another intermediate, Meta Ib, in the Meta I to Meta II transition.[18]

From electron paramagnetic resonance (EPR) measurements of spin-labeled rhodopsin (in the dark) and Meta II (in the light), the rigid-body rotation of TM6 was suggested in the formation of the Meta II state.[8] The motion of TM6 in the photoactivation cascade of rhodopsin was also demonstrated by zinc cross-linking of histidines.[26] Formation of the Meta II state requires PSB deprotonation.[27] Neutralization of the PSB renders TM3 mobile enough to leave TM7. An important proton transfer process is concomitantly required at the intracellular site for the activation of transducin.[28] The highly conserved Glu134, located at the intracellular site for TM3, appears to be responsible for the protonation by transferring the carboxylic acid side chain from a polar to a nonpolar environment.[29]

12.2 STRUCTURAL MODELS OF THE PHOTOINTERMEDIATES, BATHO, LUMI, AND META I

The photochemical isomerization of the retinylidene chromophore is accompanied by a structural change of the protein moiety.[24] The extraordinarily rapid photoisomerization (~200 fs) at a low temperature (77 K) leaves most of the opsin structure unaffected.[30] Restrained molecular dynamics simulations of the isomerization of the

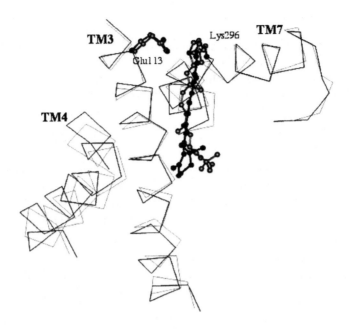

FIGURE 12.1 Motion of TM3 and TM4 in the Lumi model (black). The rhodopsin structure is shown in gray.

chromophore provided a candidate structure for the Batho chromophore in the crystal structure of opsin. The Batho chromophore showed a characteristically twisted double bond at C11-C12, with a negative dihedral angle (−148°) for C10-C11-C12-C13, whereas the cyclohexenyl moiety remained in the original binding cleft.[31]

The twisted high-energy conformation of double bonds of the polyene portion relaxed into an all-*trans* form by the outward swing of TM3. The concomitant conformational change of TM4 yielded a swing of the N-terminal end of TM4 (Figure 12.1). The flip of the cyclohexenyl group resulted in ~40° rotation of the 9-methyl group about the axis of the C9-Nζ moiety from the chromophore structure of rhodopsin. The PSB proton (Hζ) rotated out of the hydrogen-bonding distance, and the polyene moiety of the model lined up perpendicularly to a putative membrane plane, directing the 9- and 13-methyl groups toward the extracellular site (Figure 12.2). The dislocation of the PSB proton is consistent with the disappearance of the hydrogen bond acceptor for the PSB proton in Lumi.[32,33]

A further swing of the C-terminus of TM3 enabled the chromophore to rotate about 90° from that of the Lumi model. The polyene plane lies parallel to a putative membrane plane, and the PSB proton (Hζ) was reoriented to the carboxylate oxygen of Glu113 within a hydrogen bond distance (Figure 12.3).

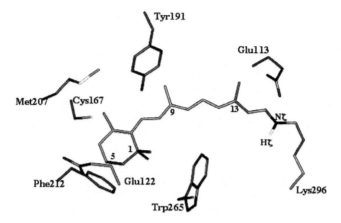

FIGURE 12.2 Conformation of the chromophore in the Lumi model (lateral view). Carbon atoms are gray; oxygen and nitrogen atoms are black; hydrogen and sulfur atoms are white.

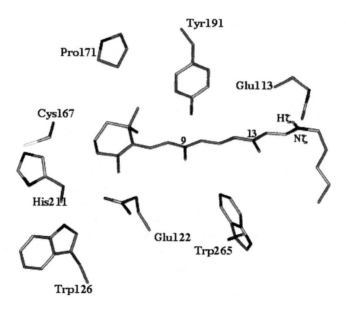

FIGURE 12.3 Conformation of the chromophore in the Meta I model. (View from the extracellular site.)

PROTOCOL 12-1: GENERAL PROCEDURE FOR MODELING THE PHOTOINTERMEDIATE STRUCTURES

The crystal structure of rhodopsin (PDB code: 1L9H)[15] was used as the starting structure for modeling the rhodopsin photointermediates. The photointermediate structures were then used as templates for modeling the GPCR structures.

Molecular dynamics calculations were performed at 300 K with a cut-off distance of 8.5 Å and a distance-dependent dielectric constant. The conformation was sampled every 1 ps with a time step of 1 fs for 100 ps, using CVFF parameters in Discover 3 (version 2000, Molecular Simulations Inc., San Diego, CA). The entire structure was energy-minimized until the final root-mean square deviation (rmsd) was less than 0.1 kcal/mol/Å, unless otherwise indicated.

In the rigid-body motions of the transmembrane helices, the C-terminal end of TM3 swung outward every 0.2 Å, pivoting on the highly conserved Cys110 residue of the N-terminal end of TM3. Interhelical Cα-Cα distances between TM2 and TM3 were maintained above 4.5 Å during the motion of TM3. The minimum interhelical Cα-Cα distance was estimated from interhelical distances of crystal structures of membrane proteins (Y. Oyama and M. Ishiguro, unpublished). The steric interactions between TM3 and TM4 caused by the motion of TM3 were eliminated by the swing of the N-terminal end of TM4 toward TM5, minimizing the structural energy. The chromophore structure was optimized using molecular dynamics calculations within the chromophore-binding site (residues within 10 Å from the chromophore). The intracellular pore generated by the outward motion of TM3 was filled with water molecules using the Assembly module in Insight II (Molecular Simulations, Inc.). The entire structure was then energy minimized.

PROTOCOL 12-2: MODEL OF BATHO, LUMI, AND META I STRUCTURES

The protein moiety was fixed, and only the chromophore was isomerized in the chromophore-binding pocket of the protein moiety. This was achieved by setting the parameter for the *trans* configuration at the C11-12 double bond in the molecular dynamics calculations.[34] The molecular dynamics calculations were performed at 300 K with a time step of 1 fs for 1 ps, sampling every 10 fs. Energy minimization of the chromophore structures was performed to give the Batho model.

The conformations of TM3 for the Lumi model were generated by swinging the C-terminal end of TM3 every 0.2 Å by 1.4 Å from the Batho structural model. A TM4 region (148–173) was minimized for every protein structure in order to eliminate interactions with TM3. The C-terminal end of TM3 was further swung every 0.2 Å by 2.0 Å from the Lumi structural model, followed by structure minimization of the TM4 region. The chromophore structure was optimized using the molecular dynamics/minimization procedure. The pore formed at the intracellular site of the Lumi and Meta I models was filled with water molecules. Final structural models were energy minimized, and the conformation of the chromophore was optimized using the molecular dynamics/minimization procedure.

12.3 STRUCTURAL MODELS OF PHOTOINTERMEDIATES, META Ib AND I_{380}

Meta Ib, an intermediate in the Meta I to Meta II transition, binds an inactive form of transducin, maintaining the PSB within a hydrogen-bond distance of its counterion, Glu113.[19] A further turn of the C-terminal end of TM3 enables Arg135 to hydrogen bond with Glu134 and Glu247 at a maximum distance between TM3 and TM6 (Figure 12.4). The N-terminal end of TM4 was concomitantly swung toward TM5 at a fairly large distance. The large conformational changes of TM3 and TM4 would cause a considerable conformational change of the second intracellular loop, which would then be recognized by transducin. A weak activation of transducin by wild-type opsin indicates that wild-type opsin binds transducin, eliciting its activity.[35] The structure of wild-type opsin is presumably analogous to that of Meta Ib.

At physiological temperatures, Lumi rapidly equilibrates with Meta I_{380},[25] which is a neutralized form of the Schiff base, as estimated from its absorption maximum (380 nm). Neutralization of the Schiff base would render TM3 highly mobile, thereby enabling a further outward swing of TM3. The motion of TM3 provoked a large gap between TM3 and TM5 at the intracellular site. Thus, the N-terminal moiety of TM4 rearranged to fit into the space. The large motion of TM3 transferred Glu134 of the highly conserved ERY triplet on TM3 to the hydrophobic lipid phase (Figure 12.5). This hydrophobic environment stabilized the protonated Glu134 residue and enabled Arg135 to switch the hydrogen bond from Glu134 to Glu247. A Glu113Gln mutant has an analogous structure to the Meta I_{380} model, because the 11-*cis* retinylidene chromophore shows the deprotonated form at its absorption maximum (380 nm).

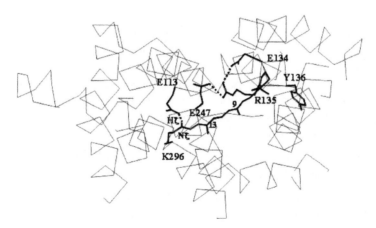

FIGURE 12.4 Hydrogen-bond networks in the Meta Ib model. (View from the intracellular site.)

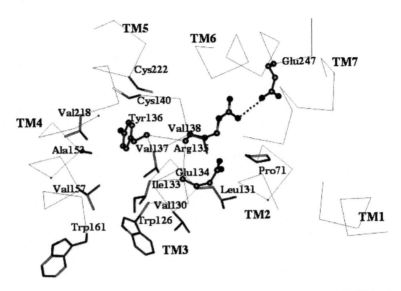

FIGURE 12.5 Residues neighboring the ERY triplet in the Meta I$_{380}$ model. (View from the intracellular site.)

PROTOCOL 12-3: MODEL META Ib AND I$_{380}$ STRUCTURES

The C-terminal end of TM3 swung within a hydrogen-bond distance of Arg135 on TM3 and Glu247 on TM6, maintaining the salt bridge between Glu134 and Arg135. The N-terminal end of TM4 swung toward TM5, pivoting on the C-terminal end of TM4 and eliminating the collision with TM3.

The conformation of TM3 for the model of Meta I$_{380}$ was generated by swinging the C-terminal end of TM3 6.6 Å from the Lumi model. The N-terminal end of TM4 then swung, pivoting on the C-terminal end of TM4, and made interhelical contact with TM5. The N-terminal end of TM4 filled the space between TM3 and TM5 generated by the movement of TM3. After filling the pore formed at the intracellular site with water molecules, the conformation of the chromophore was optimized using the molecular dynamics/minimization procedure.

12.4 STRUCTURAL MODEL OF FULLY ACTIVATED PHOTOINTERMEDIATE, META II

Large structural changes can be accompanied by rigid-body movements of transmembrane helices during the formation of Meta II.[8,9,36–41] The Meta I$_{380}$ model has a wide-open pore enclosed by seven transmembrane segments at the intracellular site. Hence, in order to restore the hydrophobic interactions at the protein interior, the wide-open pore of Meta I$_{380}$ at the intracellular site becomes compact during the Meta I$_{380}$ to II transition by translation of TM6 toward TM3. However, TM5 interferes with the inward translation of TM6 by sterically interacting with the extracellular moiety of TM6, which is kinked at the highly conserved Pro267 residue. This

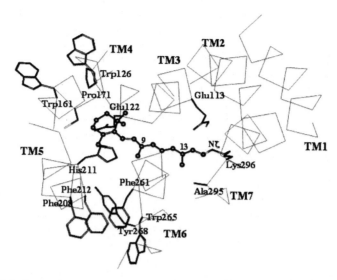

FIGURE 12.6 Aromatic residues in the Meta II model. (View from the extracellular site.)

unfavorable steric interaction may be avoided by clockwise rotation (viewed from the intracellular site) of TM6 about its helical axis. TM6 is most loosely associated with other transmembrane helices in rhodopsin.[42,43] Thus, the structure of TM6 kinked at Pro267 is thought to play an indispensable role in Meta II formation.[44]

Clockwise rotation of TM6 about the axis of the N-terminal moiety and a concomitant inward translation of TM6 provided the Meta II model (Figure 12.6). The rigid-body rotation of TM6 provokes a considerable structural change at the third extracellular and intracellular loops, considerably changing the chromophore-binding surface.

PROTOCOL 12-4: MODEL OF META II STRUCTURE

Clockwise rotation (view from intracellular site) of TM6 in Meta I_{380} about the axis of the N-terminal helix (Lys245 through Cys264) every 10° by 100° generated intermediate structures. TM6 was subsequently translated toward TM3 until a van der Waals contact was generated with TM3 and TM5, inducing steric collisions with TM5 and TM7. During the motion of TM6, the Cα-Cα distances to TM7 were maintained at greater than 4.5 Å. The initial transformed structure was energy-minimized after filling the pore with water molecules in the intracellular site. The structure, including the chromophore, was optimized using the molecular dynamics/minimization procedure.

12.5 FUNCTIONAL STRUCTURES OF GPCRS

Diffusible ligands for GPCRs of the rhodopsin family function as inverse agonists, antagonists, partial agonists, and full agonists. Recent PWR measurements in opioid and β-adrenergic receptors showed that each functionally different ligand binds a different receptor structure.[11,45] These findings imply that different structures are required for the construction of complex structural models of ligands with different functions in the same receptor. In the photointermediates of rhodopsin as well as photointermediate-related mutants, Meta I could be correlated to the inverse agonist-bound structure because it does not bind transducin and is thus totally inactive. Opsin (or Meta Ib) binds transducin and elicits a weak activity. Thus, it can be correlated to the antagonist-bound structure. Moreover, the Glu113Gln mutant (or Meta I_{380}) was correlated to the partial agonist-bound structure because it showed a high, but not full, efficacy in transducin activation. On the other hand, Meta II is the fully active form of the photoactivated rhodopsin and is expected to have a full agonist-bound structure.

In Meta II, an ionized form (Meta IIa, inactive) of Glu134 of the ERY triplet at the intracellular site of TM3 is in equilibrium with the protonated form (Meta IIb, active) at the cytoplasmic site.[28] The outward swing of the C-terminal end of TM3 in the Meta II model transferred the Glu134 residue from a polar to a nonpolar environment. The protonation of Glu134 provokes a conformational change in Arg135, facilitating the GDP–GTP exchange of G proteins (G protein activation).[18] Provided that the outward motion of TM3 determines the equilibrium rate, a larger tilt of TM3 affords a higher ratio of the protonated form of Glu134 to the deprotonated form. Namely, the fully activated form (Meta II-like) of the GPCR is thought to predominate in the protonated state of the Asp residue of the D(E)RY triplet at the C-terminal end of TM3, whereas the ionized form of the Asp residue is thought to predominate in a physiologically inactive (Meta Ib-like) structure. In the case of a highly, but not fully, active structure (partially active form, Meta I_{380}-like), the protonated form is an intermediate in the equilibrium reaction. On the other hand, the fully inactive form (Meta I-like) would not exhibit an equilibrium reaction. This is thought to be because it does not bind G protein (Scheme 12.2). Thus, the scheme including four distinct arrangements of the transmembrane segments has been suggested. In this proposed scheme, each arrangement consists of two states: the ionized and protonated forms of the Asp residue in the DRY triplet. The inverse agonist-bound structure, however, consists of a single (inactive) state.

Thus, the three-dimensional structural models of Meta I, Meta Ib, Meta I_{380}, and Meta II[18] were used to construct the structural models of the putative inverse agonist-, antagonist-, partial agonist-, and full agonist-bound forms of GPCRs of the rhodopsin family.[46]

PROTOCOL 12-5: MODEL OF RECEPTOR STRUCTURES AND LIGAND–RECEPTOR COMPLEX STRUCTURES

The multiple sequence alignment of GPCRs with the rhodopsin sequence was obtained using the Homology module installed within Insight II (2000 version, Molecular Simulations, Inc.). Deletions and insertions of amino acid residues in the transmembrane regions were not observed because the transmembrane regions are well conserved. Thus, the insertions and deletions at the extracellular site were in the loops. Moieties longer than the intracellular loops of the rhodopsin photointermediate

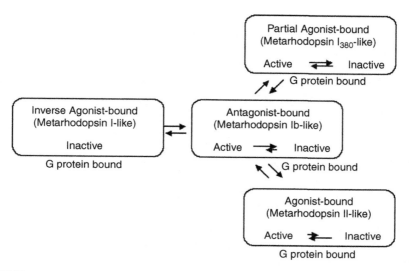

SCHEME 12.2

models were deleted. The replacement of side chains was carried out using the Homology module in Insight II, and the deleted and inserted portions were modeled by identifying appropriate peptide conformations picked up from the protein structural database. The initial structural models contained several crushed conformations of the side chains. These conformations were eliminated by searching side-chain conformations from the side-chain conformation database, and then the entire structures were optimized by energy minimization. Molecular dynamics calculations were subsequently performed for the backbone amides and side chains at 300 K.

The ligands were manually docked into the ligand-binding cleft of the corresponding receptors, guided by a salt bridge (~2.9 Å) between the cationic amine and the carboxylate oxygen atom of the conserved Asp residue in TM3. Severe steric hindrances between receptor residues and ligands were eliminated by rotating the side chains of residues. The initial complex model was energy minimized. The minimized complex structures were then optimized using the molecular dynamics/minimization procedure without distance constraints between the ligands and the receptors within the ligand-binding site (residues within 10 Å from ligands). The lowest-energy structure was selected as an energy-refined complex model.

12.6 RECEPTOR–LIGAND COMPLEX MODELS FOR MUSCARINIC ACETYLCHOLINE RECEPTOR

Acetylcholine (**1**, Figure 12.7), docked into the ligand-binding cleft of the fully activated form of the M_2 receptor models, favored the gauche conformation at the Cb-O bond (70°) in the binding cleft of the model structure. The quaternary cationic group formed a salt bridge with Asp103 in TM3, while the carbonyl oxygen of the acetyl group and the ester oxygen formed hydrogen bonds to Tyr403 in TM6 and Ser107 in TM3, respectively (Figure 12.8). In addition, Thr190 in TM5 was hydrogen

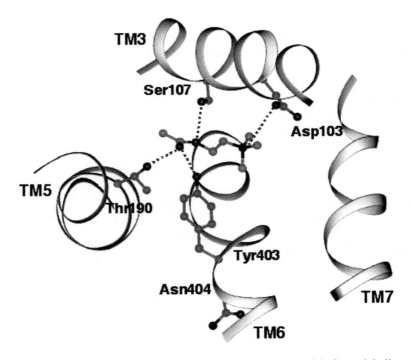

1   **2**

3   **4**   **5**

FIGURE 12.7 Chemical structures of ligands.

FIGURE 12.8 (See color insert following page 240) Complex model of acetylcholine at the binding cleft of the fully active form of the M_2 receptor models. (View from the extracellular site.) Transmembrane helical regions (TM) at the binding clefts are shown with gray ribbon. Hydrogen bonds are indicated with dotted lines. Oxygen and nitrogen atoms are black; carbon atoms are gray.

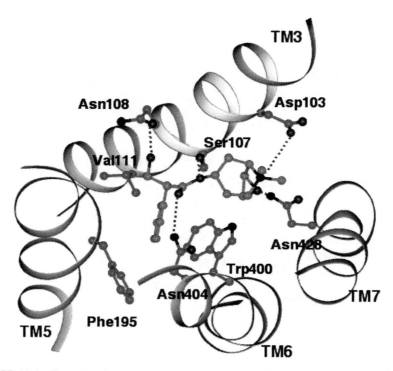

FIGURE 12.9 (**See color insert**) Complex model of N-methylscopolamine at the binding cleft of the physiologically inactive form of the M_2 receptor models.

bonded to the acetyl group. These hydrogen-bonding interactions were consistent with findings in previous reports that Thr190 and Tyr403 play critical roles in agonist binding[6,7,47] and that Asp103 binds the cationic moiety of acetylcholine.[48] The rigid-body rotation of TM6[8,18] enabled Tyr403 to form a network of hydrogen bonds between the full agonist and Thr190. The hydrogen-bond network in the complex model appears to be particularly important to the stabilization of the rotated conformation of TM6.

The ethanol amine moiety was bound at the rather narrow cleft enclosed by TM3 through TM6. Within the complex model, the introduction of the methyl group at the Cβ-position provoked severe steric interactions with TM6. Thus, metacholine favored a gauche conformation at the Cα-Cβ bond. The gauche conformation is consistent with the results observed by the transferred NOE measurements on metacholine interacting with the muscarinic acetylcholine receptor.[49]

N-methylscopolamine (2), an antagonist to the M_2 receptor, formed hydrogen bonds with Ser107 and Asn404 at the ester group, as well as Asp103 at the quaternary amine in the binding cleft of the physiologically inactive form of the M_2 receptor models. On the other hand, Tyr403 was not involved in antagonist binding in the physiologically inactive form of the M_2 receptor model (Figure 12.9). Coincidentally, the M_2 antagonists interact with Asp103 in TM3 and Asn404 in TM6 of the M_2 receptor, whereas Tyr403 does not appear to contribute to antagonist binding.[7,47,50]

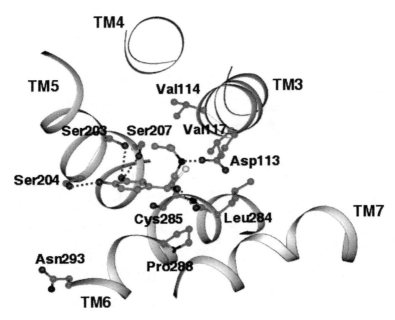

FIGURE 12.10 (**See color insert**) Complex model of *R*-isoproterenol at the binding cleft of the fully active form of the β₂ receptor models.

12.7 THE LIGAND-BINDING MODES OF ADRENERGIC β₂ RECEPTOR

A full agonist *R*-isoproterenol (**3**) formed a salt bridge between Asp113 in TM3 and the cationic amine, and a characteristic hydrogen bond between the β-hydroxyl group and the backbone carbonyl group of Leu284, which lies at the kink site of TM6 in the fully active form of the β₂ adrenergic receptor models (Figure 12.10). On the other hand, its enantiomeric isomer, *S*-isoproterenol, a partial agonist,[51] did not interact properly with the backbone carbonyl group of the fully active form of the receptor models. Furthermore, modification of the β-hydroxyl group of the full agonists to deoxy, methyl, and methoxyl groups converts the derivatives to partial agonists,[52] because these modifications are thought to break the hydrogen bond formed with the backbone carbonyl. The present complex model shows a clear contrast with the β₂-adrenergic receptor–ligand complex models constructed by *de novo* methods, which predicted the direct interaction between Asn293 and the β-hydroxyl group of agonists.[53,54] However, Asn293 may not be involved in the direct hydrogen-bond interaction in agonist binding.[51,52]

The complex model suggested that the *para*- and *meta*-hydroxyl groups of *R*-isoproterenol (**3**) bind at Ser204 and Ser207, respectively. Although the *meta*- and *para*-hydroxyl groups interact with the three serine residues (Ser203, 204, and 207) in TM5,[55–57] mutation of one of the serine residues or removal of one of the hydroxyl groups of the catechol moiety results in a reduction of not only the affinity but also

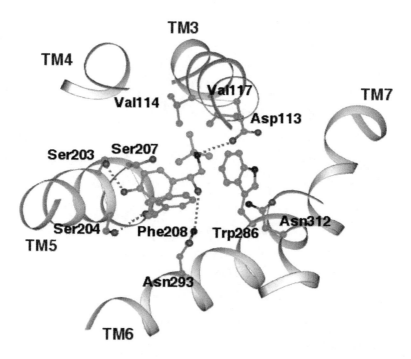

FIGURE 12.11 (See color insert) Complex model of salbutamol at the binding cleft of the partially active form of the β_2 receptor models.

the efficacy of the receptor activation.[56] Thus, the mutational experiments may indicate that the *meta-* and *para*-hydroxyl groups interact with Ser204 and Ser207 in the partially active form of the receptor, respectively. The specific catechol hydroxyl group in the full agonist that interacts with Ser204 or Ser207 in TM5 of the fully activated form of the receptor remains unknown. The binding of the bulky *tert*-butyl group of salbutamol (**4**), a typical partial agonist, at the conserved Asp113 residue in TM3 necessitated a wide space around Asp113 in the partially active form of the receptor models. The *para*-hydroxyl and *meta*-hydroxymethyl groups were directed toward Ser203 and Ser204 in the complex model, respectively (Figure 12.11). This finding is in agreement with the previous finding that Ser204 but not Ser207 is involved in the ligand recognition.[58]

The binding of the catechol moiety to the serine residues in TM5 resulted in the -hydroxyl group of the full agonist **3** to interact with the backbone carbonyl of TM6 in the fully active form of the β_2 receptor models, and *vice versa*. Although propranolol (**5**), an inverse agonist, has an N-isopropyl ethanolamine moiety with the same configuration at the β-carbon as *R*-isoproterenol (**3**), the bulky hydrophobic naphthoxymethyl group would not allow the β-hydroxyl group to form a hydrogen bond with the backbone carbonyl of TM6 (Figure 12.12).

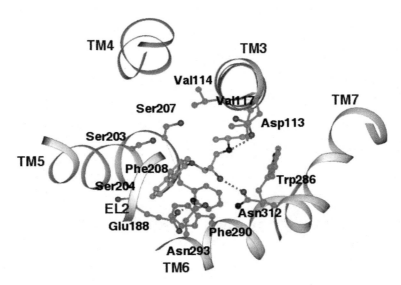

FIGURE 12.12 (See color insert) Complex model of propranolol at the binding cleft of the fully inactive form of the β₂ receptor models.

12.8 SUMMARY

The structural models of the rhodopsin photointermediates suggest putative roles of the highly conserved residues in the structural changes observed in the rhodopsin photocascade. In particular, conformational changes are possible without disrupting the conformation of the disulfide bond between Cys110 and Cys187. Furthermore, the kinked structure of TM6 at Pro267 is essential for the rigid-body rotation of TM6 in the formation of the fully active Meta II. The electrostatic change at the extracellular site (neutralization of PSB) caused by the photochemical isomerization of the 11-*cis* retinylidene chromophore was conveyed to the intracellular surface through the displacement of Glu134 on TM3 from polar to nonpolar environments. Considering the activated and inactivated states that correspond to the protonated and deprotonated forms of the highly conserved Asp (Glu) residue in the D(E)RY triplet at the intracellular site of TM3, the multiple two-state structure model is expected to be applicable to ligand recognition in GPCRs of the rhodopsin family.

The muscarinic acetylcholine and -adrenergic receptor–ligand complex models suggest that the ligands select the receptor structure according to their function (inverse agonist, antagonist, partial agonist, and full agonist). The partial agonists, in particular, are thought to bind a receptor structure that differs from the full agonist-bound receptor structure. This proposal is in agreement with the recent finding that the partial agonist-bound structure of the -adrenergic receptor is distinct from the full agonist-bound structure.[11]

ABBREVIATIONS

Batho	bathorhodopsin
CVFF	constant valence force field
EPR	electron paramagnetic resonance
GDP	guanosinediphosphate
GTP	guanosinetriphosphate
Lumi	lumirhodopsin
M$_2$	muscarinic acetylcholine receptor 2
Meta I	metarhodopsin I
Meta Ib	metarhodopsin Ib
Meta I$_{380}$	metarhodopsin I$_{380}$
Meta II	metarhodopsin II
PSB	protonated Schiff base
PWR	plasmon waveguide resonance
rmsd	root-mean square deviation
TM	transmembrane segment

REFERENCES

1. Strader, C.D. et al., Structure and function of G protein-coupled receptors, *Ann. Rev. Biochem.*, 63, 101, 1994.
2. Scheer, A. and Cotecchia, S., Constitutively active G protein-coupled receptors: potential mechanisms of receptor activation, *J. Recept. Signal Transduct. Res.*, 17, 57, 1997.
3. Robinson, P.R. et al., Constitutively active mutants of rhodopsin, *Neuron*, 9, 719, 1992.
4. Strader, C.D., Conserved aspartic acid residues 79 and 113 of the β-adrenergic receptor have different roles in receptor function, *J. Biol. Chem.*, 263, 10,267, 1988.
5. Monnot, C. et al., Polar residues in the transmembrane domains of the type 1 angiotensin II receptor are required for binding and coupling. Reconstitution of the binding site by co-expression of two deficient mutants, *J. Biol. Chem.*, 271, 1507, 1996.
6. Wess, J., Gdula, D., and Brann, M.R., Site-directed mutagenesis of the M$_3$ muscarinic receptor: identification of a series of threonine and tyrosine residues involved in agonist but not antagonist binding, *EMBO J.*, 10. 3729, 1991.
7. Heitz, F. et al., Site-directed mutagenesis of the putative human muscarinic M$_2$ receptor binding site, *Eur. J. Pharmacol.*, 380, 183, 1999.
8. Farrens, D.L. et al., Requirement of rigid-body motion of transmembrane helices for light activation of rhodopsin, *Science*, 274, 768, 1996.
9. Resek, J.F. et al., Formation of the meta II photointermediate is accompanied by conformational changes in the cytoplasmic surface of rhodopsin, *Biochemistry*, 32, 12,025, 1993.
10. Ghanouni, P. et al., Functionally different agonists induce distinct conformations in the G protein coupling domain of the β$_2$ adrenergic receptor, *J. Biol. Chem.*, 276, 24,433, 2001.

11. Devanathan, S. et al., Plasmon-waveguide resonance studies of ligand binding to the human β_2-adrenergic receptor, *Biochemistry*, 43, 3280, 2004.

12. Sharma, S.K. et al., Transformation of a κ-opioid receptor antagonist to a κ-agonist by transfer of a guanidinium group from the 5'- to 6'-position of naltrindole, *J. Med. Chem.*, 44, 2073, 2001.

13. Palczewski, K. et al., Crystal structure of rhodopsin: a G protein-coupled receptor, *Science*, 289, 739, 2000.

14. Teller, D.C. et al., Advances in determination of a high-resolution three-dimensional structure of rhodopsin, a model of G-protein-coupled receptors (GPCRs), *Biochemistry*, 40, 7761, 2001.

15. Okada, T. et al., Functional role of internal water molecules in rhodopsin revealed by X-ray crystallography, *Proc. Natl. Acad. Sci. USA*, 99, 5982, 2002.

16. Khorana, H.G., Rhodopsin, photoreceptor of the rod cell. An emerging pattern for structure and function, *J. Biol. Chem.*, 267, 1, 1992.

17. Sakmar, T.P., Rhodopsin: a prototypical G protein-coupled receptor, *Prog. Nucleic Acid Res.*, 59, 1, 1998.

18. Ishiguro, M., Oyama, Y., and Hirano, T., Structural models of the photointermediates in the rhodopsin photocascade, lumirhodopsin, metarhodopsin I, and metarhodopsin II, *ChemBioChem.*, 5, 298, 2004.

19. Tachibanaki, S. et al., Identification of a new intermediate state that binds but not activates transducin in the bleaching process of bovine rhodopsin, *FEBS Lett.*, 425, 126, 1998.

20. Tachibanaki, S. et al., Presence of two rhodopsin intermediates responsible for transducin activation, *Biochemistry*, 36, 14,173, 1997.

21. Acharya, S. and Karnik, S.S., Modulation of GDP release from transducin by the conserved Glu134-Arg135 sequence in rhodopsin, *J. Biol. Chem.*, 271, 25,406, 1996.

22. Rando, R.R., Polyenes and vision, *Chem. Biol.*, 3, 255, 1996.

23. Palings, I. et al., Assignment of fingerprint vibrations in the resonance Raman spectra of rhodopsin, isorhodopsin, and bathorhodopsin: implications for chromophore structure and environment, *Biochemistry*, 26, 2544, 1987.

24. Borhan, B. et al., Movement of retinal along the visual transduction path, *Science*, 288, 2209, 2000.

25. Thorgeirsson, T.E. et al., Effects of temperature on rhodopsin photointermediates from lumirhodopsin to metarhodopsin II, *Biochemistry*, 32, 13,861, 1993.

26. Sheikh, S.P. et al., Rhodopsin activation blocked by metal-ion-binding sites linking transmembrane helices C and F, *Nature*, 383, 347, 1996.

27. Longstaff, C., Calhoon, R.D., and Rando, R.R., Deprotonation of the Schiff base of rhodopsin is obligate in the activation of the G protein, *Proc. Natl. Acad. Sci. USA*, 83, 4209, 1986.

28. Arnis, S. and Hofmann, K.P., Two different forms of metarhodopsin II: Schiff base deprotonation precedes proton uptake and signaling state, *Proc. Natl. Acad. Sci. USA*, 90, 7849, 1993.

29. Fahmy, K., Sakmar, T.P., and Siebert, F., Transducin-dependent protonation of glutamic acid 134 in rhodopsin, *Biochemistry*, 39, 10,607, 2000.

30. Kandori, H. and Maeda, A., FTIR spectroscopy reveals microscopic structural changes of the protein around the rhodopsin chromophore upon photoisomerization, *Biochemistry*, 34, 14,220, 1995.

31. Ishiguro, M., Hirano, T., and Oyama, Y., Modelling of photointermediates suggests a mechanism of the flip of the β-ionone moiety of the retinylidene chromophore in the rhodopsin photocascade, *ChemBioChem.*, 4, 228, 2003.

32. Pan, D. and Mathies, R.A., Chromophore structure in lumirhodopsin and metarhodopsin I by time-resolved resonance Raman microchip spectroscopy, *Biochemistry*, 40, 7929, 2001.

33. Ganter, U.M., Gartner, W., and Siebert, F., Rhodopsin-lumirhodopsin phototransition of bovine rhodopsin investigated by Fourier transform infrared difference spectroscopy, *Biochemistry*, 27, 7480, 1988.

34. Ishiguro, M., A mechanism of primary photoactivation reactions of rhodopsin: modeling of the intermediates in the rhodopsin photocycle, *J. Am. Chem. Soc.*, 122, 444, 2000.

35. Acharya, S. and Karnik, S.S., Modulation of GDP release from transducin by the conserved Glu134-Arg135 sequence in rhodopsin, *J. Biol. Chem.*, 271, 25,406, 1996.

36. Farahbakhsh, Z.T. et al., Mapping light-dependent structural changes in the cytoplasmic loop connecting helices C and D in rhodopsin: a site-directed spin labeling study, *Biochemistry*, 34, 8812, 1995.

37. Kim, J.-M. et al., Structure and function in rhodopsin: rhodopsin mutants with a neutral amino acid at E134 have a partially activated conformation in the dark state, *Proc. Natl. Acad. Sci. USA*, 94, 14,273, 1997.

38. Altenbach, C. et al., Structural features and light-dependent changes in the cytoplasmic interhelical E-F loop region of rhodopsin: a site-directed spin-labeling study, *Biochemistry*, 35, 12,470, 1996.

39. Altenbach, C. et al., Structure and function in rhodopsin: mapping light-dependent changes in distance between residue 65 in helix TM1 and residues in the sequence 306-319 at the cytoplasmic end of helix TM7 and in helix H8, *Biochemistry*, 40, 15,483, 2001.

40. Altenbach, C. et al., Structure and function in rhodopsin: mapping light-dependent changes in distance between residue 316 in helix 8 and residues in the sequence 60-75, covering the cytoplasmic end of helices TM1 and TM2 and their connection loop CL1, *Biochemistry*, 40, 15,493, 2001.

41. Altenbach, C. et al., Structural features and light-dependent changes in the sequence 59-75 connecting helices I and II in rhodopsin: a site-directed spin-labeling study, *Biochemistry*, 38, 7945, 1999.

42. Filipek, S. et al., G protein-coupled receptor rhodopsin: a prospectus, *Annu. Rev. Physiol.*, 65, 851, 2003.

43. Okada, T. et al., Activation of rhodopsin: new insights from structural and biochemical studies, *Trends Biochem. Sci.*, 26, 318, 2001.

44. Nakayama, T.A. and Khorana, H.G., Mapping of the amino acids in membrane-embedded helices that interact with the retinal chromophore in bovine rhodopsin, *J. Biol. Chem.*, 266, 4269, 1991.

45. Alves, I.D. et al., Direct observation of G-protein binding to the human delta-opioid receptor using plasmon-waveguide resonance spectroscopy, *J. Biol. Chem.*, 278, 48,890, 2003.

46. Ishiguro, M., Ligand-binding modes in cationic biogenic amine receptors, *ChemBioChem*, 5, 1210, 2004.

47. Vogel, W.K., Sheehan, D.M., and Schimerlik, M.I., Site-directed mutagenesis on the M_2 muscarinic acetylcholine receptor: the significance of Tyr403 in the binding of agonists and functional coupling, *Mol. Pharmacol.*, 52, 1087, 1997.

48. Page, K.M. et al., The functional role of the binding site aspartate in muscarinic acetylcholine receptors, probed by site-directed mutagenesis, *Eur. J. Pharmacol.*, 289, 429, 1995.

49. Furukawa, H. et al., Conformation of ligands bound to the muscarinic acetylcholine receptor, *Mol. Pharmacol.*, 62, 778, 2002.

50. Hou, X. et al., Influence of monovalent cations on the binding of a charged and an uncharged ('carbo'-)muscarinic antagonist to muscarinic receptors, *Br. J. Pharmacol.*, 117, 955, 1996.

51. Wieland, K. et al., Involvement of Asn-293 in stereospecific agonist recognition and in activation of the β_2-adrenergic receptor, *Proc. Natl. Acad. Sci. USA*, 93, 9276, 1996.

52. Zuurmond, H.M. et al., Study of interaction between agonists and asn293 in helix VI of human beta(2)-adrenergic receptor, *Mol. Pharmacol.*, 56, 909, 1999.

53. Furse, K.E. and Lybrand, T.P., Three-dimensional models for β-adrenergic receptor complexes with agonists and antagonists, *J. Med. Chem.*, 46, 4450, 2003.

54. Freddolino, P.L. et al., Predicted 3D structure for the human β_2 adrenergic receptor and its binding site for agonists and antagonists, *Proc. Natl. Acad. Sci. USA*, 101, 2736, 2004.

55. Strader, C.D. et al., Identification of residues required for ligand binding to the β-adrenergic receptor, *Proc. Natl. Acad. Sci. USA*, 84, 4384, 1987.

56. Strader, C.D. et al., Identification of two serine residues involved in agonist activation of the β-adrenergic receptor, *J. Biol. Chem.*, 264, 13,572, 1989.

57. Sato, T. et al., Ser203 as well as Ser204 and Ser207 in fifth transmembrane domain of the human β_2-adrenoceptor contributes to agonist binding and receptor activation, *Br. J. Pharmacol.*, 128, 272, 1999.

58. Kikkawa, H. et al., Differential contribution of two serine residues of wild type and constitutively active β_2-adrenoceptors to the interaction with β_2-selective agonists, *Br. J. Pharmacol.*, 121, 1059, 1997.

Index

(Note: Numbers in italics indicate figures and tables.)